Design and Analysis of Connections in Steel Structures

Design and Analysis of Connections in Steel Structures

Fundamentals and Examples

Alfredo Boracchini

Author

Alfredo Boracchini, P. E.
info@steeldesign.info

Cover
Detail of a Moment Connection
in a Composite Building Structure
("InterPuls spa" Building, Reggio Emilia,
Italy)
Photo: Alfredo Boracchini

■ All books published by Ernst & Sohn are carefully produced. Nevertheless, authors, editors, and publisher do not warrant the information contained in these books, including this book, to be free of errors. Readers are advised to keep in mind that statements, data, illustrations, procedural details or other items may inadvertently be inaccurate.

Library of Congress Card No.:
applied for

British Library Cataloguing-in-Publication Data
A catalogue record for this book is available from the British Library.

Bibliographic information published by the Deutsche Nationalbibliothek
The Deutsche Nationalbibliothek lists this publication in the Deutsche Nationalbibliografie; detailed bibliographic data are available on the Internet at <http://dnb.d-nb.de>.

© 2018 Wilhelm Ernst & Sohn, Verlag für Architektur und technische Wissenschaften GmbH & Co. KG, Rotherstraße 21, 10245 Berlin, Germany

All rights reserved (including those of translation into other languages). No part of this book may be reproduced in any form – by photoprinting, microfilm, or any other means – nor transmitted or translated into a machine language without written permission from the publishers. Registered names, trademarks, etc. used in this book, even when not specifically marked as such, are not to be considered unprotected by law.

Coverdesign Sophie Bleifuß, Berlin, Germany
Typesetting SPi Global, Chennai, India
Printing and Binding CPI Group (UK) Ltd, Croydon, CR0 4YY

Print ISBN: 978-3-433-03122-3
ePDF ISBN: 978-3-433-60606-3
ePub ISBN: 978-3-433-60607-0
oBook ISBN: 978-3-433-60605-6

Printed on acid-free paper.

To my mom Alda

Preface

Structural Steel Connection Design is an engineering manual directed toward the engineering audience. The first section provides an introduction to key concepts, then progresses to provide a more in-depth description for the design of structural steel connections.

A correct approach to connection design is fundamental in order to have a safe and economically sound building. Therefore, this book will attempt to explain how to set up connections within the main calculation model, choose the types of connections, check them (limit states to be considered), and utilize everything in practice.

The focal point of the book is not to closely follow and explain one specific standard; rather the aim is to treat connections generally speaking and to understand the main concepts and how to apply them. This means that, even though Eurocode (EC) and the American Institute of Steel Construction (AISC) are the most referenced standards, other international norms will be mentioned and discussed. This helps to understand that connection design is not an exact science and that numerous approaches can be viable.

Type by type, connection by connection, detailed examples will be provided to help perform a full analysis for each limit state.

An excellent software tool (SCS – Steel Connection Studio) will be illustrated and used as an aid to assist in the comprehension of connection design. The software can be downloaded for free at www.steelconnectionstudio.com or at www.scs.pe and can be installed as a demo (trial) version (limitations about printing, saving, member sizes, and reporting), see "Software Downloads and its Limitations" (page xxiv). A professional full version can also be purchased online but the demo version is enough to reproduce the examples in the book.

The book will also try to deliver some practical suggestions for the professional engineer: how to talk about bracings to the architect, how to interact with fabricators showing an understanding of erection and fabrication, and much more.

Many countries have a deeper engineering culture about concrete structures than steel structures. This manual therefore aims to illustrate to engineers that do not design steel structures daily, some concepts that will facilitate and make their design of connections for steel structures more efficient. This will be done using a practical, rather than a theoretical, approach.

Design of steel structures can become tricky when it is about stability (buckling) and joints: this second fundamental aspect of steel constructions, which is crucial for economic performance, will be examined in detail.

The text, figures, charts, formulas, and examples have been prepared and reported with maximum care in order to help the engineer better understand and set up his or her own calculations for structural steel connections. However, it is possible that the book contains errors and omissions, and therefore readers are encouraged to have standards at hand as their primary reference. No responsibility is accepted and taken for the application of concepts explained in the manual: the engineer must prepare and perform any analysis and design under his or her complete competence, responsibility, and liability.

For a list of errors and omissions found in the book and their corrections, please check www.steeldesign.info.

Finally, please use www.steeldesign.info to send comments, suggestions, criticisms, and opinions. The author thanks you in advance.

April 2018

Alfredo Boracchini
Reggio Emilia

About the Author

Alfredo Boracchini is a Professional Engineer in Italy, Canada, and some states of the United States. His professional experience is mainly in steel structures that he has designed and calculated for many applications and in various parts of the world. He is an active member in some international steel associations and the owner of an engineering firm with offices in Europe, Asia, and America. This allowed him to collect extensive international experience in the field of steel connection design that he shares in this manuscript with other engineers interested in this field.

Contents

Acknowledgments *xxi*
List of Abbreviations *xxiii*

1	**Fundamental Concepts of Joints in Design of Steel Structures** *1*	
1.1	Pin Connections and Moment Resisting Connections *1*	
1.1.1	Safety, Performance, and Costs *1*	
1.1.2	Lateral Load Resisting System *2*	
1.1.3	Pins and Fully Restrained Joints in the Analysis Model *7*	
1.2	Plastic Hinge *8*	
1.2.1	Base Plates *9*	
1.2.2	Trusses *11*	
	References *12*	
2	**Fundamental Concepts of the Behavior of Steel Connections** *13*	
2.1	Joint Classifications *13*	
2.2	Forces in the Calculation Model and for the Connection *14*	
2.3	Actions Proportional to Stiffness *17*	
2.4	Ductility *18*	
2.5	Load Path *19*	
2.6	Ignorance of the Load Path *20*	
2.7	Additional Restraints *21*	
2.8	Methods to Define Ultimate Limit States in Joints *21*	
2.9	Bolt Resistance *22*	
2.10	Yield Line *22*	
2.11	Eccentric Joints *22*	
2.12	Economy, Repetitiveness, and Simplicity *22*	
2.13	Man-hours and Material Weight *23*	
2.14	Diffusion Angles *23*	
2.15	Bolt Pretensioning and Effects on Resistance *24*	
2.15.1	Is Resistance Affected by Pretensioning? *24*	
2.15.2	Is Pretensioning Necessary? *24*	
2.15.3	Which Pretensioning Method Should Be Used? *25*	

2.16	Transfer Forces	*25*
2.17	Behavior of a Bolted Shear Connection	*25*
2.18	Behavior of Bolted Joints Under Tension	*27*
	References	*29*

3	**Limit States for Connection Components**	*31*
3.1	Deformation Capacity (Rotation) and Stiffness	*31*
3.1.1	Rotational Stiffness	*32*
3.2	Inelastic Deformation due to Bolt Hole Clearance	*33*
3.3	Bolt Shear Failure	*34*
3.3.1	Threads Inside the Shear Plane	*35*
3.3.2	Number of Shear Planes	*37*
3.3.3	Packing Plates	*37*
3.3.4	Long Joints	*38*
3.3.5	Anchor Bolts	*39*
3.3.6	Stiffness Coefficient	*39*
3.4	Bolt Tension Failure	*40*
3.4.1	Countersunk Bolts	*41*
3.4.2	Stiffness Coefficient	*41*
3.5	Bolt Failure in Combined Shear and Tension	*42*
3.6	Slip-Resistant Bolted Connections	*42*
3.6.1	Combined Shear and Tension	*44*
3.7	Bolt Bearing and Bolt Tearing	*44*
3.7.1	Countersunk Bolts	*49*
3.7.2	Stiffness Coefficients	*49*
3.8	Block Shear (or Block Tearing)	*49*
3.9	Failure of Welds	*52*
3.9.1	Weld Calculation Procedures	*54*
3.9.1.1	Directional Method	*54*
3.9.1.2	Simplified Method	*57*
3.9.2	Tack Welding (Intermittent Fillet Welds)	*58*
3.9.3	Eccentricity	*58*
3.9.4	Fillet Weld Groups	*58*
3.9.5	Welding Methods	*60*
3.9.6	Inspections	*60*
3.9.6.1	Visual Testing	*60*
3.9.6.2	Penetrant Testing	*60*
3.9.6.3	Magnetic Particle Testing	*60*
3.9.6.4	Radiographic Testing	*60*
3.9.6.5	Ultrasonic Testing	*61*
3.10	T-stub, Prying Action	*61*
3.10.1	T-stub with Prying Action	*62*
3.10.2	Possible Simplified Approach According to AISC	*64*
3.10.3	Backing Plates	*65*
3.10.4	Length Limit for Prying Forces and T-stub without Prying	*66*
3.10.5	T-stub Design Procedure for Various "Components" According to Eurocode	*67*

3.10.5.1	Column Flange *67*	
3.10.5.2	End Plate *71*	
3.10.5.3	Angle Flange Cleat *71*	
3.10.6	T-stub Design Procedure for Various "Components" According to the "Green Book" *71*	
3.10.6.1	ℓ_{eff} for Equivalent T-stubs for Bolt Row Acting Alone *74*	
3.10.6.2	ℓ_{eff} to Consider for a Bolt Row Acting Alone *77*	
3.10.6.3	ℓ_{eff} to Consider for Bolt Rows Acting in Group *79*	
3.10.6.4	Examples of ℓ_{eff} for Bolts in a Group *80*	
3.10.7	T-stub for Bolts Outside the Beam Flanges *81*	
3.10.8	Stiffness Coefficient *81*	
3.11	Punching *82*	
3.12	Equivalent Systems *82*	
3.13	Web Panel Shear *82*	
3.13.1	Stiffness Coefficient *84*	
3.14	Web in Transverse Compression *84*	
3.14.1	Transformation Parameter β *86*	
3.14.2	Formulas for Other Local Buckling Limit States *87*	
3.14.3	Stiffness Coefficient *88*	
3.14.4	T-stub in Compression *88*	
3.15	Web in Transverse Tension *88*	
3.15.1	Stiffness Coefficient *89*	
3.16	Flange and Web in Compression *89*	
3.17	Beam Web in Tension *89*	
3.18	Plate Resistance *90*	
3.18.1	Material Properties *90*	
3.18.2	Tension *90*	
3.18.2.1	Staggered Bolts *92*	
3.18.3	Compression *92*	
3.18.4	Shear *92*	
3.18.5	Bending *93*	
3.18.6	Design for Combined Forces *93*	
3.18.7	Whitmore Section *93*	
3.19	Reduced Section of Connected Profiles *93*	
3.19.1	Shear Lag *95*	
3.20	Local Capacity *99*	
3.21	Buckling of Connecting Plates *100*	
3.21.1	Gusset Plate Buckling *100*	
3.21.2	Fin Plate (Shear Tab) Buckling *101*	
3.22	Structural Integrity (and Tie Force) *103*	
3.23	Ductility *105*	
3.24	Plate Lamellar Tearing *106*	
3.25	Other Limit States in Connections with Sheets and Cold-formed Steel Sections *108*	
3.26	Fatigue *108*	
3.27	Limit States of Other Materials in the Connection *109*	
	References *109*	

4 Connection Types: Analysis and Calculation Examples *113*

- 4.1 Common Symbols *113*
- 4.1.1 Materials *113*
- 4.1.2 Design Forces *113*
- 4.1.3 Bolts *113*
- 4.1.4 Geometric Characteristics of Plates and Profiles *114*
- 4.2 Eccentrically Loaded Bolt Group: Eccentricity in the Plane of the Faying Surface *115*
- 4.2.1 Elastic Method *115*
- 4.2.1.1 Example of Eccentricity Calculated with Elastic Method *116*
- 4.2.2 Instantaneous Center-of-Rotation Method *118*
- 4.2.2.1 Example of Eccentricity Calculated with the Instantaneous Center-of-Rotation Method *119*
- 4.3 Eccentrically Loaded Bolt Group: Eccentricity Normal to the Plane of the Faying Surface *120*
- 4.3.1 Neutral Axis at Center of Gravity *121*
- 4.3.1.1 Example of Eccentricity Normal to Plane Calculated with Neutral Axis at Center-of-Gravity Method *122*
- 4.3.2 Neutral Axis Not at Center of Gravity *123*
- 4.3.2.1 Example of Eccentricity Normal to Plane Calculated with Neutral Axis not at Center-of-Gravity Method *124*
- 4.4 Base Plate with Cast Anchor Bolts *125*
- 4.4.1 Plate Thickness *125*
- 4.4.1.1 AISC Method *125*
- 4.4.1.2 Eurocode Method *130*
- 4.4.2 Contact Pressure *135*
- 4.4.2.1 AISC Method *135*
- 4.4.2.2 Eurocode Method *136*
- 4.4.3 Anchor Bolts in Tension *139*
- 4.4.3.1 AISC Method *139*
- 4.4.3.2 Eurocode Method *140*
- 4.4.3.3 Other Notes *141*
- 4.4.4 Welding *142*
- 4.4.5 Shear Resistance *142*
- 4.4.5.1 Friction *142*
- 4.4.5.2 Anchor Bolts in Shear *143*
- 4.4.5.3 Shear Lugs *144*
- 4.4.6 Rotational Stiffness *144*
- 4.4.7 Measures to Improve Ductility *145*
- 4.4.8 Practical Details and Other Notes *145*
- 4.4.9 Fully Restrained Schematization of Column Base Detail *148*
- 4.4.10 Example of Base Plate Design According to Eurocode *149*
- 4.4.10.1 Uplift and Moment *149*
- 4.4.10.2 Shear *152*
- 4.4.10.3 Welding *153*
- 4.4.10.4 Joint Stiffness *153*

4.4.10.5	Comparison with AISC Method for SLU1	*153*
4.5	Chemical or Mechanical Anchor Bolts	*153*
4.6	Fin Plate/Shear Tab	*154*
4.6.1	Choices and Possible Variants	*155*
4.6.1.1	Pin Position	*155*
4.6.1.2	Location of Plate Welded to Primary Member	*156*
4.6.1.3	Notches (Copes) in Secondary Member	*157*
4.6.1.4	Reinforcing Beam Web	*158*
4.6.2	Limit States to Be Considered	*161*
4.6.3	Rotation Capacity	*161*
4.6.4	Measures to Improve Ductility	*162*
4.6.5	Measures to Improve Structural Integrity	*162*
4.6.6	Design Example According to DIN	*162*
4.6.6.1	Bolt Shear	*163*
4.6.6.2	Bearing	*165*
4.6.6.3	Block Shear	*166*
4.6.6.4	Plate Resistance	*167*
4.6.6.5	Beam Resistance	*167*
4.6.6.6	Plate Buckling	*168*
4.6.6.7	Local Check for Primary-Beam Web	*168*
4.6.6.8	Welding	*168*
4.6.6.9	Rotation Capacity	*169*
4.6.6.10	Ductility	*169*
4.6.6.11	Structural Integrity	*169*
4.7	Double-Bolted Simple Plate	*169*
4.7.1	Rotation Capacity	*170*
4.7.2	Ductility	*170*
4.7.3	Structural Integrity	*171*
4.7.4	Beam-to-Beam Example Designed According to Eurocode	*171*
4.7.4.1	Bolt Shear	*172*
4.7.4.2	Bearing	*173*
4.7.4.3	Block Shear	*174*
4.7.4.4	Plate Resistance	*174*
4.7.4.5	Beam Resistance	*174*
4.7.4.6	Plate Buckling	*174*
4.7.4.7	Primary-Beam Web Local Check	*174*
4.7.4.8	Welding, Ductility, and Structural Integrity	*174*
4.8	Shear ("Flexible") End Plate	*175*
4.8.1	Variants and Rotation Capacity	*175*
4.8.2	Limit States to be Considered	*177*
4.8.3	Rotational Stiffness	*177*
4.8.4	Ductility	*178*
4.8.5	Structural Integrity	*178*
4.8.6	Column-to-Beam Example Designed According to IS 800	*178*
4.8.6.1	Bolt Resistance	*179*
4.8.6.2	Rotation Capacity and Structural Integrity	*179*
4.8.6.3	Bearing	*180*

4.8.6.4	Block Shear	*180*
4.8.6.5	Plate Check	*180*
4.8.6.6	Beam Shear Check	*180*
4.8.6.7	Column Resistance	*180*
4.8.6.8	Welds	*181*
4.8.6.9	Conclusion	*181*
4.9	Double-Angle Connection	*181*
4.9.1	Variants	*183*
4.9.2	Limit States to Be Considered	*183*
4.9.3	Structural Integrity, Ductility, and Rotation Capacity	*183*
4.9.4	Practical Advice	*183*
4.9.5	Beam-to-Beam Example Designed According to AISC	*184*
4.10	Connections in Trusses	*186*
4.10.1	Intermediate Connections for Compression Members	*186*
4.11	Horizontal End Plate Leaning on a Column	*188*
4.11.1	Limit States to be Considered	*189*
4.12	Rigid End Plate	*189*
4.12.1	Column Web Panel Shear	*191*
4.12.2	Lever Arm	*191*
4.12.3	Stiffeners	*192*
4.12.4	Supplementary Web Plate Check	*193*
4.12.5	Check for Column Stiffeners in Compression Zone	*193*
4.12.6	Check for Column Stiffeners in Tension Zone	*195*
4.12.7	Check of Column Diagonal Stiffener for Panel Shear	*196*
4.12.8	Shear Due to Vertical Forces	*196*
4.12.9	Design with Haunches	*196*
4.12.10	Beam-to-Beam Connections	*196*
4.12.11	BS Provisions	*197*
4.12.12	AISC Approach	*197*
4.12.13	Limit States to Be Considered	*199*
4.12.14	Rotational Stiffness	*200*
4.12.15	Simplifying the Design	*201*
4.12.16	Practical Advice	*201*
4.12.17	Structural Integrity, Ductility, and Rotation Capacity	*201*
4.12.18	Beam-to-Column End-Plate Design Example According to Eurocode	*202*
4.12.18.1	Column Flange Thickness Check for Bolt Row 1	*204*
4.12.18.2	Column Web Tension Check for Bolt Row 1	*204*
4.12.18.3	Beam End-Plate Thickness Check for Bolt Row 1	*205*
4.12.18.4	Beam Web Tension Check for Bolt Row 1	*205*
4.12.18.5	Final Resistant Value for Bolt Row 1	*205*
4.12.18.6	Column Flange Thickness Check for Bolt Row 2 Individually	*205*
4.12.18.7	Column Web Tension Check for Bolt Row 2 Individually	*206*
4.12.18.8	Beam End-Plate Thickness Check for Bolt Row 2 Individually	*206*
4.12.18.9	Beam Web Tension Check for Bolt Row 2 Individually	*206*

4.12.18.10	Column Flange Thickness Check for Bolt Row 2 in Group with Bolt Row 1	207
4.12.18.11	Column Web Tension Check for Bolt Row 2 in Group with Bolt Row 1	207
4.12.18.12	Beam End-Plate Thickness Check for Bolt Row 2 in Group with Bolt Row 1	207
4.12.18.13	Beam Web Tension Check for Bolt Row 2 in Group with Bolt Row 1	207
4.12.18.14	Final Resistant Value for Bolt Row 2	208
4.12.18.15	Vertical Shear	208
4.12.18.16	Web Panel Shear	209
4.12.18.17	Column Web Resistance to Transverse Compression	209
4.12.18.18	Stiffener Design	210
4.12.18.19	Welds	210
4.12.18.20	Rotational Stiffness	210
4.13	Splice	212
4.13.1	Calculation Model and Limit States	213
4.13.2	Structural Integrity, Ductility, and Rotation Capacity	215
4.13.3	Column Splice Design Example According to AS 4100	215
4.13.3.1	Flanges	216
4.13.3.2	Web	217
4.13.3.3	Conclusions and Final Considerations	217
4.13.3.4	Possible Alternative	217
4.14	Brace Connections	217
4.14.1	AISC Methods: UFM and KISS	220
4.14.1.1	KISS Method	222
4.14.1.2	Uniform Force Method	222
4.14.1.3	UFM Variant 1	223
4.14.1.4	UFM Variant 2	224
4.14.1.5	UFM Variant 3	225
4.14.1.6	UFM Adapted to Existing Connections	226
4.14.2	Practical Recommendations	227
4.14.3	Complex Brace Connection Example According to CSA S16	227
4.14.3.1	Friction Connection for Brace	227
4.14.3.2	Brace and Gusset Bearing	228
4.14.3.3	Block Shear	228
4.14.3.4	Channel Shear Lag	229
4.14.3.5	Whitmore Section for Tension Resistance and Buckling of Gusset Plate	229
4.14.3.6	UFM Forces	229
4.14.3.7	Gusset-to-Column Shear Tab	229
4.14.3.8	Gusset-to-Beam Weld	229
4.14.3.9	Beam-to-Column Shear Tab	229
4.14.3.10	Ductility and Structural Integrity	230
4.15	Seated Connection	230
4.16	Connections for Girts and Purlins	233
4.17	Welded Hollow-Section Joints	236

4.18	Connections in Composite (Steel–Concrete) Structures 236
4.19	Joints with Bolts and Welds Working in Parallel 236
4.20	Expansion Joints 237
4.21	Perfect Hinges 238
4.22	Rollers 239
4.23	Rivets 240
4.24	Seismic Connections 241
4.24.1	Rigid End Plate 242
4.24.2	Braces 243
4.24.3	Eccentric Braces and "Links" 244
4.24.4	Base Plate 244
	References 246

5 Choosing the Type of Connection 249

5.1	Priority to Fabricator and Erector 249
5.2	Considerations of Pros and Cons of Some Types of Connections 249
5.3	Shop Organization 250
5.3.1	Plates or Sheets 250
5.3.2	Concept of "Handling" One Piece 250
5.4	Culture 252
	References 252

6 Practical Notes on Fabrication 253

6.1	Design Standardizations 253
6.1.1	Materials 253
6.1.2	Thicknesses 253
6.1.3	Bolt Diameters 253
6.2	Dimension of Bolt Holes 254
6.2.1	Bolt Hole Clearance in Base Plates 255
6.3	Erection 256
6.3.1	Structure Lability 256
6.3.2	Erection Sequence and Clearances 256
6.3.3	Bolt Spacing and Interferences 257
6.3.4	Positioning and Supports 257
6.3.5	Holes or Welded Plates for Handling and Lifting 258
6.4	Clearance Needed to Operate Tightening Wrenches 258
6.4.1	Double Angles in Connections 259
6.5	Bolt Spacing and Edge Distances 260
6.6	Root Radius Encroachment 260
6.7	Notches 264
6.8	Bolt Tightening and Pretensioning 265
6.8.1	Calibrated Wrench 266
6.8.2	Turn of the Nut 266
6.8.3	Direct Tension Indicators 270
6.8.4	Twist-Off Type Bolts 271
6.8.5	Hydraulic Wrenches 273

6.9	Washers 274
6.9.1	Tapered (Beveled) Washers 275
6.9.2	Vibrations 277
6.10	Dimensions of Screws, Nuts, and Washers 277
6.10.1	Depth of Bolt Heads and Nuts 277
6.10.2	Washer Width and Thickness 277
6.11	Reuse of Bolts 278
6.12	Bolt Classes 279
6.13	Shims 280
6.14	Galvanization 281
6.14.1	Tubes 281
6.14.2	Plate Welded over Profiles as Reinforcement 281
6.14.3	Base Plates 282
6.15	Other Finishes After Fabrication 282
6.16	Camber 283
6.17	Grout in Base Plates 284
6.18	Graphical Representation of Bolts and Connections 286
6.19	Field Welds 287
6.20	Skewed Joints 287
	References 291

7 Connection Examples 293

Index 355

Acknowledgments

The author is grateful to Giovanna Zanardi for her assiduous support and to Renzo Mazzali for his precious advice.

I also wish to express my gratitude to Antonella, Alda, Emma, Irma, Vera, and Lea for their love and for sharing daily with me, sometimes in spite of themselves, my passion for structural engineering.

List of Abbreviations

AISC	American Institute of Steel Construction
ASD	allowable stress design
AWS	American Welding Society
BS	British Standards
CG	center of gravity
EC	Eurocode
EN	European Standard
DIN	German Institute for Standardization
ECCS	European Convention for Constructional Steelwork
FCAW	flux-cored arc welding
FEA	finite element analysis
FEM	finite element method
GMAW	gas metal arc welding
HSS	hollow structural steel
ISO	International Organization for Standardization
KISS	keep it simple stupid
LRFD	load and resistance factor design
NDP	nationally determined parameter
NTC	Italian Standard for Constructions
OSHA	Occupational Safety and Health Administration
PL	preload
PR	partially restrained
RCSC	Research Council on Structural Connections
SAW	submerged arc welding
SCS	Steel Connection Studio
SMAW	shielded metal arc welding
TOS	top of steel
UFM	uniform force method

Software Download and its Limitations

The engineer can use the book to get familiar with SCS – Steel Connection Studio, a fantastic software tool that can be used in the design of steel connections. The software (which can be downloaded from www.steelconnectionstudio.com or www.scs.pe) works in a demo version with the following limitations:

- Only 90 consecutive days of usage are allowed after installation. Providing the code *Ej8Z4pn1* the demo version can be extended with no time limitation. Please send an email to info@scs.pe if you are interested in this offer.
- File saves are not possible.
- Commercial and academic usage is not possible.
- Maximum dimensions of members are 254 mm/10 in. in width and 361 mm/14.2 in. in depth. The sketch cannot be printed.
- The word *demo* is watermarked in the sketch.
- The maximum number of reports that can be generated is 25.

For videos, tutorials, validation examples, the manual, and additional (commercial too) information, please visit www.steelconnectionstudio.com or www.scs.pe.

1

Fundamental Concepts of Joints in Design of Steel Structures

Regarding joints, the first fundamental concept the engineer must be clear about when he or she starts to design is which connections will develop moment resistance and which can be executed as simple pin joints. To do this, it is necessary to clarify the lateral load resisting system.

1.1 Pin Connections and Moment Resisting Connections

1.1.1 Safety, Performance, and Costs

Steel structures should be safe, able to perform, and be cost-effective.

They must be safe because they act as canopies, mezzanines, buildings, skyscrapers, bridges, and much more that give shelter, protect, and be welcoming to men and women. A structural collapse is extremely dangerous and likely to cause severe harm to anyone in the surrounding area.

Structures must also effectively serve their commercial purpose while efficiently and comfortably (for the users) maintaining their design features over time. These are the basic notions of serviceability limit state design specifying that, just as a nonlimiting example, deformations will not damage secondary structures or that excessive vibrations will not make users uncomfortable.

Poor performance might also decrease the structure's value and harm the property owner.

Simultaneously, the market logic requires that the structural system be economically sound and cost-effective when compared to alternatives using different materials and design. Being economically sound is a complex matter that must take into account many factors in the building design. However, the engineer must make the structure as cost-effective as possible without compromising safety and performance. The service and expertise that engineers are expected to deliver should include reducing costs while maintaining high standards of functionality and protection.

For the principles stated, the design of connections is a focal point and it must be well defined in the engineer's mind from the commencement of the project.

1.1.2 Lateral Load Resisting System

The choice of connections is related to the choice of the lateral load resisting system.

Taking a closer look at this key point, we consider these initial hypotheses: that the structure geometry is defined, that steel will be used as structural material, and that the design loads are provided. This means that the engineer can set up the analysis model with the finite element software available. However, before building the model wireframe, the engineer must have a clear vision of the lateral resisting system(s). This choice influences costs and architectural restraints.

Lateral load resisting systems can be diverse and variously combined among themselves. Each horizontal direction can have its own system, one that may be different from the other direction.

The basic lateral resisting systems (Figure 1.1) are as follows:

- Braces (bracings)
- Moment connections (portals)
- Base rigid restraints (cantilever columns or inverted pendulum)
- Connection to an existing structure or another ad hoc structure built with different materials (say a concrete staircase, masonry or concrete walls, etc.).

The structural engineer attentive to fabrication logics usually tries to adopt bracings as this will deliver maximum cost performance. The main advantages of using braces are as follows:

- The structure is easily sized against horizontal forces (mainly wind and earthquakes) allowing less weight for beams and, most of all, columns (braces take care of lateral forces and the column can work only in compression).
- Connections can roughly be just in shear or axial action and so are light and economic.
- Lateral deflection control is excellent.
- Seismic response is good (given that the necessary detailing is provided).

At the same time bracing has some disadvantages:

- It laterally obstructs the transit, limiting windows or gates.
- The architect or the owner might not like it for esthetic reasons.

This last problem might be solved by "highlighting" the braces and assigning architectural importance to them. Some famous examples can be found, such as landmark skyscrapers (Figure 1.2) and more "ordinary" buildings (Figure 1.3), where the architect was able to create an interesting contrast with materials that nicely emphasize the braces.

The problem of transit obstruction is usually bypassed by choosing one specific bay for braces, if possible. This is done either in the middle or at the end of the building system. Horizontal braces are implemented to bring forces to the localized braces. (This book does not discuss the layout of horizontal braces. Rather it discusses one of their main functions, beyond limiting flexural torsional buckling of beams, that is, to connect unbraced bays to braced ones.)

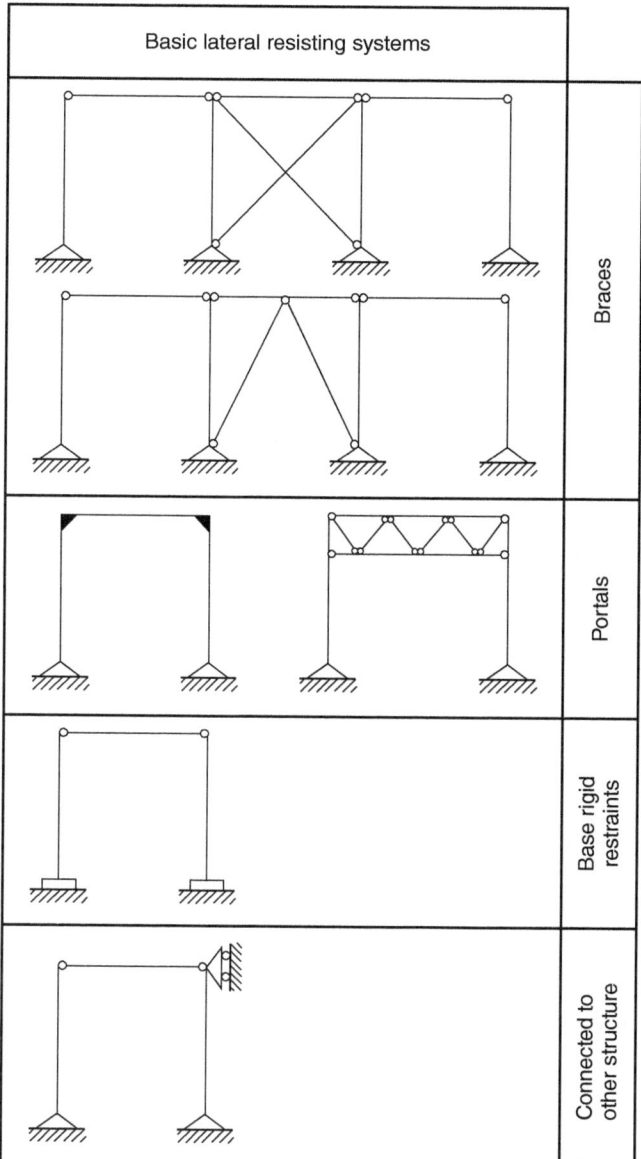

Figure 1.1 Lateral load resisting systems.

Another method to limit the obstruction in the space occupied by the braces is to adapt their geometry to the challenges of architectural restraints using different schemes and shapes (V, inverted-V, X, K, Y, and more).

Having given the many advantages of using braces and the importance of informing the owner and the other players about this solution in order to have it approved, in many situations it is not possible to use braces, especially in both directions. As a consequence, it is necessary to use portals or base rigid

Figure 1.2 Braces emphasized esthetically in the John Hancock Tower of Chicago. Source: From Wikipedia; photo courtesy of "Akadavid", 2008.

connections or a combination of them, if not different additional schemes such as shear steel walls or other concrete or composite systems that are outside the scope of this book.

The main advantage of using portals and rigid bases is what made braces undesirable; that is, there are no obstacles in fully exploiting all the space of the bays. In addition, moment resisting systems (by the way, it is not trivial to underline that a system made by trusses and columns is a specific case of a portal) have the following advantages:

- Possible savings (at the expense of the dimension and cost of the columns) in beam depth since the moment connection allows a better exploitation of the beam strength along the full length.
- A more "convincing" look of the columns that, being heavier, seem safer.
- Pin (hinge) connections at the base, then savings in foundation work (larger even compared to braces, which could give an uplift and require more expensive tension details and some "ballasting" of the plinths).
- Reasonable seismic resistance (if the necessary detailing is followed).

Figure 1.3 Valorization of internal braces (InterPuls, Reggio Emilia, Italy).

Disadvantages of portals might be the following:

- Moment connections are required and they are usually complex and more expensive.
- Additional encumbrance may be provided by the beam-to-column connection (net height at the eaves is impacted and this could make it mandatory to raise the whole structure); also, the obstruction given by trusses is similarly and evidently large.
- On average, the weight per unit of area will worsen.
- Lateral deflections should be checked carefully.
- Buckling length of columns worsens.

A lateral resisting system having the columns rigidly connected to the base may have the following benefits:

- No obstructed bays, as already mentioned.
- Larger columns inspiring more confidence in the safety of the building.

The following are some of the disadvantages:

- Expensive foundation work required: large plinths, piles likely mandatory
- Lateral deflections to check (but usually better than portals)
- More material (steel) necessary to build the structure
- Longer buckling length of columns
- Poor seismic performance (for example, the American Society of Civil Engineers (ASCE) basically bans this system for buildings if the area is highly seismic (see Ref. [1] for more precise information)).

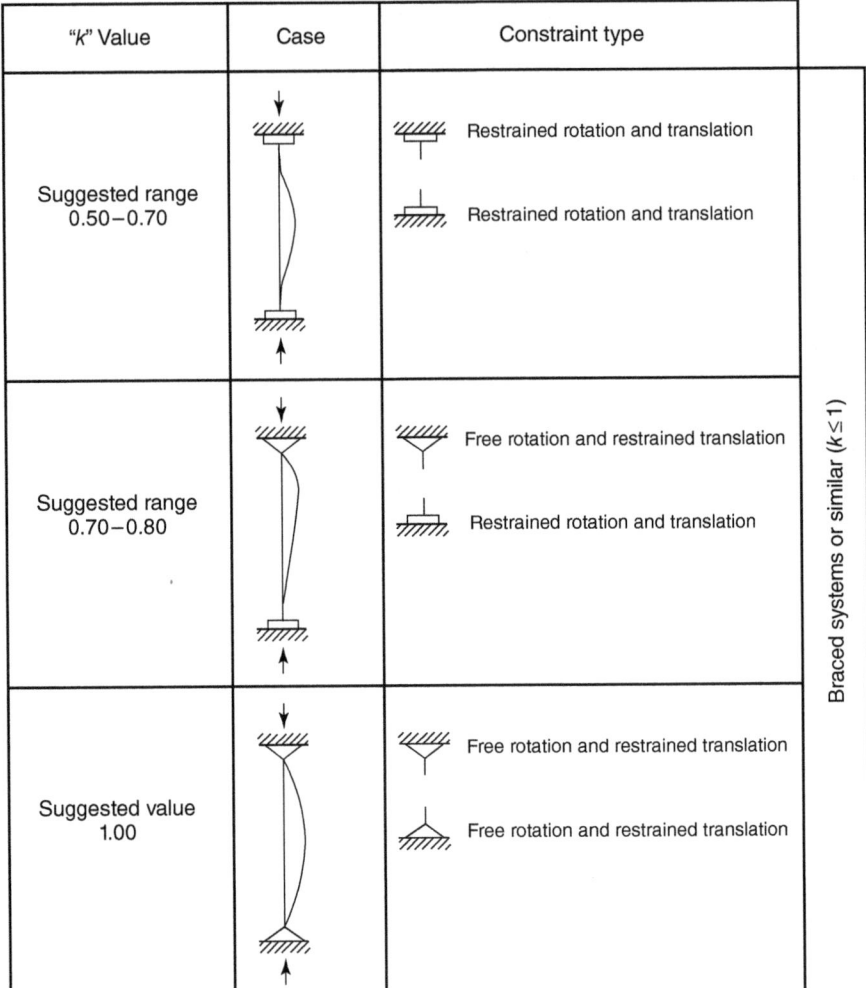

Figure 1.4 Buckling length coefficients (effective length factors) for braced systems.

Sometimes, to solve the problems of lateral deflections and column buckling length (see Figures 1.4 and 1.5 for reference values), both systems are contemporarily adopted.

As Figure 1.5 shows, the buckling length of columns is two times the physical length in each system when taken by itself, but it goes back to almost unity (braced systems have 1) when used in a combined system.

Every situation is different and braces are not always the best option. For example, if the structure has large bays (beyond 20 m, or 60 ft) and that direction already uses trusses in its architectural layout, it is already a moment resisting system that can be exploited as a lateral resisting system. Braces can be used only in the orthogonal direction, effectively restraining the weak side of the columns.

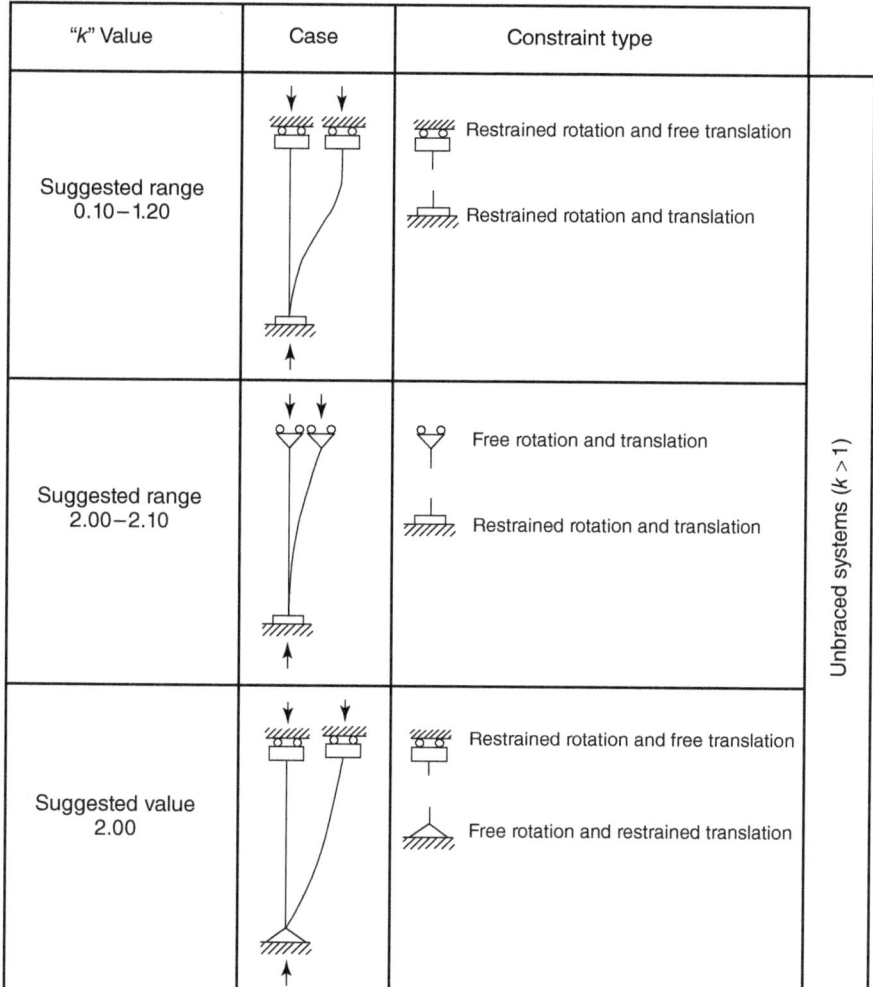

Figure 1.5 Buckling length coefficients (effective length factors) for unbraced systems.

An engineer with a clear understanding of a lateral load resisting system will correctly prioritize and evaluate, in any situation, the benefits of each option, choosing the best method with regards to economy, safety, and performance.

1.1.3 Pins and Fully Restrained Joints in the Analysis Model

As described, the designer must choose the lateral load resisting system, in agreement with the architect and the owner, before setting up the analysis model.

It is important to underline that the matter should not be considered to the owner in terms that are too technical, that is, the problem should not be introduced as a choice of lateral load resisting system. Rather, the designer should talk about this from an architectural perspective, where braces can be placed and about the economic and performance benefits that braces can bring. Where bracing is not accepted, for esthetic or other reasons, the engineer must think about the alternatives previously illustrated.

Only at this point will the engineer know where in the design model to put *fully restrained joints* and where it is possible to unrestrain beams and to consider their connections as *pins (hinges)*.

Not taking into account possible decisions of having beams in continuity (therefore fully restrained) to help deflections and the final weight, all the connections that are not necessary for global stability (that is, to the lateral resisting system when it is a portal or an inverted pendulum) should be considered as pins. This is conservative and helps the project budget.

If an engineer who is not familiar with structural steel develops a model without careful consideration of the lateral load resisting system and the connections among members, it could severely impact the project. If the entire model has rigid connections, the structure could be underdesigned and unstable if the joints are not correctly dimensioned and realized as fully restrained. Also, in the event that the joints are correctly fabricated as rigid, the competitive price of a similar fully restrained system with complex and labor-intensive connections is suspicious.

To summarize, the correct order to follow during the design stage is as follows:

- Choose the lateral resisting system(s).
- Model as fully restrained the joints that are strictly necessary for this purpose (overall stability).
- Model as pins (hinges) all the other connections.
- Design the structure.
- Decide if stiffening some joints (from pin to fully restrained) can be beneficial to the total weight or deflections.
- Design the connections.

If following Eurocode (EC), the additional steps are:

- Calculate joint rigidity.
- Check if the assumptions in the model are consistent with the results of joint rigidity (pin or fully restrained joints).
- If necessary, update the calculation model; if needed, use joint springs in the model to simulate exact rigidity (semirigid joints).
- When necessary, rerun the analysis.

According to the classical elastic method, it is not required to check connection rigidity because the experience of the engineer is enough to assess this. However, some standards (EC primarily) have started to ask for an analytical check of this component.

1.2 Plastic Hinge

In contrast to reinforced concrete where simple supports and fully restrained connections are more easily understandable because the physical connection is similar to the ideal, this concept is less intuitive with regards to steel.

If a concrete construction has a beam leaning on a girder (a true simple support), in constructions made of steel the connections that are considered as hinges (pins) might not be immediately recognizable as such to designers unfamiliar with the material.

Hinge/pin connections are neither real pins nor simple supports. They are initially able to resist bending moments, more or less relevant in absolute value. The engineer must indeed learn that the experience (in the sense of the history of structural engineering) and the ductility of the material makes this kind of connection representable as a pin, and years of structural steel buildings have shown that this approach is both sound and reliable. This also means that it is not conservative to assign calculation moments to these kinds of connections, especially if those bending moments are essential to the overall stability of the structure. In fact, the steel is ductile and the material will become plastic when the yield limit is reached and there are no brittle or buckling behaviors. This means that the connection will develop into a hinge, thus redistributing forces. This explains why some joints that do not look like pin connections are represented as such in the design practice.

As mentioned in the previous section, intermediate behavior (semirigid) is discussed, for example in EC and AISC (partially restrained (PR) joints), but it is crucial that the designer comprehends the plastic hinge concept that has been used for many years in steel construction. With this in mind, we take a closer look at two typical examples of welded connections in trusses and base plates.

1.2.1 Base Plates

Base plates can be represented as either hinges or fully restrained joints.

It is not necessary to physically realize a "real pin" to represent a base plate as a pin connection. This method was used several years ago, as seen in the example of the Milan railway station in Figure 1.6. Nowadays, it is not considered necessary to put, for example, only one row of anchor bolts in order to have a pin because even configurations like the ones illustrated in Figure 1.7 have enough ductility to be considered as hinges: any yield due to an initial bending moment will make the connection evolve to a plastic state, similar to a hinge. This means that it is conservative and conventional to consider the joint as a pin (and the bending moment that can be resisted at least initially is an additional benefit). Many books, including [2], agree on this concept, explicitly articulating on the subject. French standards partially disagree since they take into account Yvon Lescouarc'h's publications [3, 4], which set limits for the representations of base connections as pins. Another important concept that seems to give credit to [2] is that the plate-to-column systems (with stiffeners in case) and the base plate-to-concrete systems are always stiffer than the concrete-to-soil systems. Therefore, the behavior of the joint will depend on how the foundation is designed and realized: if an initial moment creates any settlement in the foundation, the whole connection system will behave like a hinge since the locally low stiffness will activate a more rigid lateral load resisting system.

In other words, it is the engineer's choice whether the base joint is considered as a pin or a fully restrained connection. To arrive at a decision, he or she will

10 | *1 Fundamental Concepts of Joints in Design of Steel Structures*

Figure 1.6 Column bases at the Milan Central Railway Station. Source: Picture courtesy of Massimiliano Manzini.

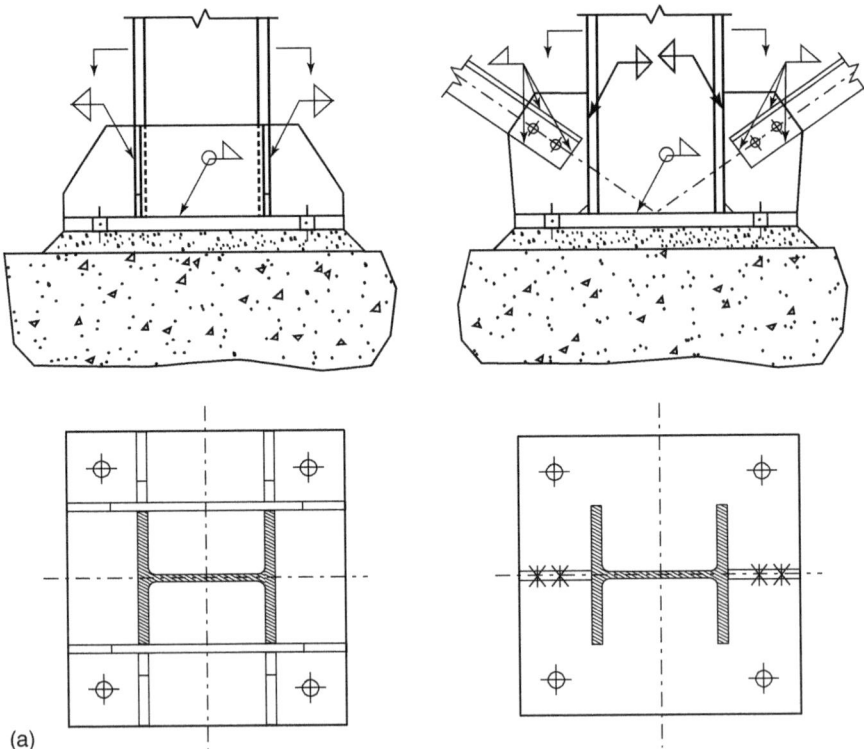

(a)

Figure 1.7 Base plate configurations that can be considered (last one excluded) as either a pin or a fully restrained connection. Source: Taken from [2].

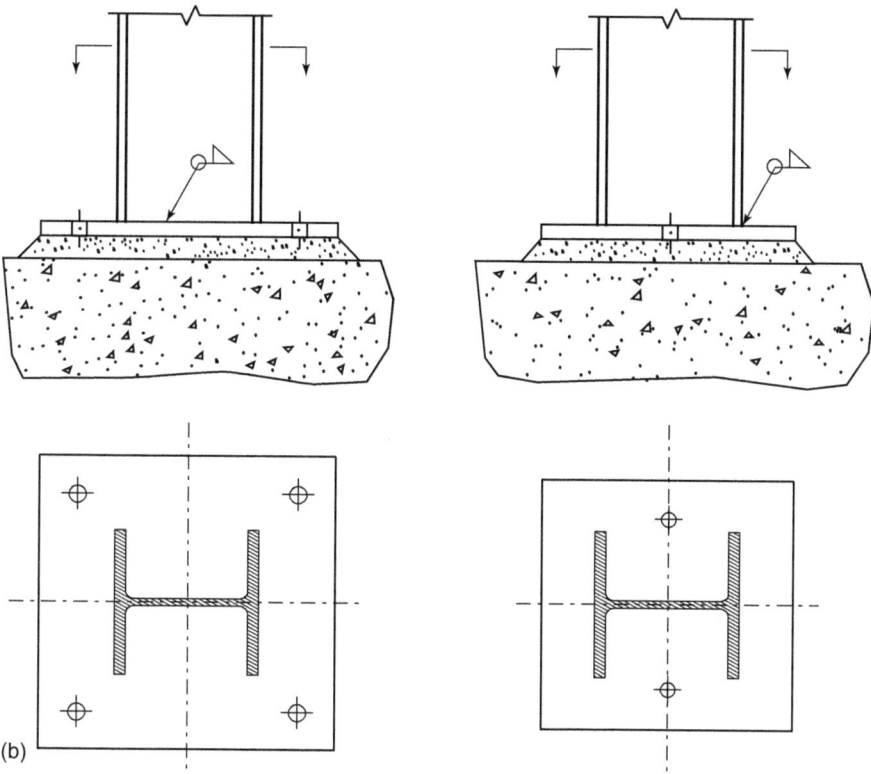

Figure 1.7 (Continued)

consider the lateral resisting system, the importance of lowering lateral displacements or adding hyperstatic restraints to better resist design forces (therefore saving some material but adding labor), and the characteristics of the foundation system and the soil (foundation costs are heavily impacted if rigid restraints have to be adopted).

The designer must carefully evaluate special situations: the project may be about designing a mezzanine inside an existing building/warehouse, leaning on the existing slab that should not be modified (due to either costs or possible delays in production); if a bending moment threatens to shear punch the concrete slab, it is certainly advisable to realize the connection with only a row of anchor bolts or without stiffening details in order to avoid any considerable moment, even if only initial.

1.2.2 Trusses

Truss connections are normally considered as pinned, even when welded (and the effective length factor taken as 1). The reason is that a plastic hinge will form: Even if the connection is initially rigid and able to resist non-negligible moments, it becomes a hinge after the material yields.

The history of steel construction confirms this method (and structural scheme) as conservative if the effective length factor is not taken less than 1 (which is the correct coefficient when the plastic hinges are in place).

References

1 American Society of Civil Engineers (ASCE) (2010). *Minimum Design Loads for Buildings and Other Structures*, ASCE/SEI 7–10. Reston, VA: ASCE Standard.
2 Ballio, G. and Mazzolani, F. (1983). *Theory and Design of Steel Structures*. London: Taylor & Francis.
3 Lescouarc'h, Y. (1982). *Le pied de poteaux articulés en acier*: CTICM. www.cticm.org (accessed 22 January 2018).
4 Lescouarc'h, Y. (1998). *Le pied de poteaux encastrés en acier*: CTICM. www.cticm.org (accessed 22 January 2018).

2

Fundamental Concepts of the Behavior of Steel Connections

The focus of this book, and in particular this chapter, is on the mechanisms that rule connections, instead of a series of on complicated calculation expressions.

Therefore, this chapter will not present formulas; rather the formulas will be provided in Chapter 3 and, to a lesser extent, in Chapter 4. Here, we will discuss ideas and concepts, some of which are not always clear and intuitive to a designer not very experienced with steel.

2.1 Joint Classifications

There are different possible classifications of connection types, a few of which are presented in the following discussion.

First, it is possible to divide them according to the rotational stiffness and the consequent capacity of transferring a bending moment. Under this connotation, joints can be hinges (pins), rigid (fully restrained) connections, and semirigid joints (intermediate behavior), as illustrated in Figure 2.1.

On the other hand, if we want to make the distinction according to strength (usually this is done for plastic analysis), the connection can be defined as partial-strength capacity (in the sense it will resist the calculated actions but not the largest actions that the connected member could transmit) or full-strength capacity if the joint is designed to resist the maximum force the connected member can carry (even if amplified, where, for seismic applications, the material yield value is larger than the minimum required by the standards and, therefore, actions could be bigger than the nominal values).

Another possible distinction from a seismic point of view and even, generally speaking, to reach a good performance design is with regard to ductility. This assessment is done by checking all the limit states to see if the governing limit state is ductile or nonductile. From an experimental point of view, ductility is measured by the deformation (usually rotation) capacity of a joint before its collapse.

Many other ways of sorting joints are possible, such as welding or bolting, or depending on the kind of connected members (e.g. beam-to-column, beam-to-beam, base joints). See Table 2.1.

Design and Analysis of Connections in Steel Structures: Fundamentals and Examples,
First Edition. Alfredo Boracchini.
© 2018 Ernst & Sohn Verlag GmbH & Co. KG. Published 2018 by Ernst & Sohn Verlag GmbH & Co. KG.

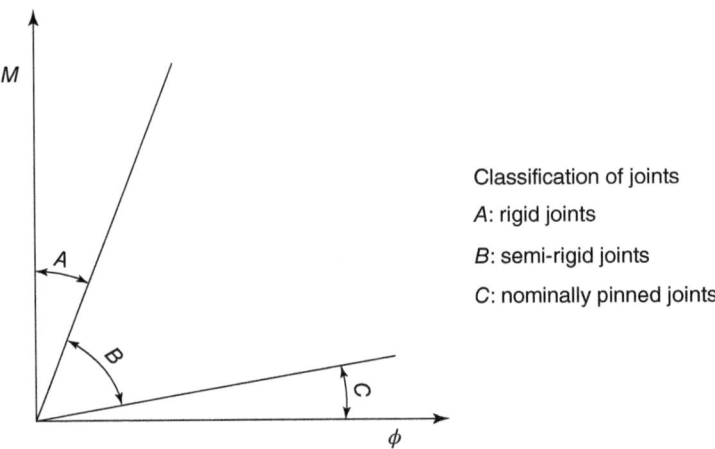

Figure 2.1 Moment-rotation diagram defines the joint behavior.

Table 2.1 Eurocode joint sorting.

Analysis method		Joint classification	
Elastic	Nominally pinned	Rigid	Semirigid
Rigid-plastic	Nominally pinned	Full strength	Partial strength
Elasto-plastic	Nominally pinned	Rigid and full strength	Semirigid and partial strength
			Semirigid and complete strength
			Rigid and partial strength
Joint modeling	Simple	Continuous	Semicontinuous

Source: Taken from Ref. [1]

2.2 Forces in the Calculation Model and for the Connection

Usually the engineer takes loads from his or her finite element analysis (FEA) (or finite element method (FEM)) and applies them to the connection. Actually some important considerations should be made because, according to the scheme assumed in connection design, forces can change.

Let us consider the possibility that the FEM model will connect joints by default along their axes. Although commercial software is available, which allows the connection location to be changed, since this process can be time-consuming, it is not done in many situations. Furthermore, it is not always necessary since the joint position can be left along the axes if the engineer takes this into consideration and applies a modified value to the connection design.

2.2 Forces in the Calculation Model and for the Connection

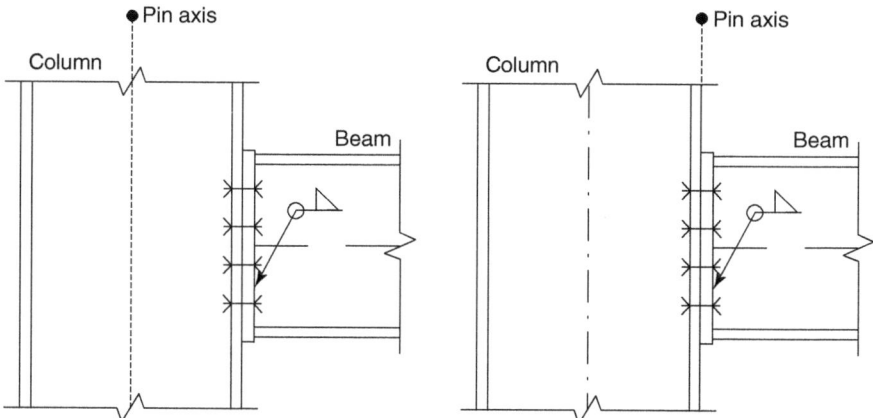

Figure 2.2 Possible locations for the axis of the connection.

For many connections, the engineer has the option of considering the location of the hinge in different positions. While perhaps initially not the case, a change can make a substantial difference in the design of bolts, plates, welds, and even columns and beams because local additional moments could be added (or removed) due to the assumed location of the pin.

Let us consider, as an example, a flexible end plate (which can be considered as a hinge) connected to a column flange. The exact pin location can be taken at any of the following positions (Figure 2.2):

1. On the column axis
2. At the contact point between the flange and the plate.

The assumption has notable consequences: in the second case, the column has an additional moment due to the joint eccentricity while in the first situation a non-negligible bending moment will stress the plates and bolts. Also, for congruity in the second case, the beam could be calculated on a reduced span, that is, instead of the distance between column axes, the distance between column flanges.

This important concept is valid for every type of connection and must be carefully evaluated in each instance by the engineer: sometimes there are small differences and assumptions are inconsequential but the changes are often meaningful.

Let us also notice that the two cases mentioned above are the ones usually referred to because one minimizes the eccentricity on the column (making it zero) and the other minimizes the actions on the bolt. However, there are an infinite number of possible cases and they are all acceptable as long as equilibrium is respected and the joint has enough ductility to go into the plastic state.

In other situations, there are more than two possible basic configurations. Consider, for example, a beam that is connected to a column by double angles bolted on both the beam web and the column flange.

Here there are three possible basic schemes (Figure 2.3), two as seen in the previous example of the end plate on the column flange and one on the axis of the bolt group connecting the beam.

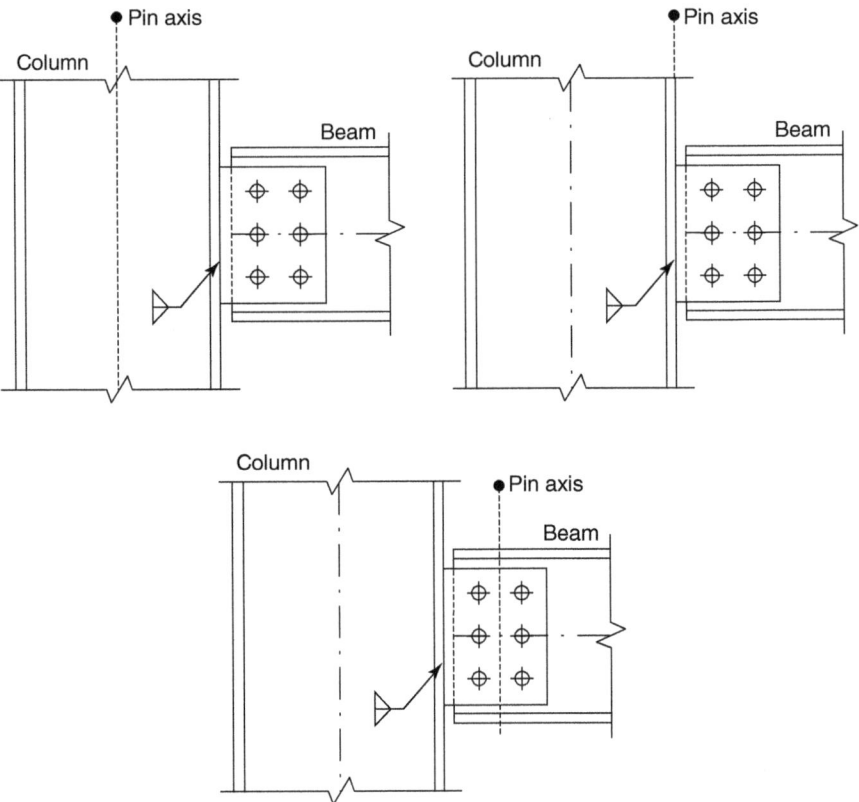

Figure 2.3 Possible locations of the joint axis.

Another example is a brace connection to a column and beam (see Section 4.14). Especially in the United States, different methods are presented to the engineering community and they are all valid: Some of them will minimize, roughly speaking, the bolts (uniform force method (UFM)), others the welds (L weld method), and others will give a very conservative design (KISS (keep it simple stupid) method).

At this point it is important to stress that any situation can be chosen if the balance of forces is correctly imposed. The engineer will likely choose the one that will allow reducing actions over the "preferred" element. As discussed, the actions on the connection elements can be reduced to the expenses of actions on the column, or the opposite can be done (according to preferences). For example, let us say that ¾-in. bolts were used in all the connections of a structure and that dimensioning one kind of connection to the column axis will make the connection bolts 1 in. as a minimum. At this point, if possible, the engineer could try to move the axis of the connection to reduce the forces on the bolts due to eccentricity. The engineer can therefore choose the optimum configuration as long as the calculations for each limit state are consistent with the hypothesis taken.

If there is no particular element design to be minimized, the engineer will likely choose the location of the hinge as suggested by his or her common sense.

This location is where the rotation capacity seems larger and the load path looks near the "real" path. Again, though, this is not the only way; it is just one of the possibilities. The real distribution of forces is the one that will maximize the load, in other words the one that can withstand the maximum actions.

An example given in [2] will help in understanding this idea: if some weight is supported by three steel bars, intuition says that the load will be uniformly divided. However, if the central bar bears no load, a possible solution for this design problem is to divide the load between the two external bars, which balances the distribution of weight and, therefore, can be deemed as acceptable. Assigning some percentage of the resisted load to the central bar, we can also obtain an acceptable design, due to steel ductility. However, it is true that the solution that will maximize the allowable load is the one where all bars take the same part of the load, and in fact this is the closest to the real solution and yet, once again, it is not the only one available.

We will see in the following chapters (in particular, Chapter 4) that, in each kind of joint, the engineer can choose among different possibilities in order to optimize the design but, not to be overlooked, it is crucial to remember that the global equilibrium must be respected and consistent while fragile mechanisms must be avoided.

2.3 Actions Proportional to Stiffness

The actions will distribute according to stiffness. This is true globally not only for the structural system but also locally for the connections. The basic concept is about parallel springs with different stiffness that share the force according to their stiffness (Figure 2.4). Structurally speaking, although there are other factors to consider, stiffness is fundamental.

This concept will help solve problems in nonordinary connections and will also help in understanding the response of complex systems. Let us say that forces "chase" stiffness.

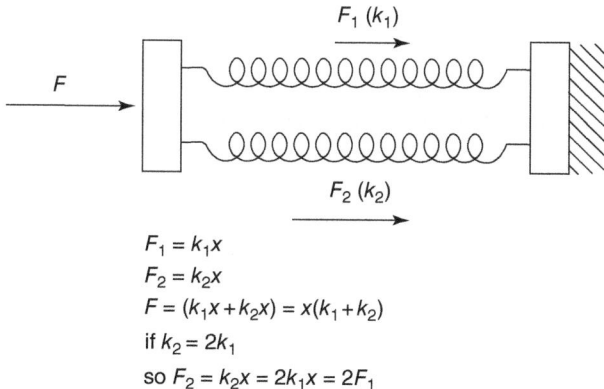

$F_1 = k_1 x$
$F_2 = k_2 x$
$F = (k_1 x + k_2 x) = x(k_1 + k_2)$
if $k_2 = 2k_1$
so $F_2 = k_2 x = 2k_1 x = 2F_1$

Figure 2.4 Actions on springs are uniformly proportional to their own stiffness.

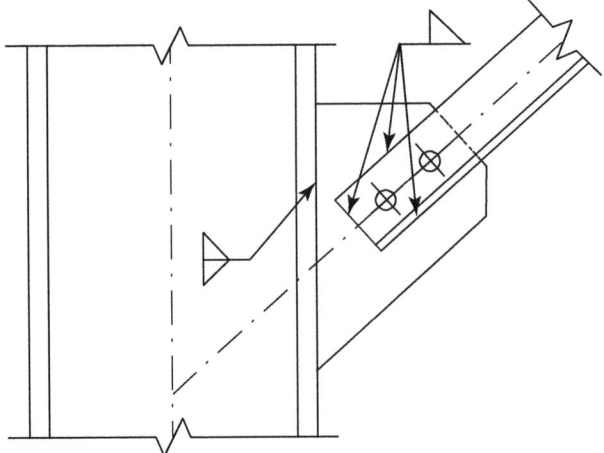

Figure 2.5 Connection with both welds and bolts sharing the load.

A possible example is a connection between plates that is realized with both bolts and welds (Figure 2.5). It is not feasible to divide the force between the welds and bolts because welds have much larger stiffness than bolts (unless they are designed for friction), and, therefore, the load will act primarily over the welds. If the weld breaks (nonductile), the bolts will support the load but, also in this case, the whole action must be resisted. The result is that a design that divides the forces might not work and could eventually have serious consequences (e.g. see Section 4.19 for some exceptions).

The notion of forces chasing rigidity is also effective in understanding why, once one connection evolves toward plasticity (forming, e.g. a plastic hinge), the later forces will redistribute to stiffer connections, allowing a distribution (ductility) that makes the steel a special material.

2.4 Ductility

Ductility is an important concept in a discussion of steel structures, even though it is dispensable.

Ductility allows for better use of steel resources, exploiting the plastic behavior (recall the example of the three bars in Section 2.2). It is however important to remember that a "fragile" design can also satisfy standards and yield acceptable results, usually at the cost of using more material and with lower chances of absorbing "extra" forces that are not well evaluated during the design phase. Using a seismic design example can confirm this: a ductile design (large response modification coefficient R or behavior factor q, depending on the standard name for it) will allow for consistently lower design forces compared to a nonductile design that does not have to worry about ductility. However, even a nonductile design, if correctly sized, is acceptable. Generally, it is well known that it is good if a ductile limit state governs the design so that a redistribution of forces is possible if the foreseen values are exceeded.

In other words, ductility is beneficial, as opposed to fragility. The latter has a collapsing mechanism that interrupts the transfer of loads, ending in collapse. In contrast, ductility will permit a loss in stiffness without breaking the transfer of forces, allowing other elements, if available (redundant system), to step in and collect additional forces. Thus, ductility can be seen as rotational capacity, meaning that joints with good rotational capacities and, more commonly, deformations are likely identifiable as ductile. The Steel Construction Institute [3] defines a rotation of 0.02–0.03 rad before collapse as the limit for connection ductility.

It is also crucial to remember which limit states display nonductile behavior:

- Shear bolt rupture
- Weld rupture
- Rupture (not yielding) of the net section for shear or tension
- Block shear (also called block tearing).

Additional comments about these as well as other limit states are given in Chapters 3 and 4.

2.5 Load Path

How does the load spread from one element to another? In other words, what is the path of the load when transferring from a secondary member to a primary member? It is important to consider this because it can uncover the possible problems a joint can experience.

It can help the engineer to think about the force as a "fluid," representable with arrows, which "flows" from one member (element) to the next in order to eventually bring all forces to the ground.

Consideration of the load path is fundamental in the design of connections and to understand the behavior of the joint, so as to avoid forgetting some local checks and thus prevent dangerous blunders. The example in Figure 2.6 shows how the axial force will converge and concentrate into the plate, representing how the secondary beam web is not allowed to transmit high axial forces with

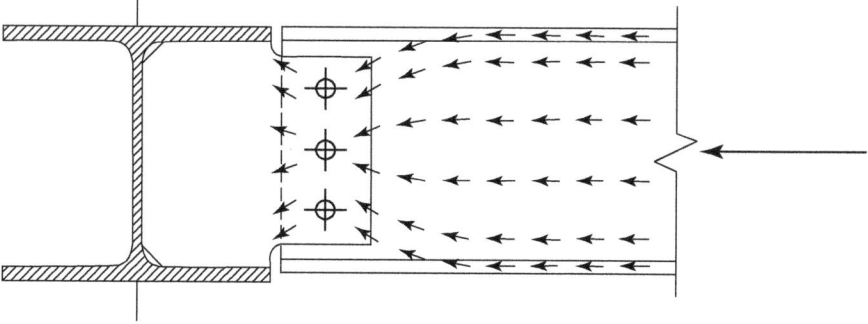

Figure 2.6 Path of the load.

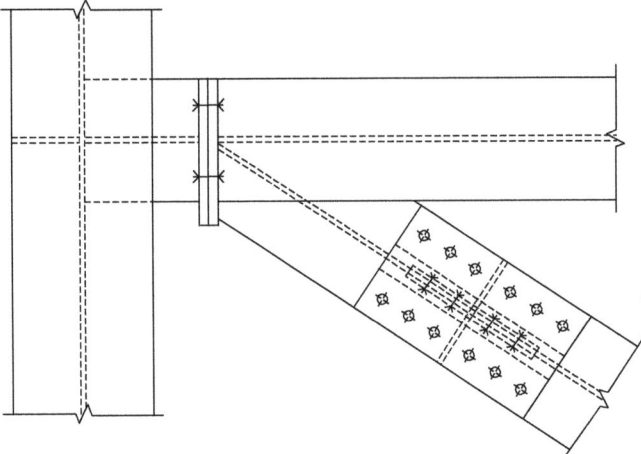

Figure 2.7 Wrong connection because the load path concept was missed.

such a connection (an additional reduction coefficient will have to be inserted because of shear lag, as per Section 3.19.1).

Let us also notice that the more classical simple connections might have single and well-defined load paths, in contrast to the multiple paths in more complex connections.

Following the load path can avoid serious design errors, as in Figure 2.7: The brace is connected at full strength but the forces in the brace must pass through the beam to get into the column, which does not look to be designed correctly with the inclusion of brace forces (and it would even include a remarkable eccentricity).

2.6 Ignorance of the Load Path

The structural engineer should remember that the load path is only an assumption and thus he or she does not know exactly the way the force is transmitted.

As engineer Brown wrote [4]: "Structural engineering is the art of molding materials we do not wholly understand into shapes we cannot precisely analyze, so as to withstand forces we cannot really assess, in such a way that the community at large has no reason to suspect the extent of our ignorance."

Beyond the scheme with hinges and fully restrained connections that, as we have already discussed, is not real but just a good hypothesis for calculations, many other simplifying assumptions are considered for analysis, such as isotropy, perfect elasticity (or perfect plasticity), no residual stresses, and no stress concentrations, just to mention the most important. Let us also remember that the forces we use in our FEMs are based on mere guesses.

Especially for steel structures it is well known that commonly used elastic theories would be wrong without consideration of plasticity (see Ref. [5] for advanced considerations about this topic) and important phenomena such as residual stresses and geometric imperfections are neglected when counting on plasticity resources. In other words, experience has shown that these simplified theories are conservative, the reason being that they implicitly rely on the ductility of the steel. If our materials were glass or ceramics, the theories would be seriously flawed and no longer applicable.

Additional elements that impact load paths (yet are normally neglected) are roof or floor components – slabs, sheets, panels, and grating – all of which can also transfer horizontal and vertical loads. Although they are usually ignored in order to be conservative, they do influence load paths.

It is appropriate to reiterate that the assumed load path is just one of the options within the many possibilities available; each is acceptable if equilibrium and congruity are maintained.

The "real" load path is, as already stated, the one that will maximize the load capacity of the structure. It is the artistic prowess performed by the engineer to get as close as possible, which helps to arrive at cost-effective solutions.

2.7 Additional Restraints

It is a corollary of plasticity theorems (the lower bound theorem in particular) that structural safety will not get worse if restraints are added.

This is another confirmation of the concepts expressed in Chapter 1: Considering some connections as pins, even though they are realized in a way that some moments can be resisted, is conservative as long as the joint can go into plasticity without instability or fragile breaks. This "check" can be empirically verified by the practice that says, after hundreds of years of steel structures, some connections can be considered as hinges even though considerable bending moments might develop at some stages (see Section 1.2.1 for an example).

When strictly applying EC design methods, though, stiffness assumptions should be analytically checked and, when necessary, springs should be included in the FEM.

2.8 Methods to Define Ultimate Limit States in Joints

Even though this is not helpful in "every day" practice, it is important to know that the theories at the root of joint calculation models are obtained by bringing joints to collapse. Some local yielding of a joint component (if, as seen, it is not coming with instability or fragile problems) is not necessarily an ultimate limit state and connection behavior is quite complex due to stress concentrations, strain hardening, effective widths, and force redistributions. Therefore, only experimental models and/or properly calibrated FEMs can help engineers to acquire the theories and formulas to help in the calculation. In short, only after thorough testing can the formulas discussed in this chapter (e.g. in the next section on bolt resistance) be defined.

2.9 Bolt Resistance

A bolt must not be considered a beam in bending since its "span" has the same order of magnitude as the diameter and, therefore, there is no relevant deformation in the elastic range. A bolt is hence commonly designed considering only shear, taking as a limit a value proportional to the ultimate resistance as determined by tests. This value is approximately 0.5–0.6, varying slightly according to standards and material classes.

2.10 Yield Line

The yield line method mentioned in many standards is based on the principle of virtual work, which is useful in structural steel engineering to define formulas for plate yield. A good explanation of the method is in [6], but a more detailed analysis dedicated to concrete structures is found in [7] since the method was originally kicked off in the 1950s and 1960s to calculate reinforced concrete slabs.

2.11 Eccentric Joints

Attention must be given to connections where eccentricity generates local moments to be withstood by the joint itself and/or the connecting members.

Let us differentiate between two different cases. In the first situation, the eccentricity is a routine element of connections, for example, when there is eccentricity in the bolt group of a shear tab because the connection axis was taken on the primary member. In the second case, the eccentricity and its moment can occur unexpectedly because of poor design. Those bending moments are parasitical and come out of inattentive design or detailing, for example, in trusses or bracings where the detailer or the fabricator has not been correctly briefed about the possible problem.

If the design sketches represent the neutral axis and the detailers are correctly instructed, this category of parasitic eccentricities can be avoided.

If the eccentricity is unavoidable (due to fabrication, erection, or other reasons), it is important that the detailer informs the engineer of all eccentricities present in order to make the necessary recalculations. This could mean using larger bolts, stiffeners, or thicker plates or making a change in the profile sizes.

2.12 Economy, Repetitiveness, and Simplicity

As a basic concept, it is true that connection economy comes from simple and repetitive connections.

Some studies [2, 8] define the impact of connections as 30–50% of the total cost of a structure (while the weight of the connections is usually less than 15% when compared to the global weight). This clarifies just how important the connection design is, not only from a safety point of view, but also with regard to economic competitiveness.

2.13 Man-hours and Material Weight

If the statement in the previous paragraph is true, implying that simplicity must be embraced, the question remains then how to make decisions when a simple design can make a structure heavier?

The engineering community is almost unanimous in recommending simplicity, eliminating, for example, stiffeners (e.g. as in end plates and base plates) even though this usually means using more materials (that is, thicker plates).

However, several fabricators prefer to weld stiffeners and save on plate thickness, possibly because they already have something in-house or just to use more uniform thicknesses. Another situation that could make the fabricator opt for more labor is if the company is not busy, so as not to lay workers off, the choice of saving on material and using more labor could prove cost-effective.

The advice is consequently to discuss the matter with the fabricator and choose accordingly.

2.14 Diffusion Angles

As a basic concept, a 45° force distribution is applicable in several situations, as in Figure 2.8, which illustrates a cantilever or a similar system. See Ref. [9] for a demonstration of how to calculate the angle (the exact value would be 48°).

There are some exceptions, among which the following should be deemed notable:

- A tension load on a flat plate (used to find the Whitmore section) which is considered distributing at 30° on each side (see Section 3.18.7).
- An action (usually from an end plate) that spreads into the column flange and then to web, as in Figure 2.9, which can be taken to distribute on each side with a 2.5 : 1 ratio, that is, with a 68° angle (like the stated assumption in [1, 10]).

Figure 2.8 Effective width of a cantilevered plate loaded by a concentrated force.

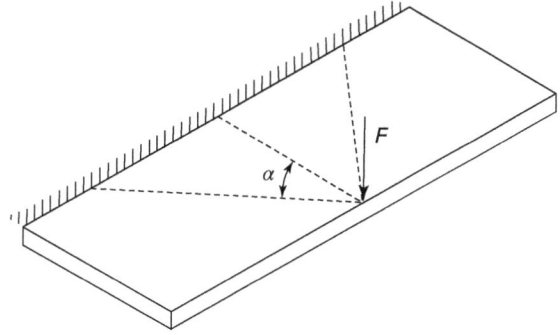

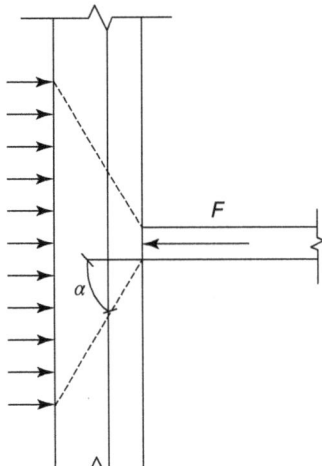

Figure 2.9 Force diffusion from a beam flange (right-hand side) into a column (left-hand side).

2.15 Bolt Pretensioning and Effects on Resistance

There are at least three questions to answer when talking about bolt pretensioning because the common perception is sometimes opposite to what standards, tests, and books report.

2.15.1 Is Resistance Affected by Pretensioning?

It is important to note that in high-resistance bolts pretensioning, only negligibly, modifies their resistance.

This is documented for example in [11]. The evidence provided is experimental: bolts pretensioned beyond the recommended limit and bolts that are not pretensioned have a similar collapse load, with differences that are less than 10%. Pretensioning therefore has a marginal effect over collapse load.

2.15.2 Is Pretensioning Necessary?

Pretensioning comes at a cost because one method must be applied (see Section 6.8), an inspection later disposed, and, as we just saw, it can negatively impact, even if only marginally so, the performance of the bolt. For all these reasons, it should be prescribed only when necessary.

There are US guidelines (Ref. [10] and others) that recommend use of pretensioning in the following situations:

- Bolts that work in tension (to avoid separation of parts) or combined tension and shear
- Slip-critical connections, for example, when there are slots and oversize bolts
- Meaningful bolt load reversals
- Fatigue

- Connection between columns and braces in tall buildings (i.e. about 120 ft, or 40 m) and column splices in buildings with remarkable height-to-width ratios
- Connections supporting cranes with capacity over 5 tons.

Eurocode 1993-1-8 requires pretensioning for categories B, C (shear with friction resistance respectively for serviceability and ultimate limit states), and E (pretensioned joints working in tension).

This EC adds an additional note pertaining to this stating that, when pretensioning is not strictly required, it can be specified because of durability. If pretensioning is not considered in design checks (it can occur if there are friction-resistant connections), it becomes a personal decision whether to prescribe it or not. Thus, even though there might be some performance improvements (at a cost), pretensioning is not strictly necessary.

2.15.3 Which Pretensioning Method Should Be Used?

Section 6.8 highlights the fact that using a torque wrench is not the only method and, while contrary to common perception, it is not the most trustworthy and simple of options, just one of the many with its own disadvantages.

2.16 Transfer Forces

If several members are connected to the same joint, the transfer forces for the connection are not commonplace. For example, if a beam frames into a column but there are also other beams or braces on the other side of the column, it is not clear how forces divide and, for example, what is the shear taken by the column. This means the engineer should check how forces from the calculation model should be vector summed. The problem is that, sometimes, depending on the local bidding regulations, the model is not available because the connection designer is not the designer of the structure. The construction design drawings issued to fabricators that design connections should provide the forces case by case rather than (as it occurs frequently) simply giving the maximum and minimum (absolute) values or it could result in incorrect transfer forces (see some interesting examples in Appendix D of [12]) in addition to making the design too conservative and uneconomic.

Another similar but different concept is illustrated in Figure 2.10. It shows two situations that are likely modeled equally when using the FEM software but the forces to consider in the joint design are actually quite different. Analogous situations can occur in alternate cases too, especially if bracings are involved (see Section 4.14).

2.17 Behavior of a Bolted Shear Connection

Figure 2.11 sketches the standard behavior of a shear joint:
- Curve *a* represents an elastic deformation where friction is the resisting force of the bolts.

26 | *2 Fundamental Concepts of the Behavior of Steel Connections*

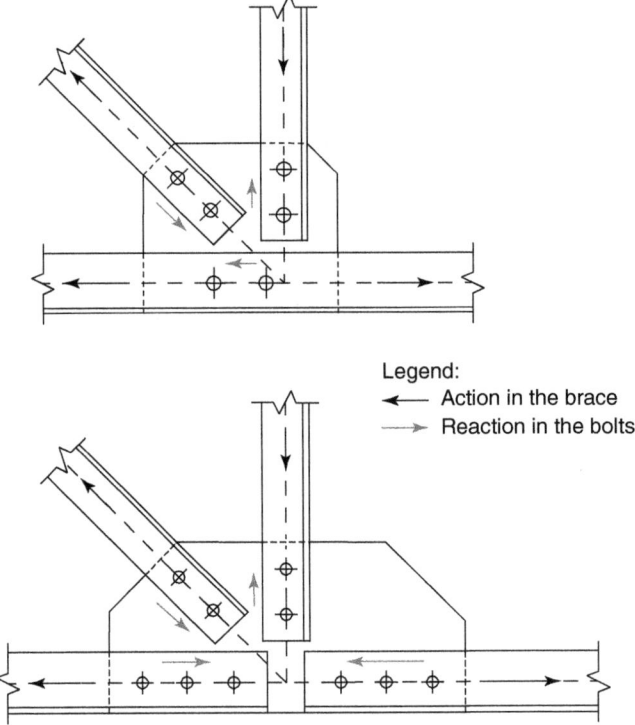

Legend:
⟵ Action in the brace
⟶ Reaction in the bolts

Figure 2.10 Joint load versus joint configuration.

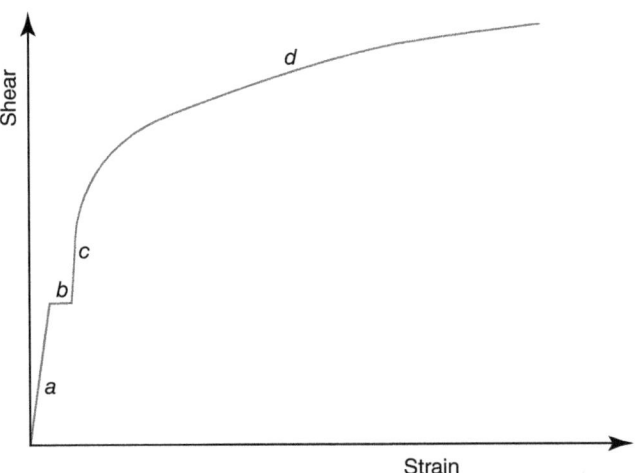

Figure 2.11 Behavior of a shear connection.

- In *b* there is a slight but definite displacement due to sliding (the acting force goes over the resisting friction).
- In *c* the resisting response by the bolts is by true contact, resulting in the parts deforming elastically.
- The net area (then the gross) of the plates (parts) in the connection eventually yields, progressing into *d*, where the deformation is plastic, until the joint collapses in any of the components (shear, tear-out, block shear, etc.).

Looking at the diagram, it becomes clear that if the bolts break in *c* before plates yield in *d*, the joint is fragile because the plastic part in *d* might allow a redistribution of forces due to a change in stiffness (the forces would chase stiffer parts, similar to what we saw in Section 2.3).

A plate that is too "strong" could consequently be counterproductive (example of resistance hierarchy) for ductility because the bolt shear should not be the lower (governing) limit state. As the graph shows, it is advisable that limit states like plate yielding or bearing step in before the bolts collapse, since those ductile limit states allow the joint to develop the plasticity necessary to lower the stiffness and redistribute the forces to other parts of the structure.

2.18 Behavior of Bolted Joints Under Tension

Section 2.15 discusses a few advantages pertinent to pretensioning if the joint is working in tension. One advantage is that pretensioning makes sure there is no separation unless loads go well beyond service loads.

As sketched in Figure 2.12a, the applied preload (PL) will generate a pressure (ppl) that keeps the parts under strict contact, creating a plate compression and a tension on the bolt that will elongate it. The force *P* (Figure 2.12b) will, therefore, only lower the pressure originally given by the preload, yet it will not separate

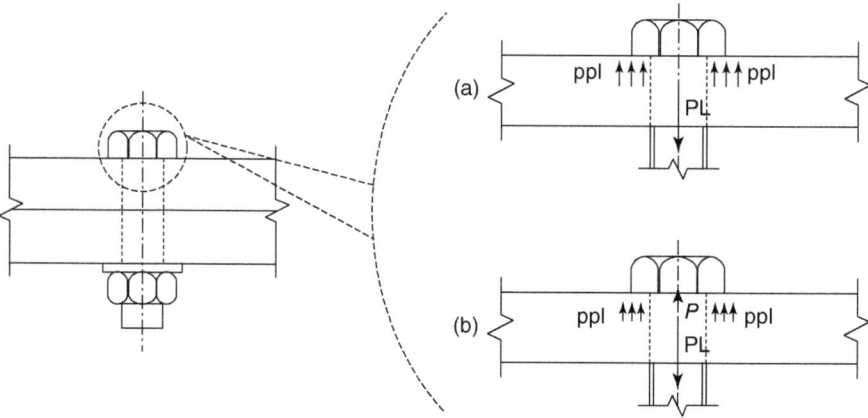

Figure 2.12 Preload effects.

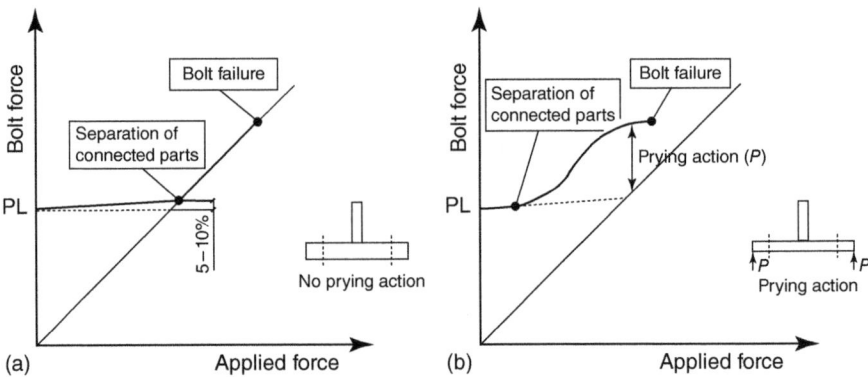

Figure 2.13 Bolted connections in tension.

the parts and significantly increase the bolt tension so long as P approximately reaches the preload (at least in the rigid-plate case). The exact value when parts separate depends on the prying action that, as we will see in Chapter 3, is based mainly on the connecting plate thickness. However, if the plate is "flexible," the force will bend the plate and pry the bolt loose, causing premature detachment.

The two cases in Figure 2.13 summarize the behavior when dealing with bolted connections in tension.

Empirically, when there is a separation of connected parts as illustrated in case Figure 2.13a with no prying action, the load on the bolts is about 5–10% higher than the preload and the stress for the bolt does not have to add the prestress because it is in fact almost independent. For a more detailed discussion of this the reader is referred to [13].

Notice that even in the prying action case illustrated in Figure 2.14 the preload does not influence the collapse load (as already mentioned in Section 2.15).

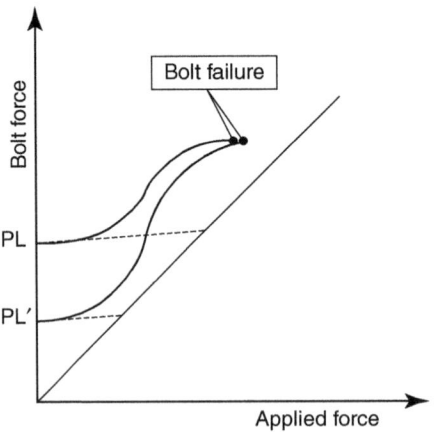

Figure 2.14 Different preload diagrams in a tension connection with prying action.

References

1 CEN (2005). *Eurocode 3: Design of Steel Structures – Part 1–8: Design of Joints*, EN 1993-1-8: 2005. Brussels: CEN.
2 Muir, L.S. and Thornton, W.A. (2009). *Practical Connection Design for Economical Steel Structures*. Chicago, IL: American Institute of Steel Construction.
3 The Steel Construction Institute (SCI), The British Constructional Steelwork Association (BCSA) (1995). *Joints in Steel Construction: Moment Connections*. Ascot, UK: SCI, BCSA.
4 Brown, E.H. (1967). *Structural Analysis*, vol. 1. New York: Wiley.
5 Massonet, C. and Save, M. (1993). *Calcolo plastico a rottura nelle costruzioni*. Milano: Città Studi.
6 Murray, T.M. and Sumner, E.A. (2003). *Extended End Plate Moment Connections – Seismic and Wind Applications*, 2e, AISC Design Guide 4. Chicago, IL: American Institute of Steel Construction.
7 Kenedy, G. and Goodchild, C. (2003). *Practical Yield Line Design*. Cambridge: British Cement Association, Reinforced Concrete Council.
8 The Steel Construction Institute (SCI), The British Constructional Steelwork Association (BCSA) (2002). *Joints in Steel Construction: Simple Connections*. Ascot, UK: SCI, BCSA.
9 Ballio, G. and Mazzolani, F. (1983). *Theory and Design of Steel Structures*. London: Taylor & Francis.
10 American Institute of Steen Construction (AISC) (2011). *Steel Construction Manual*, 14e. Chicago, IL: AISC.
11 Kulak, G. (2002). *High Strength Bolts, a Primer for Engineers*, AISC Design Guide 17. Chicago, IL: American Institute of Steel Construction.
12 Muir, L.S. and Thornton, W.A. (2014). *Vertical Bracing Connections – Analysis and Design*, AISC Design Guide 29. Chicago, IL: American Institute of Steel Construction.
13 Bickford, J.H. (1995). *An Introduction to the Design and Behavior of Bolted Joints*, 3e. Boca Raton, FL: CRC Press.

3

Limit States for Connection Components

It is essential in structural steel design to check each single-limit state in connections or, in other words, to check the limit states of all the components of a joint: bolts, welds, plates, and profiles (if modified to fit the connection, i.e. if notched).

The connections are indeed made by components and each of them needs to be proven for strength, stability, deformation, and anything else that gives satisfactory performance and safety to the structural system.

3.1 Deformation Capacity (Rotation) and Stiffness

Current standards set limits dividing connections between pins and rigid connections, even providing instructions to deal with semirigid joints. Also, rotation capacity can be verified following those guidelines and this confirmation becomes important for plastic design (e.g. see Ref. [1]).

However, these methods (outlined for only some kinds of connections and with important limitations that cannot be overlooked) are difficult to apply, which means that the normal industry practice has not fully incorporated them yet. Engineers still mainly use the faster (and safer, as proven from years of structural steel constructions) and, therefore, more inexpensive and simple division of rigid and pin joints, as explained elsewhere in this book, which aspires to be "practical" and informative about the common "real" habits of design firms. This does not negate the need for engineers to subsequently make sure that the connection deformation capacity is consistent with the design assumptions.

For the typical joints described in this book, the instructions here given, joint by joint, coupled with member displacements (in other words, joint rotation) in the standard limits (which should be by rule) supposedly guarantee that the deformation capacity is acceptable.

For different nonstandard situations, this aspect should be evaluated case by case, possibly with more sophisticated tools.

Considering that important standards like Eurocode (EC) demand an analytical check of the rotational stiffness, the engineer should become more familiar with this approach, as discussed in the next section.

Connection stiffness is indeed an important parameter, not just for the connection, but to correctly assess forces in the global structural model (displacements and actions change according to joint stiffness).

Design and Analysis of Connections in Steel Structures: Fundamentals and Examples,
First Edition. Alfredo Boracchini.
© 2018 Ernst & Sohn Verlag GmbH & Co. KG. Published 2018 by Ernst & Sohn Verlag GmbH & Co. KG.

3.1.1 Rotational Stiffness

Eurocode provides guidance on rotational stiffness evaluation for several types of joints, but their practical application is rather laborious and is limited by hypotheses that are not always verifiable. For example, the general formula from [1] is based on the premise that the design value of the axial force N_{Ed} in the connected member does not exceed 5% of the plastic resistance of its cross-section; in addition, the formula only holds true for joints connecting H or I sections ([1] does not provide information for other types of profiles). The fulfillment of these conditions allows applying the formula

$$S_j = \frac{Ez^2}{\mu \sum_i \frac{1}{k_i}}$$

where S_j is the rotational stiffness of the joint formed by the various i components, k_i is the stiffness coefficient for each i component, z is the lever arm (it will change with the type of joint; additional information can be found in Chapter 4). The stiffness ratio μ can be evaluated as 1 if $M_{j,Ed} \leq 0.66 M_{j,Rd}$; otherwise it can be obtained as

$$\mu = \left(\frac{1.5 M_{j,Ed}}{M_{j,Rd}}\right)^{\psi}$$

in which the coefficient ψ is 3.1 for bolted angle flange cleats and 2.7 for all the remaining cases (welded, bolted end plate, base plate).

It is important to remember that the initial rotational stiffness $S_{j,ini}$ of the joint is given by the same expression for S_j setting $\mu = 1$.

For the entry of S_j in the global analysis model, refer to the indications in [1]. In the case of elastic global analysis, its value can be simplified as $S_{j,ini}/\eta$, where η is the stiffness modification coefficient as found in Table 3.1.

A joint may be classified as nominally pinned, rigid, or semirigid according to its rotational stiffness by comparing its initial rotational stiffness $S_{j,ini}$ with the value

$$k \frac{EI_b}{L_b}$$

where I is the second moment of area (if the subscript is b it concerns a beam, if the subscript is c it concerns a column) and L is the length (to be more precise it

Table 3.1 Stiffness modification coefficient H according to Eurocode.

Type of connection	Beam-to-column joints	Other types of joints (beam-to-beam joints, beam splices, column base joints)
Welded	2	3
Bolted end plate	2	3
Bolted flange cleats	2	3.5
Base plates	—	3

corresponds to the story height for columns and to the span, taken as center to center between columns, for beams). Regarding the coefficient k, the value 0.5 is taken as the upper limit to define a joint as nominally pinned. Therefore, if

$$S_{j,\text{ini}} \leq 0.5 \frac{EI_b}{L_b}$$

the joint has to be considered nominally pinned.

The lower bound for the classification as a rigid joint depends on the presence or absence of a bracing system that can reduce the horizontal displacement by at least 80%. If there is such a bracing system, k (k_b in EC) is equal to 8; otherwise it is necessary to evaluate the ratio

$$\frac{I_b/L_b}{I_c/L_c}$$

If the ratio is less than 0.1, the joint should be classified as semirigid. If the ratio is ≥ 0.1, k is taken as 25 and the joints classified as rigid will be the ones with $S_{j,\text{ini}}$ greater than the value calculated as shown.

For an in-depth analysis of the matter, the reader is referred to various texts (e.g. Refs. [2–4]).

3.2 Inelastic Deformation due to Bolt Hole Clearance

In addition to the discussion in the previous section, a possible effect of connection design on structure deformations might be given by bolt hole clearances that could lead to undesired inelastic deflections.

A typical example is a fully bolted truss (i.e. with members of the truss web also bolted) in which the accumulated tolerance of the holes (with respect to the bolts) may lead to a permanent deflection already above the initial reference limit.

This deflection is inelastic since it is not recovered elastically when the structural element is unloaded. Unless special precautions are adopted when tightening the bolts, the deflection appears due to the natural position assumed by members because of their weight. Even if special care is dedicated to tightening the bolt under its own weight (e.g. bolting it to the ground in a horizontal position and then raising the entire truss assembly), work stresses that go beyond the friction limit could still cause a problem.

A 10-m-(33-ft)-long truss with 10 diagonal struts, 10 vertical struts, and horizontal members also divided into 10 parts, with bolt hole clearances of 2 mm (1/16 in.), would likely collect up to 3–4 mm (1/8 in.) of inelastic deformation for each joint (see Figure 3.1 and notice the 4-mm axis-to-axis distance reached between the left and right struts) and, therefore, a total $10 \times 4\,\text{mm} = 40\,\text{mm}$ (1.5 in.), which is 1/250 of the span with only the weight of the truss.

One possible applicable solution to prevent this is by prescribing the necessary precamber in the shop drawings. Another possibility as already mentioned is to pre-erect some parts on the ground where there is no gravity force deforming the structure.

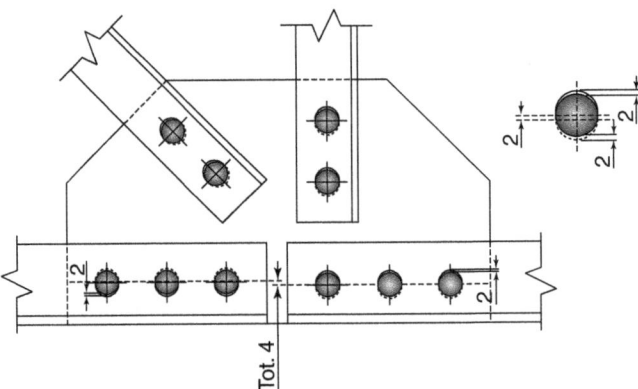

Figure 3.1 Maximum possible "permanent" deformation: each plate adds 1 mm, which means 1 + 1 mm in the left horizontal member and an additional 1 + 1 mm on the right.

3.3 Bolt Shear Failure

The shear failure of the bolt is one of the most intuitive limit conditions.

It is necessary to highlight that the shear failure of the bolt is a brittle limit state, and therefore it is not desirable that it control the design (i.e. it should not be the first critical limit state for the connection).

The discussion in Section 2.9 is depicted in Tables 3.2 and 3.3.

The AISC defines (Table 3.4) the shear resistance–ultimate axial resistance ratio as 0.625 for AISC bolts [5], but it is reduced by multiplying it by a factor of 0.9 (it was 0.8 until the 13th version of the manual [5]) to account for the nonuniform stress distribution (the first rows of bolts in both directions are more stressed); to this end see also Section 3.3.4.

According to EC, the design shear resistance (per shear plane) $F_{v,Rd}$ of a bolt is

$$\frac{\alpha_v f_{ub} A}{\gamma_{M2}}$$

Table 3.2 Values for class of bolts – Eurocode 3, it is a so-called nationally determined parameter (NDP) and it may vary by country.

Class	f_{yb} (N mm^{-2})	f_{ub} (N mm^{-2})	α_v	γ_{M2}
4.6	240	400	0.6	1.25
4.8	320	400	0.5	1.25
5.6	300	500	0.6	1.25
5.8	400	500	0.5	1.25
6.8	480	600	0.5	1.25
8.8	640	800	0.6	1.25
10.9	900	1000	0.5	1.25

Table 3.3 Values for class of bolts – DIN 18800.

Class	$f_{y,b,k}$ (N mm^{-2})	$f_{u,b,k}$ (N mm^{-2})	α_a	γ_M
4.6	240	400	0.6	1.1
5.6	300	500	0.6	1.1
8.8	640	800	0.6	1.1
10.9	900	1000	0.55	1.1

Table 3.4 Values for class of bolts – AISC (revisited for X-type bolts).

Class	F_{nv} (N mm^{-2})	F_u (N mm^{-2})	F_{nv}/F_u	Φ
A325	465	827	0.563 ($\approx 0.625 \times 0.9$)	0.75
A490	585	1040	0.563 ($\approx 0.625 \times 0.9$)	0.75

where A is the reference area of the bolt (see Section 3.3.1) and α_v is the coefficient shown in Table 3.2. If the shear plane passes through the unthreaded portion of the bolt, α_v remains 0.6 independent of the bolt class.

The DIN (German Institute for Standardization) has (see Table 3.3) a similar equation ($f_{u,b,k}$ corresponds to f_{ub}):

$$\frac{\alpha_a f_{u,b,k} A}{\gamma_M}$$

The AISC nominal resistance R_n (which has to be multiplied by Φ) for bolts with the threaded portion external to the shear plane (type X bolts in the United States, where X stands for eXcluded) is defined as

$$F_{nv} A_b$$

where A_b is the nominal area of the bolt and F_{nv} is the maximum shear stress the reference material can bear. If the threaded portion is in the shear plane (type N bolts in the United States, where N stands for iNcluded), the value has to be divided by 1.25.

See Table 3.7 for some results (maximum design shear) for metric bolts.

The concept of bolt shear failure is clear and intuitive while two concepts just mentioned are less intuitive and need to be discussed: failure also depends on the number of resistant sections and the presence of the bolt thread inside the shear plane. More detail of those concepts will be provided in the following sections.

3.3.1 Threads Inside the Shear Plane

If the shear plane goes through the threaded portion of the bolt, then the net resistant area must be used when checking shear. If, instead, the bolt is only partially threaded and the threads stop before the shear plane (part of the available literature suggests that, as a more precautionary step, the point at which the thread should stop is at the washer), the whole gross cross-sectional area can be considered, with a sensible increase of the resistant area (around 20–35% is gained, depending on the diameter of the bolt; see Table 3.5). The AISC (Table 3.6), as seen earlier, instead considers a standard reduction if the threads are included in the shear plane by dividing the nominal area by 1.25.

Many fabricators/erectors utilize fully threaded bolts, and therefore it is advisable to use the net area as a precautionary first step. However, the usage of bolts with the right thread length (matched with the competence to give to erectors, who have to be mindful of this important detail) can lead to savings, especially in

Table 3.5 Area for standard metric bolt.

Metric bolt (mm)	M10	M12	M14	M16	M18	M20	M22	M24
d: mm (in.)	10 (0.39)	12 (0.47)	14 (0.55)	16 (0.63)	18 (0.71)	20 (0.79)	22 (0.87)	24 (0.94)
A: mm^2 (in.^{2})	78 (0.12)	113 (0.18)	154 (0.24)	201 (0.31)	254 (0.39)	314 (0.49)	380 (0.59)	452 (0.70)
A_s: mm^2 (in.^{2})	58 (0.09)	84 (0.13)	115 (0.18)	157 (0.24)	192 (0.30)	245 (0.38)	303 (0.47)	353 (0.55)
Metric bolt (mm)	M27	M30	M33	M36	M39	M42	M45	M48
d: mm (in.)	27 (1.06)	30 (1.18)	33 (1.30)	36 (1.42)	39 (1.54)	42 (1.65)	45 (1.77)	48 (1.89)
A: mm^2 (in.^{2})	573 (0.89)	707 (1.10)	855 (1.33)	1018 (1.58)	1195 (1.85)	1385 (2.15)	1590 (2.46)	1810 (2.81)
A_s: mm^2 (in.^{2})	459 (0.71)	581 (0.90)	694 (1.08)	817 (1.27)	976 (1.51)	1120 (1.74)	1310 (2.03)	1470 (2.28)

Note: A is the nominal area and A_s is the net area of the threaded portion.

Table 3.6 Area for standard imperial bolts.

Imperial bolt (in.)	½	⁵/₈	¾	⁷/₈	1	1⁷/₈	1¼	1³/₈	1½
d: in. (mm)	0.5 (12.7)	0.625 (15.9)	0.75 (19.1)	0.875 (22.2)	1 (25.4)	1.125 (28.6)	1.25 (31.8)	1.375 (34.9)	1.5 (38.1)
A_b: in.^{2} (mm^2)	0.20 (127)	0.31 (198)	0.44 (285)	0.60 (388)	0.79 (507)	0.99 (641)	1.23 (792)	1.48 (958)	1.77 (1140)
A_{net}: in.^{2} (mm^2)	0.16 (101)	0.24 (158)	0.35 (228)	0.48 (310)	0.63 (405)	0.80 (513)	0.98 (633)	1.19 (766)	1.41 (912)

Note: A_b is the nominal area.

3.3 Bolt Shear Failure

the case of heavy joints (e.g. for full-capacity bolted connections of braces when seismic ductility is required).

3.3.2 Number of Shear Planes

The engineer must have a clear understanding that a connection on two resisting sections will halve the shear stress of the bolt (Figure 3.2).

As for the previous point (threads included or excluded), this notion can be used to contain the shear stress of bolts.

German DIN (now superseded but here referenced because it is still used and it sometimes provides interesting advice) supplies an interesting indication regarding the potential problem for a bolt in double shear but with only one shear plane crossing the thread. DIN recommends evaluating the two resistances separately, then summing them and comparing the result with the applied load. Also, Australian standard AS 4100 [6] and Indian standard IS 800 [7] embrace this approach.

3.3.3 Packing Plates

According to EC, when bolts transmitting load in shear pass through packings (Figure 3.3) of total thickness t_p greater than one-third of the nominal diameter d, the design shear resistance $F_{v,Rd}$ should be reduced by multiplying it by a reduction factor $\beta_p \leq 1$ defined as

$$\beta_p = \frac{9d}{8d + 3t_p}$$

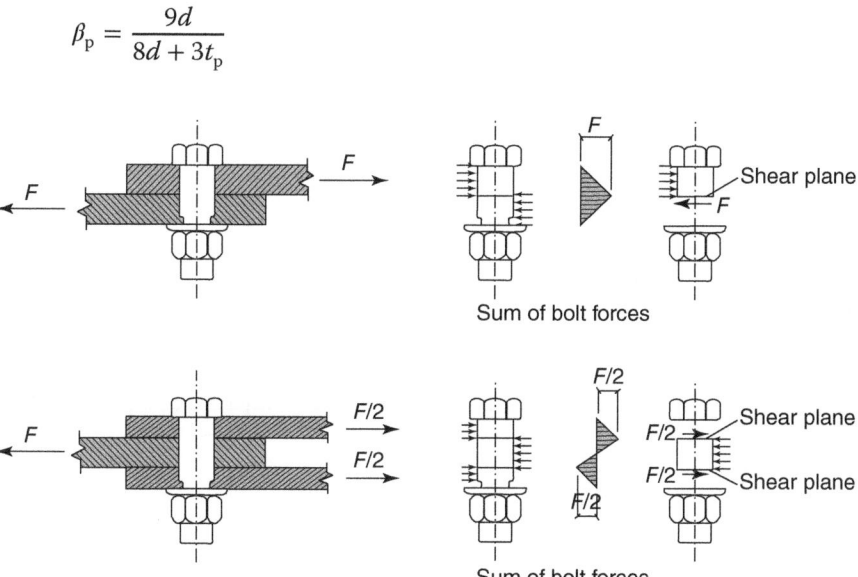

Figure 3.2 Shear planes and effects on bolts.

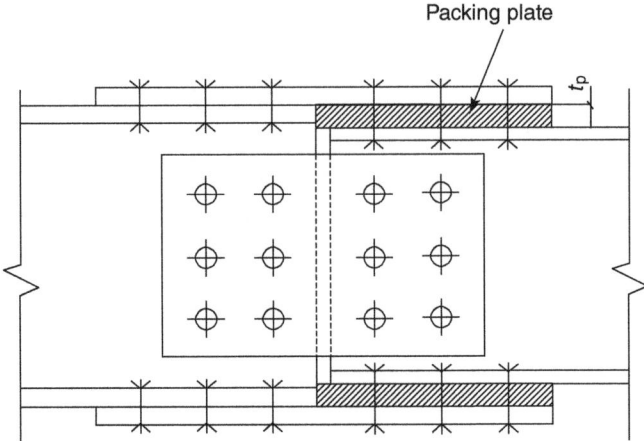

Figure 3.3 Packing plates in a splice connection.

For double-shear connections with packings on both sides of the splice, t_p should be taken as the thickness of the thicker plate.

According to [8], packing plates should not exceed a quantity of 3 and should not have a thickness less than 2 mm.

3.3.4 Long Joints

Eurocode says that when the distance L_j between the centers of the end fasteners in a joint, measured in the direction of the force transfer (Figure 3.4), is more than 15 times the diameter, the design shear resistance $F_{v,Rd}$ of all the fasteners should be reduced by multiplying it by a factor β_{Lf} ($0.75 \leq \beta_{Lf} \leq 1$) given by

$$\beta_{Lf} = 1 - \frac{L_j - 15d}{200d}$$

This means that for a "standard" bolt pitch equal to three times the hole diameter, the reduction is applied (although with an initially negligible factor) even for five to six rows of bolts.

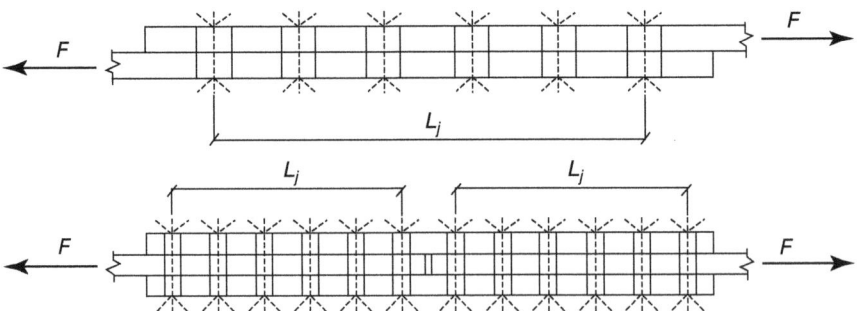

Figure 3.4 Long joints.

Table 3.7 Maximum design shear ($\gamma_{M2} = 1.25$) for metric bolts.

Bolt: Class 8.8, type N	M10	M12	M14	M16	M18	M20	M22	M24
$F_{v,Rd}$, EC: kN (kips)	22 (4.9)	32 (7.2)	44 (9.9)	60 (13.5)	74 (16.6)	94 (21.1)	116 (26.1)	136 (30.6)
$V_{a,Rd}$, DIN: kN (kips)	25 (5.6)	37 (8.3)	50 (11.2)	69 (15.5)	84 (18.9)	107 (24.1)	132 (29.7)	154 (34.6)
ΦR_n, AISC: kN (kips)	21 (4.7)	31 (7.0)	42 (9.4)	54 (12.1)	69 (15.5)	85 (19.1)	103 (23.2)	122 (27.4)
Bolt: Class 8.8, type N	M27	M30	M33	M36	M39	M42	M45	M48
$F_{v,Rd}$, EC: kN (kips)	176 (39.6)	223 (50.1)	266 (59.8)	314 (70.6)	375 (84.3)	430 (96.7)	503 (113)	564 (127)
$V_{a,Rd}$, DIN: kN (kips)	200 (45)	254 (57.1)	303 (68.1)	357 (80.3)	426 (95.8)	489 (110)	572 (129)	641 (144)
ΦR_n, AISC: kN (kips)	155 (34.8)	191 (42.9)	231 (51.9)	275 (61.8)	323 (72.6)	374 (84.1)	429 (96.4)	489 (110)

Note: Unless additional coefficients apply, as explained in the text, for threads included in the shear plane, that is "N" type according to the US definition. Not recommended for structural applications.

It has been experimentally verified that end fasteners are subject to greater stress than internal fasteners and, therefore, are the first to fail, causing a domino collapse of the other rows. AISC, as previously mentioned (see Table 3.7), accounts for this effect in the definition of the nominal shear stress–ultimate stress ratio (the 0.9 factor given in the introduction of Section 3.3) and evaluates long joints subject to additional reduction factors only if longer than 38 in. (965 mm, whereby it is applied a further reduction of 16.7%; previously it was 50 in. with a reduction of 20%).

3.3.5 Anchor Bolts

For the evaluation of anchor bolt shear resistance please refer to Section 4.4 for base plates.

3.3.6 Stiffness Coefficient

According to EC, slip-resistant bolts in shear have an infinite stiffness coefficient (at the concerned load level), or

$$k_{11}(\text{or } k_{17}) = \frac{16 n_b d^2 f_{ub}}{E d_{M16}}$$

Some symbols have been discussed already while d_{M16} is the nominal diameter of an M16 bolt and n_b is the number of bolt rows in shear. Using millimeters as the

3.4 Bolt Tension Failure

The tension failure of bolts is a fundamental check for several types of joints, in particular for the so-called T-stubs, whose most typical example can be found within the end plate moment connection.

The check of the resistant net area of the bolt (independent of the presence of a partially threaded shank) takes place in relation to simple tension as well as to tension due to moments, both possibly increased by prying action (see Section 3.10).

Additional safety factors are provided for this limit state with regard to the possible local parasitic bending moments usually accompanying the tension loading of fasteners, which explains why the actual resistance is considerably less than the material resistance.

The tension failure of the bolt is not as brittle as the shear failure since there is elongation of the bolt prior to the failure. This effect can allow a rotation of the connection with a resulting redistribution (if possible) of forces.

In EC, the design tension resistance per bolt is defined as

$$\frac{k_2 f_{ub} A_s}{\gamma_{M2}}$$

where A_s is the net area of the bolt and $k_2 = 0.9$ (except for countersunk bolts; see Section 3.4.1). For the definition of the other values, depending on the class in which they belong, see Tables 3.2–3.6.

DIN recommends the lesser of the two values given by

$$\frac{A_{Sch} f_{y,b,k}}{1.1 \gamma_M}$$

and

$$\frac{A_{Sp} f_{u,b,k}}{1.25 \gamma_M}$$

with A_{Sp} being the net area and A_{Sch} the nominal area.

According to AISC, the nominal tensile resistance R_n (for A325 and A490) is

$$F_{nt} A_b = 0.75 F_u A_b$$

where, A_b is the nominal area of the bolt, and thus the factor 0.75 accounts for the approximate ratio of the effective area of the threaded portion to the area of the shank (the result then has to be multiplied by Φ in order to obtain the value to be compared with the design actions).

See Table 3.8 for a ready reference for class 8.8 bolt capacities. Consider that M10 should not be used for structural design and that currently the recommended types for minor size are M12, M16, M20, and M24.

3.4 Bolt Tension Failure

Table 3.8 Design maximum tensile strength for metric bolts.

Bolt: Class 8.8	M10	M12	M14	M16	M18	M20	M22	M24
$F_{t,Rd}$, EC: kN (kips)	33 (7.4)	48 (10.8)	66 (14.8)	90 (20.2)	111 (25)	141 (31.7)	175 (39.3)	203 (45.6)
N_{Rd}, DIN: kN (kips)	34 (7.6)	49 (11)	67 (15.1)	91 (20.5)	112 (25.2)	143 (32.1)	176 (39.6)	205 (46.1)
ΦR_n, AISC: kN (kips)	35 (7.9)	51 (11.5)	69 (15.5)	90 (20.2)	114 (25.6)	141 (31.7)	171 (38.4)	203 (45.6)
Bolt: Class 8.8	M27	M30	M33	M36	M39	M42	M45	M48
$F_{t,Rd}$, EC: kN (kips)	264 (59.3)	335 (75.3)	400 (89.9)	471 (106)	562 (126)	645 (145)	755 (170)	847 (190)
N_{Rd}, DIN: kN (kips)	267 (60)	338 (76)	404 (90.8)	475 (107)	568 (128)	652 (147)	762 (171)	855 (192)
ΦR_n, AISC: kN (kips)	258 (58)	318 (71.5)	385 (86.6)	458 (103)	538 (121)	623 (140)	716 (161)	815 (183)

Note: Unless additional coefficients apply, as explained in the text. Not recommended for structural applications.

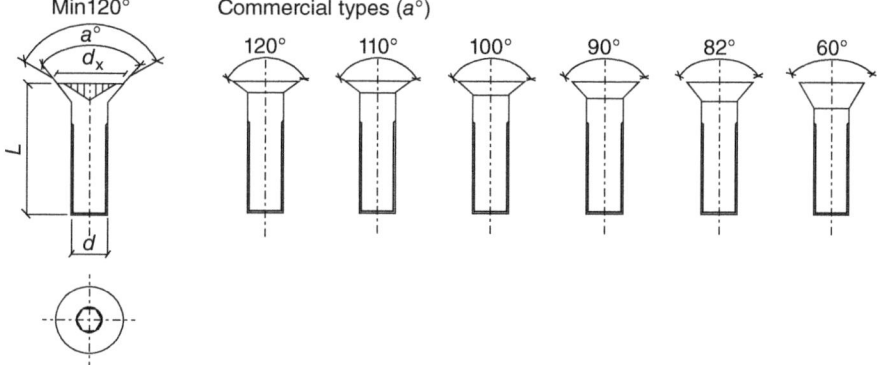

Figure 3.5 Countersunk bolts.

3.4.1 Countersunk Bolts

Eurocode reduces the tension resistance of countersunk bolts (Figure 3.5) by approximately 30% considering that the factor k_2 is 0.63 instead of 0.9 (further considerations should be made with regard to the angle and depth of the bolt head).

3.4.2 Stiffness Coefficient

The stiffness coefficient of a single row of bolts in tension should be determined, according to EC, by

$$k_{10} = \frac{1.6 A_s}{L_b}$$

where L_b is the bolt elongation length, taken as equal to the grip length (total thickness of material and washers), plus half the sum of the depth of the bolt head and the depth of the nut; A_s is the net area of the bolt as seen previously.

3.5 Bolt Failure in Combined Shear and Tension

In the case of combined actions, EC prescribes the formula

$$\frac{F_{v,Ed}}{F_{v,Rd}} + \frac{F_{t,Ed}}{1.4 F_{t,Rd}} \leq 1$$

where the subscripts v and t stand for shear and tension, respectively, R for resistance, and E for force (d design).

DIN also requires the equation

$$\left(\frac{N}{N_{Rd}}\right)^2 + \left(\frac{V_a}{V_{a,Rd}}\right)^2 \leq 1$$

with comparable meaning of the symbols (ratio between applied and resistant forces).

For AISC, in combined shear and tension, the equation to apply for the nominal resistance R_n is

$$A_b \left(1.3 F_{nt} - \frac{f_v}{\Phi F_{nv}} F_{nt} \right) \leq A_b F_{nt}$$

wherein the only new term is f_v and it represents the shear stress. In other words, the ratio $f_v/\Phi F_{nv}$ is the shear usage ratio that consequently lowers the available tensile resistance. The second term ($A_b F_{nt}$) shows that the resistance cannot be greater than the case where only tension is included. Summing up the given information, the shear resistance is the first to be verified, and subsequently tension will be checked decreasing the available resistance as indicated in the expression. It is necessary to report that, in order to verify combined tension and shear, Ref. [9] shows an expression comparable to the DIN equation and this method is expressly accepted by [10].

For combined actions in which the shear force is resisted by friction, see the next section.

3.6 Slip-Resistant Bolted Connections

The engineer might decide to design the shear connection as slip resistant, which means the friction between the contact surfaces of the connected elements (proportional to the compressive force that is the preload of the bolt) will resist the design forces, rather than the bolt shank by contact.

The design of a slip-resistant connection can be realized at the serviceability limit state (category B in [1]) or at the ultimate limit state (category C according to EC).

3.6 Slip-Resistant Bolted Connections

In addition to the bearing resistance (to check for all the categories), category B requires the verification of the slip resistance with serviceability loads in addition to the "classic" shear resistance of the bolt.

The design slip resistance of a preloaded bolt should be taken as per [1] (EC) as

$$F_{s,Rd} = \frac{k_s n \mu}{\gamma_{M3}} F_{p,C}$$

where n represents the number of friction surfaces, μ the slip factor, k_s the coefficient given in Table 3.9, and $F_{p,C}$ the preloading force. According to EC, μ is derivable from specific tests or, in their absence, from Table 3.10; Italian standard for construction (NTC) [11] simply prescribes the value at 0.45 "in case of white metal blasted connections protected until bolt preloading" and at 0.30 for all other cases. For a category B check, $\gamma_{M3,ser}$ is used instead of γ_{M3}. The NTC names both as γ_{M3}, assigning the value 1.25 for the ultimate limit state and the value 1.1 for the serviceability limit state.

In either category B or C, bolts have to belong to classes 8.8 or 10.9 and actually all the components needed for the assembly (bolt, nut, and washer) have to comply with EN 14399 (which is divided in several parts) for a design according to EC. For bolts conforming to these standards, the preloading force used in the previous equation is taken as (all symbols familiar)

$$F_{p,C} = \frac{0.7 f_{ub} A_s}{\gamma_{M7}}$$

Table 3.9 Slip-resistant connections [1], values of k_s.

Case	k_s
Holes with standard nominal clearance	1
Oversize holes	0.85
Short slotted holes with axis perpendicular to the direction of force	0.85
Short slotted holes with axis parallel to the direction of force	0.76
Long slotted holes with axis perpendicular to the direction of force	0.7
Long slotted holes with axis parallel to the direction of force	0.63

Table 3.10 Slip-resistant design from [1, 8], values of μ in the absence of specific tests.

Friction surface class according to [8]	μ
A	0.5
B	0.4
C	0.3
D	0.2

In the case of controlled tightening, γ_{M7} can be taken as 1.

It must be noted that in category C, according to EC, the following must also be checked:

$$\sum F_{v,Ed} \leq N_{net,Rd}$$

where $\sum F_{v,Ed}$ is the design shear force acting upon the bolts (it is the sum of single contributions by each bolt) to check. For the definition of $N_{net,Rd}$ please refer to Section 3.18.2.

As previously discussed, the stiffness coefficient is evaluated as infinite for slip-resistant connections according to EC.

3.6.1 Combined Shear and Tension

If a slip-resistant connection has a tensile force $F_{t,Ed}$ applied as well, the slip resistance per bolt should be taken as

$$F_{s,Rd} = \frac{k_s n \mu}{\gamma_{M3}}(F_{p,C} - 0.8 F_{t,Ed})$$

3.7 Bolt Bearing and Bolt Tearing

Bearing of the connected parts is often the reference limit state (i.e. it rules the design) for connections subjected to tension only, for example, braces or truss elements.

Both bolt bearing and bolt tearing depend on the material, bolt diameter, plate thickness, and hole edge distance. The last two variables are certainly the most easily adjustable and in particular the increase in distance between the hole and the edge of the plate is the easiest (and cheapest) to change. The distance between the axis of the hole and the plate edge is usually designed as 1.5 times the diameter of the hole, but in the case of trusses or brace connections, the standard should be increased twice and sometimes further upgraded (up to three times the diameter; greater distances do not help the cause).

If there is more than one bolt in the direction of the load transfer, then it is necessary to evaluate the failure also with regard to the spacing of the bolts, since bearing can occur for inner bolts too and it directly depends on the spacing of the bolts.

Bolt tearing (Figure 3.6) consists in a real shear tear-out, whereas bolt bearing (Figure 3.7) is more a locally visible deformation of the material created by the contact between the bolt and the contour of the hole.

According to DIN, the bearing resistance limit for plate thickness t is obtained using the equation:

$$V_{l,Rd} = \frac{\alpha_1 f_{y,k,pl}}{\gamma_M} dt$$

where t represents the reference thickness, d the bolt diameter, $f_{y,k,pl}$ the yield strength, and α_1 a coefficient depending on the edge distance and the spacing of the bolts.

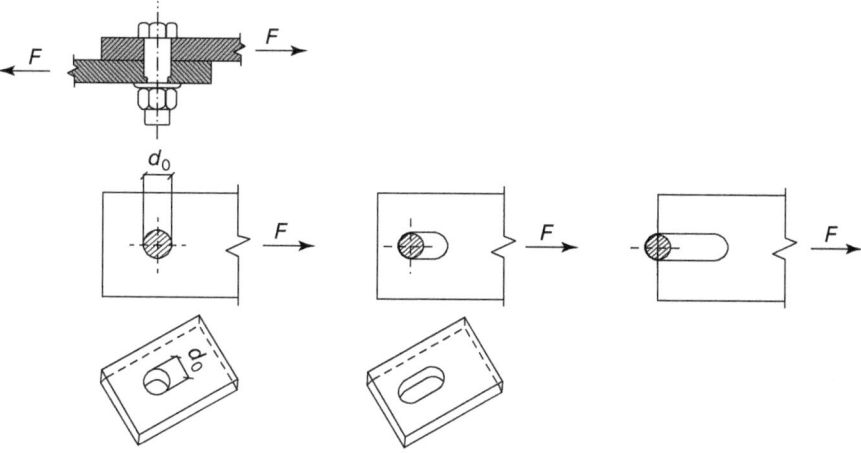

Figure 3.6 Bolt tearing.

Figure 3.7 Bolt bearing.

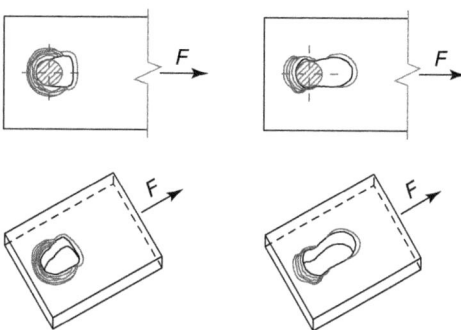

Considering the symbols in Figure 3.8 (and d_0, the diameter of the hole) and having $e_2 \geq 1.5d_0$ and $e_3 \geq 3.0d_0$, α_1 is derivable as (upper limit)

$$\alpha_1 = \min\left(\frac{1.1e_1}{d_0} - 0.30, \frac{1.08e}{d_0} - 0.77\right)$$

If $e_2 = 1.2d_0$ and $e_3 = 2.4d_0$, α_1 is given as (lower limit)

$$\alpha_1 = \min\left(\frac{0.73e_1}{d_0} - 0.20, \frac{0.72e}{d_0} - 0.51\right)$$

For intermediate values of e_2 and e_3:

$1.2d_0 < e_2 < 1.5d_0$

$2.4d_0 < e_3 < 3.0d_0$

and the optimal value for α_1 may be obtained by a linear interpolation of the previous cases (the worst between e_2 and e_3 will be taken).

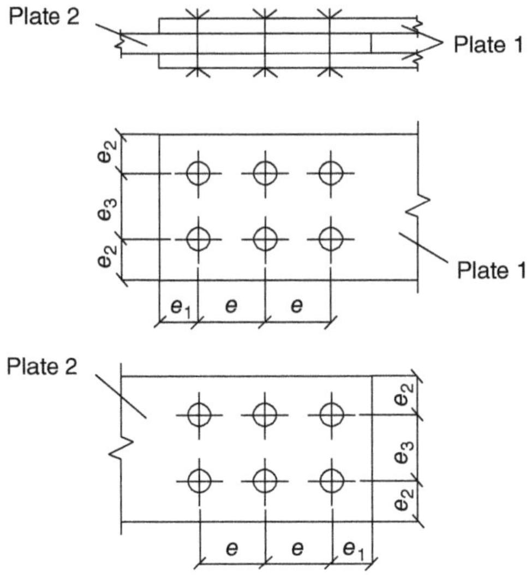

Figure 3.8 DIN symbols.

Please notice that the maximum values that can be used for e and e_1 are

$$e_1 = 3d_0$$
$$e = 3.5d_0$$

This means that greater values of end distance and spacing do not produce any further gain.

The EC method is similar but supplies coefficients for inner and end bolts, for both the direction of the load transfer and the direction perpendicular to the load. The bearing resistance of every single bolt should be compared with the limit value. Alternatively, with an easier and more operational method, it is possible to find the minimum coefficients and prudently adopt them for all the bolts.

The formula to obtain the bearing resistance is

$$F_{b,Rd} = \frac{k_1 \alpha_b f_u t d}{\gamma_{M2}}$$

with

$$\alpha_b = \min\left(1, \alpha_d, \frac{f_{ub}}{f_u}\right)$$

where f_u is the ultimate strength of the material, f_{ub} is the ultimate tensile strength of the bolt, and $F_{b,Rd}$ is the EC symbol corresponding to DIN $V_{l,Rd}$.

In the direction of load transfer,

$$\alpha_d = \frac{e_1}{3d_0}$$

for end bolts and

$$\alpha_d = \frac{p_1}{3d_0} - \frac{1}{4}$$

for inner bolts. Refer to Figure 3.9 for the meaning of the symbols.

Figure 3.9 Symbols according to EC.

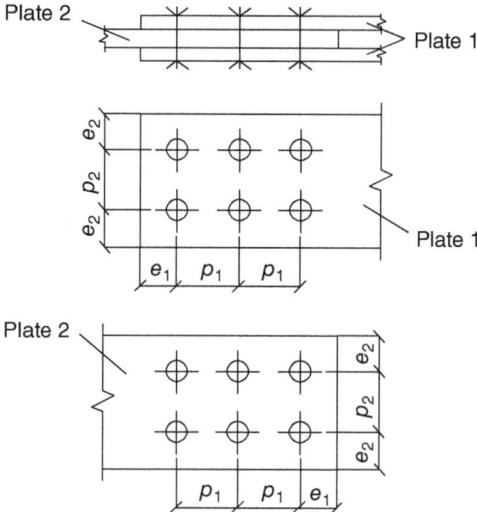

In the direction perpendicular to the load transfer,

$$k_1 = \min\left(2.5, 2.8\frac{e_2}{d_0} - 1.7\right)$$

for edge bolts and

$$k_1 = \min\left(2.5, 1.4\frac{p_2}{d_0} - 1.7\right)$$

for inner bolts. When using the equations above, the EC approach implicitly includes a check of the cross-section and application of these expressions also in compression controls this (tear-out would occur only in tension conditions and bearing too is usually critical in tension because in compression the edge distance is not a factor).

Changing the reference standard, an easy-to-recall formula for tear-out check (in case quick hand calculations are needed) is the AISC formula for R_n (multiply as usual by Φ to obtain the design value):

$$2(0.6F_u)L_c t = 1.2 F_u L_c t$$

That is, the formula considers two resisting shear sections as in Figure 3.10 ($0.6F_u$ is exactly the shear limit).

The material collapses according to the angle shown in Figure 3.11, but the "simplified" diagram shown in Figure 3.10 is quite useful when needing to remember the equation (obtained from laboratory tests).

AISC also insists on checking the resistance against the bearing strength, which is equal to

$$2.4 F_u d t$$

and the lesser value is the one that will be used as reference.

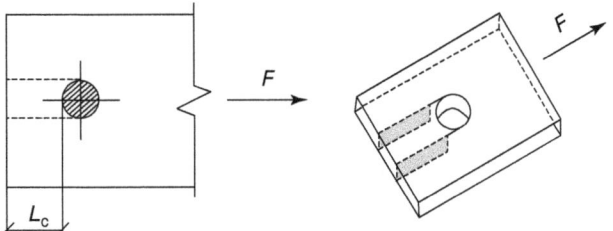

Figure 3.10 AISC representation.

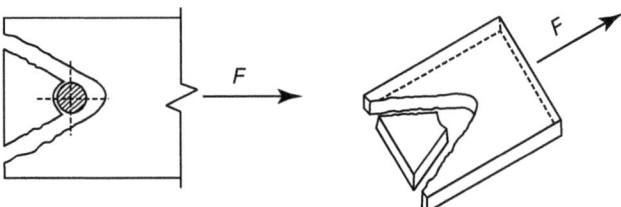

Figure 3.11 Representation of actual collapse.

To be precise, AISC prescribes the previous formulas "when the deformation at the bolt hole at service load is a design consideration" (literally from the specs). If this condition does not apply, it is possible to evaluate the strength R_n as

$$R_n = 1.5 F_u L_c t \leq 3 F_u \, dt$$

Instead, the Australian standard (see Ref. [6]) prescribes a resisting value equal to the minimum value between

$$a_e f_{up} t_p$$

and

$$3.2 d_f f_{up} t_p$$

where f_{up} is the plate ultimate strength, d_f the bolt diameter, r_p the plate thickness, and a_e the end distance from the hole (in the direction of the load transfer) plus half the diameter of the bolt.

To conclude the overview and certify how the phenomenon has various formulations, consider that, according to the old Italian standard UNI 10011 [12], the limit for the contact pressure value due to bearing (referred to the projected area of the cylindrical surface of the bolt) was

$$\alpha f_d$$

where α (maximum value 2.5) is equal to a/d with a being the end distance from the axis of the hole and d the bolt diameter.

Inserting the value of the pressure area $(t \times \pi d/2)$, the result is a bearing strength equal to

$$a \frac{f_y}{\gamma_m} t \frac{\pi}{2} \leq 2.5 d \frac{f_y}{\gamma_m} t \frac{\pi}{2}$$

The reference to the yield strength (then compensated by higher coefficients) should be noticed (typically found in older standards) for phenomena in which the ultimate strength should actually be the design reference.

Let us not forget that the verification must be done on both "sides" of the connection, usually a plate and another element (commonly the web) of the connected member. In addition, if there are two resisting sections, the double element will have the total double thickness in calculating the resistance (or half the force if this is checked against only one plate).

For a comprehensive calculation, the check must be executed in both perpendicular directions (i.e. toward both edges of the plate) since (due to eccentricity) the bolts will likely have forces in each direction.

3.7.1 Countersunk Bolts

Countersunk bolts must be checked against bearing using a reduced thickness instead of the nominal value.

Eurocode recommends reducing the reference thickness to a value equal to half the countersink depth.

3.7.2 Stiffness Coefficients

The formulas in [1] are here reported to evaluate the bearing stiffness of the components (unless the joint is designed by friction, where the stiffness is infinite):

$$k_{12} \text{ (or } k_{18}) = \frac{24 n_b k_b k_t \, d f_u}{E}$$

where

$$k_b = \min\left(1.25, \frac{e_b}{4d} + 0.5, \frac{p_b}{4d} + 0.375\right)$$

and

$$k_t = \min\left(\frac{1.5 t_j}{d_{M16}}, 2.5\right)$$

and n_b represents the number of bolt rows in shear, d_{M16} is the nominal diameter of an M16 bolt, t_j and f_u are the thickness and yield strength of the component, e_b is the distance of the row of bolts from the free edge in the force direction, and p_b is the distance between the bolts in the direction of the force; all other symbols have been previously defined.

3.8 Block Shear (or Block Tearing)

Block shear (or block tearing as it is called in EC) is a phenomenon that occurs due to the combined effects of shear and tension on a plate area in which bolts are present (see Figures 3.12 and 3.13).

The limit state is different from bearing both qualitatively (see Section 3.7) and quantitatively (see the formulas in that section). The block shear is a global type

50 | *3 Limit States for Connection Components*

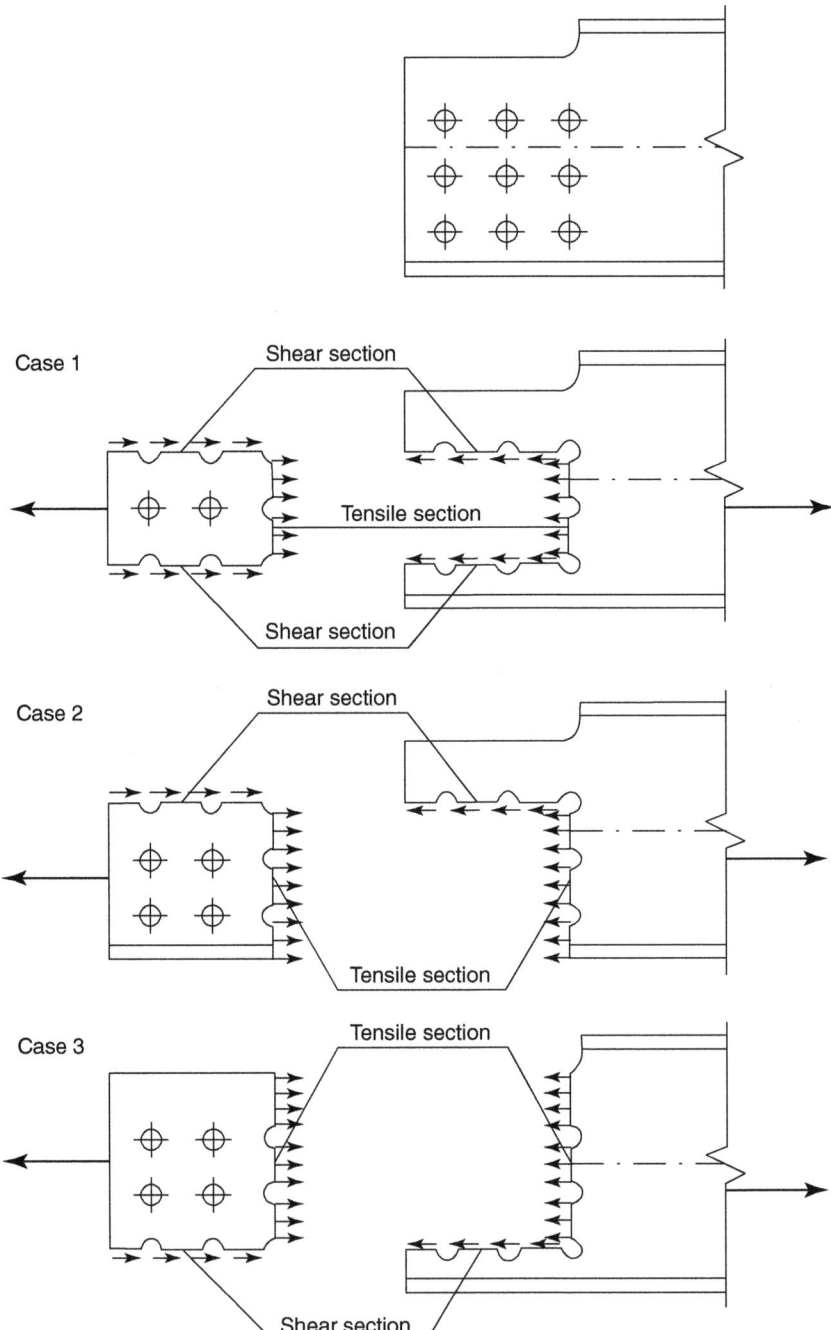

Figure 3.12 Block shear possible modes.

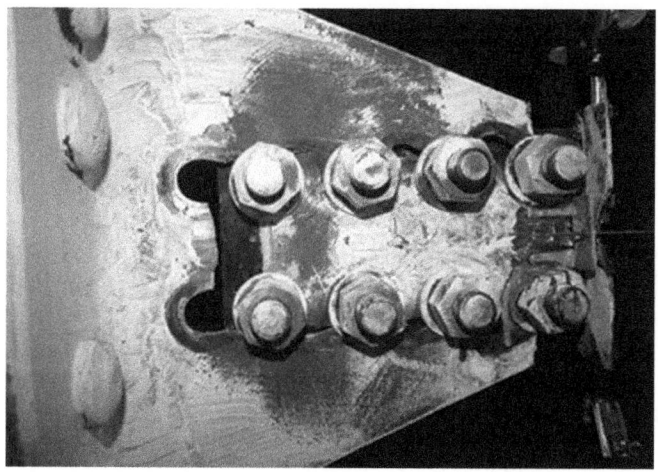

Figure 3.13 Laboratory test dramatically shows the block shear phenomenon. Source: Photo courtesy of J. Swanson and R. Leon, Georgia Tech.

of collapse over a group of bolts: one section is stressed by shear and another is perpendicularly stressed by tension, thus causing a breaking mechanism as in Figure 3.12 (the weakest of the three cases).

Eurocode offers the following formula to estimate the resistance:

$$k\frac{f_u A_{nt}}{\gamma_{M2}} + \frac{\left(f_y/\sqrt{3}\right) A_{nv}}{\gamma_{M0}}$$

where A_{nt} is the net area for tension and A_{nv} the net area for shear. Eurocode also requires halving the first term if there is eccentricity (therefore k is 0.5 if there is eccentricity, otherwise it is 1).

The AISC approach is similar (with the tension part that must be halved for a nonuniform stress) but the second term, in analogy with the plate verification, is taken as the lesser between $0.6F_u A_{nv}$ and $0.6F_y A_{gv}$ with A_{gv} the gross shear area.

Operationally, it would be necessary to check the block shear in both directions unless there is either only a simple axial force or shear with no eccentricity. This is currently poorly addressed in provisions that do not provide any special instructions on how to perform the check in those cases.

The same simplification just viewed that a 0.5 coefficient must be used with stresses that are not uniform (which could be either high or insignificant) and is coarse. It should also to be noted that the formulas for block shear according to the British Standards (BS) are quite different (see Ref. [13] in addition to the official standard [14]). Reference [13] also adds some interesting formulas to check shear and bending interaction of the beam web.

Hopefully, future regulatory updates can give the engineer more precise equations in this field.

It is then important to remember that the block shear collapse is classified as nonductile.

3.9 Failure of Welds

The plates in bolted joints are also largely connected by welds so it is necessary to check their resistance to various actions.

It is not the purpose of this text to delve into the different problems and various methods of analysis and implementation of welds. After an introduction of some important aspects to be kept in mind, only rudiments for basic verifications of fillet welds (which are what is commonly used in connections of standard steel structures because of their low cost and execution simplicity) will be given.

First of all, it must be kept in mind that welds have an extremely limited capacity of deformation, so overcoming the resistance of a weld usually activates a mechanism of brittle type, in particular if the weld is stressed transversely (in spite of the increased load that it can bear).

See Figure 3.14, which shows how the behavior is very fragile for a weld loaded at 90° (angle in relation to the weld longitudinal axis).

It is therefore common practice that, for small-to-medium carpentry jobs, plates are welded to completely restore the full resistance (two fillets with a throat of 5 mm restore a 10-mm plate with good approximation). This kind of design (full strength) also means that the checks are omitted in the calculation reports. For jobs of medium- to large-sized structures and for moment connections, the verification is required for both safety reasons and to avoid unnecessary oversizing. However, it is desirable that the welding failure is not the limit state that governs the design since it does not allow the redistribution of loads. If weld failure governs, the design structure would be considered fragile.

To ensure ductility so that it does not become too expensive, it is enough to size the sum of the throats of the weld equal to the thickness connected, which transmits the force (DIN recommends that each throat of a double fillet be taken as 0.7 times instead of 0.5 when the plate is S355 or equivalent, that is high resistance).

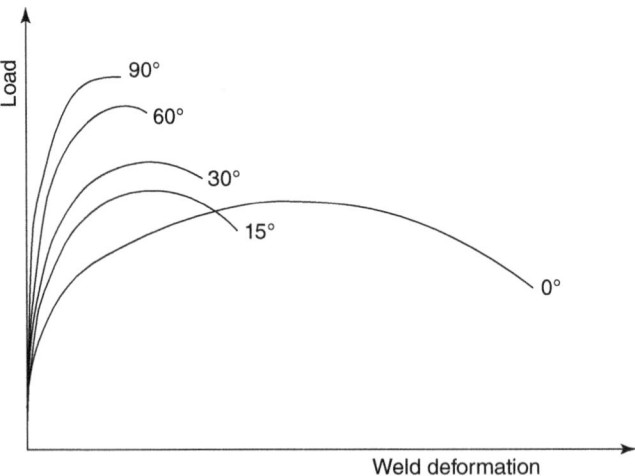

Figure 3.14 Weld deformation depending on the load angle. Source: Taken from Ref. [15].

The exact result that [16] is obtained in calculating the thickness of double fillet welds which guarantees the full strength of connected plates is, depending on the quality of the material, a throat thickness greater than 0.46 times the thickness for S235, 0.48 for S275, and 0.55 for S355. For S420 and S460 materials, each fillet must be greater than values between 0.68 and 0.74 times the thickness.

It is however interesting to note that the latest AISC indications will reduce this request by comparing the yield strength of the plate with the rupture of the weld, whereby the ductility is reached with a leg (not throat!) of $5/8$ the thickness for each of the two fillets instead of the $3/4$ required by comparing yield with yield. In fact, the AISC standards take as a reference the leg of the fillet, not the throat, and therefore the values seem to be higher (by dividing by $\sqrt{2} \approx 1.4$ the throat equivalent can be calculated). Ultimately, two throat fillets equal to 0.44 times the thickness completely reset the detail strength, ensuring the necessary ductility (which, according to AISC, even with material grade 50 is roughly equivalent to S355).

Another interesting aspect related to what was explained above as well as the economy of the welds is the fact that large welds require multiple runs (also known as "passes"). In fact, up to a throat of about 6 mm ($1/4$ in.) a single pass may suffice, but for greater thicknesses it is advisable to have multiple passes to achieve good welding quality. As Figure 3.15 illustrates, many passes are required to reach a slightly higher thickness. For example, a throat thickness of 9 mm requires about three passes, while one of 12 mm requires about five or six runs. This means that, to achieve a resistance equal to about 50% more than a 6 mm fillet, three times more labor is necessary (without considering that it is necessary to "clean" the various welds) and even five or six times the work for double strength. We conclude, then, wherever possible, that it is preferable to "stretch" the welded area with fillets that are not thick, rather than having very thick fillets of limited length.

A much more performing weld is the "full-penetration" weld, which will restore the strength of the connected elements but requires more preparation and control and therefore increasing costs for the fabrication shop (in contrast this solution is inexpensive for the engineer and would avoid any calculation with the simple full-penetration instruction).

It is noteworthy that it is typical to use V- or half-V-shaped (both on only one side) complete penetration welds up to 20-mm- ($3/4$-in.) thick plates. Beyond this limit it is convenient to use a double-V or K weld (see Figure 3.19) because, despite requiring a double preparation, it reduces the thickness of welds (consequently also the material and the labor) and guarantees a lower distortion.

A partial-penetration weld (compare Figure 3.17 with Figure 3.16) instead requires careful preparation and a calculation check based on the actual

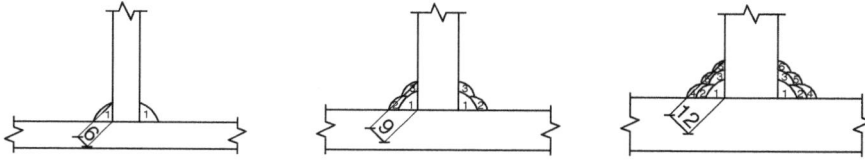

Figure 3.15 Increase of number of weld passes to have larger throats (mm).

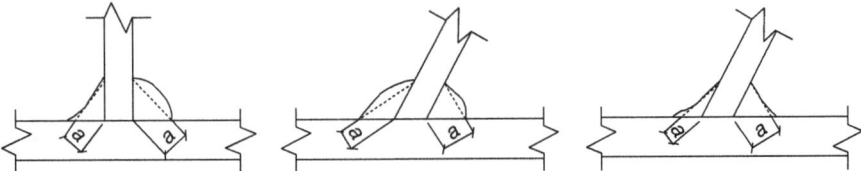

Figure 3.16 Net throat thickness of fillet welds.

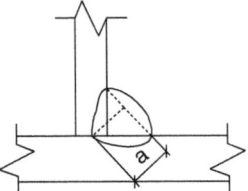

Figure 3.17 Net throat thickness of a partial-penetration weld.

thickness of the throat (similar to those for fillet welds) and it might become necessary to reduce the number of passes or when a "normal" fillet weld adding material to the connected element would interfere with some other element (a partial-penetration weld can exploit some "base material space" as a throat, therefore reducing total thickness).

When feasible, it is preferable to perform a fillet weld on both sides of the piece. A single fillet weld (that has the advantage of avoiding the rotation of the piece in the shop) can be effective for shear but not for tension in a butt joint.

About welding positions (see Figure 3.18), it must be remembered that the flat and horizontal positions are preferred to the vertical and overhead positions (where the gravity makes the operation quite cumbersome).

The welding symbols are many and various, in part depending on different geographical regions, for which the reader is referred to specialized manuals. Some of the most frequently used symbols are given in Figure 3.19. It is also significant to remember that if the indication of the weld is on one side, then the weld is on the near side according to European standards but on the far side by US standards.

As mentioned, the design checks for welds are various and will not be discussed in detail here, but some general guidelines are given below.

3.9.1 Weld Calculation Procedures

Eurocode divides the analysis according to two possible methods of calculation, the directional and simplified methods.

3.9.1.1 Directional Method

According to the directional approach, the design stress must be calculated taking tension and shear separately in both longitudinal and transverse directions, thus obtaining four values (Figure 3.20):

$\sigma_\perp$, normal stress perpendicular to the throat plane
$\sigma_\parallel$, normal stress parallel to the axis of the weld

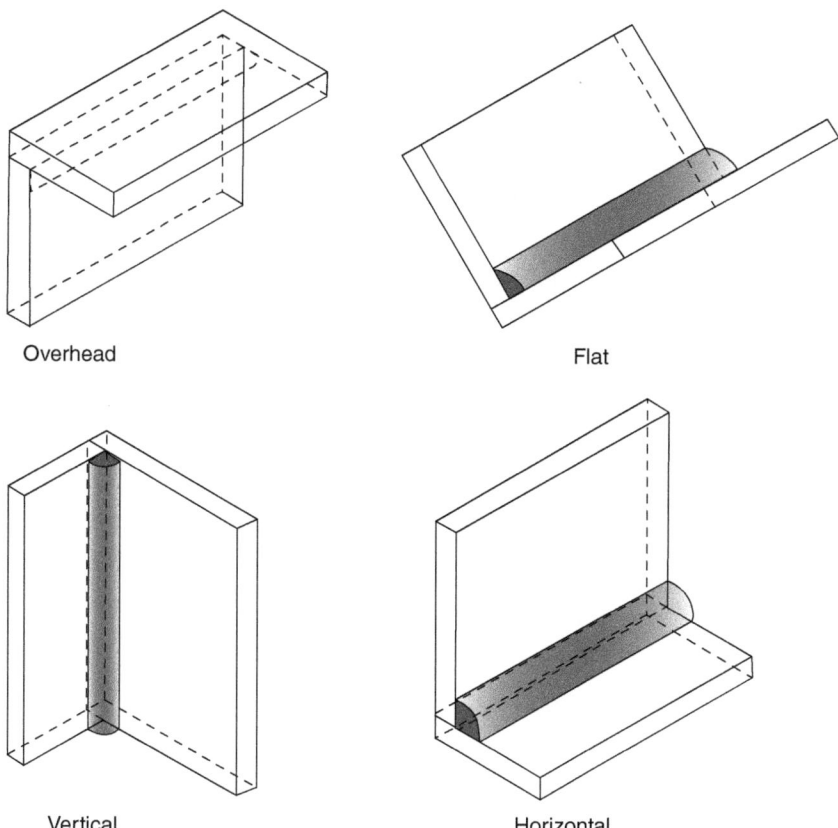

Figure 3.18 Welding positions.

$\tau_\perp$, shear (in the throat plane) perpendicular to the axis of the weld
$\tau_\parallel$, shear (in the throat plane) parallel to the axis of the weld.

The individual components are calculated by dividing the forces that act upon the weld.

It is important to remember that $\sigma_\parallel$ must not be checked in fillet welds (Ref. [17] describes all the steps that, in the 1960s, showed that this value does not affect the weld capacity). Tensions of type $\sigma_\parallel$ are, for example, present in a welded profile working in bending (in fact, fillet welds are designed for only the shear loads unless there are normal perpendicular stresses such as, say, a crane traveling on the same beam, as also explained in [17]).

The other components of a fillet weld should be composed according to Eurocode as

$$\sqrt{\sigma_\perp^2 + 3(\tau_\perp^2 + \tau_\parallel^2)}$$

to compare with the term

$$\frac{f_u}{\beta_w \gamma_{M2}}$$

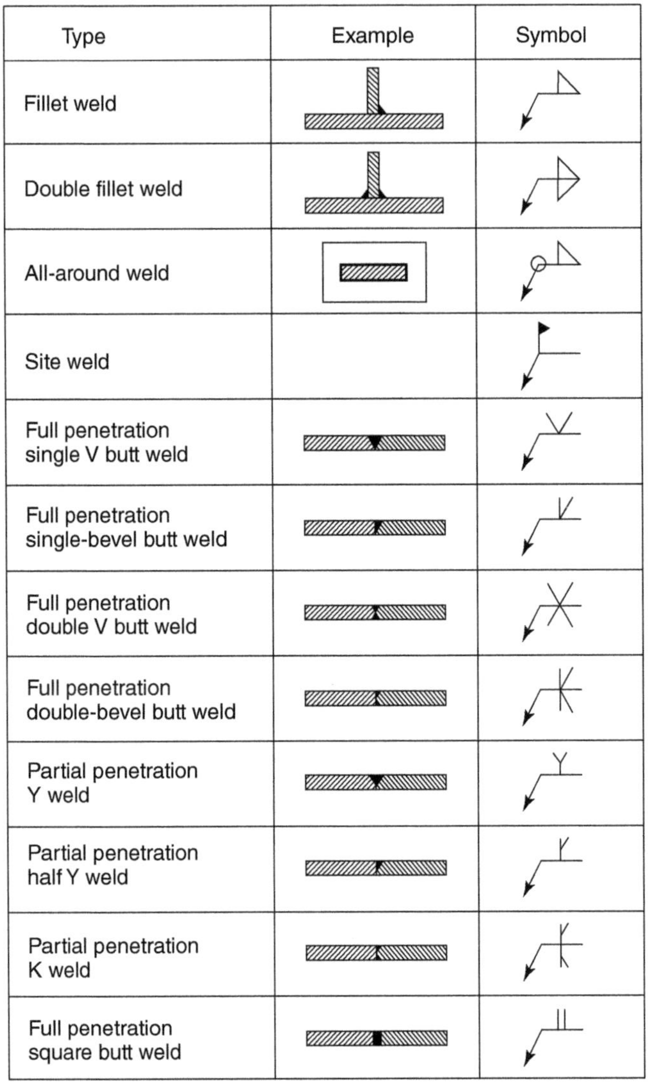

56 | *3 Limit States for Connection Components*

Type	Example	Symbol
Fillet weld		
Double fillet weld		
All-around weld		
Site weld		
Full penetration single V butt weld		
Full penetration single-bevel butt weld		
Full penetration double V butt weld		
Full penetration double-bevel butt weld		
Partial penetration Y weld		
Partial penetration half Y weld		
Partial penetration K weld		
Full penetration square butt weld		

Figure 3.19 Most common weld symbols.

where f_u is the ultimate resistance for the weakest of the connected materials and β_w is an appropriate correlation coefficient that depends on the material (again, the weakest in the connection; see Table 3.17). For commonly used materials, the value is 0.8 for S235, 0.85 for S275, and 0.9 for S355. In addition, Eurocode demands to verify that

$$\sigma^\perp \leq \frac{f_u}{\gamma_{M2}}$$

Notice that the throat area is taken in the actual position and not, as in regulations such as the Italian CNR 10011, "overturned" on one side of the fillet weld.

Figure 3.20 Stresses on the throat section of a fillet weld.

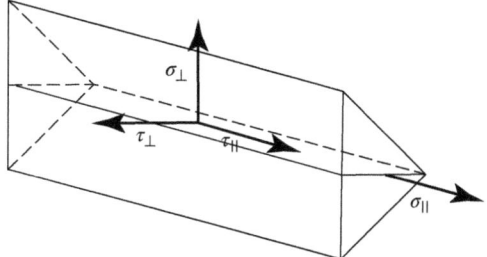

A detailed discussion can be found in [17], noting in particular that if we consider the actions on the overturned section (n is the normal stress, t the shear stress in the overturned section), the following equations are valid for fillets with identical legs:

$$\sigma_\perp = (n_\perp + t_\perp)/\sqrt{2}$$

$$\tau_\perp = (t_\perp - n_\perp)/\sqrt{2}$$

$$\tau_\parallel = t_\parallel$$

$$\sqrt{\sigma_\perp^2 + 3(\tau_\perp^2 + \tau_\parallel^2)} = \sqrt{2(n_\perp^2 + t_\perp^2) + 3\tau_\parallel^2 - 2n_\perp t_\perp}$$

The NTC (which is the local application of Eurocode) instead gives the additional opportunity to follow the path of the overturned section but considering two simplified formulas that refer to two new coefficients and to yield strength:

$$\sqrt{n_\perp^2 + t_\perp^2 + \tau_\parallel^2} \leq \beta_1 f_{yk}$$

and

$$|n_\perp| + |t_\perp| \leq \beta_2 f_{yk}$$

where β_1 and β_2, respectively, are 0.85 and 1 for S235, 0.7 and 0.85 for S275 and S355, and 0.62 and 0.75 for S420 and S460.

The DIN formulas also reference the overturned section and, by writing the inequality with the symbology used here, we have

$$\sqrt{n_\perp^2 + t_\perp^2 + \tau_\parallel^2} \leq \frac{\alpha_w f_{y,k}}{\gamma_M}$$

where $f_{y,k}$ is the yield strength and α_w is an appropriate coefficient depending on the material (see Table 3.17); in short, the formula is very similar to that prescribed by the NTC. Also note that in the DIN there is only one value for γ_M (1.1).

3.9.1.2 Simplified Method

Eurocode and NTC provide a simplified check for fillet welds demonstrating that

$$F_{w,Ed} \leq F_{w,Rd}$$

with $F_{w,Ed}$ being the design action on the fillet weld (per unit of length) and $F_{w,Rd}$ representing the weld resistance (per unit of length), computable as (using the symbols previously described)

$$\frac{a f_{tk}}{\beta \sqrt{3}\, \gamma_{M2}}$$

The AISC approach to a simplified verification is instead, writing the formula of the resistance as a function of the throat a:

$$\Phi R_n = 0.75 \cdot 0.6 F_{EXX} a l$$

with l the length and F_{EXX} the filler metal strength (usually superior to the base material strength). Since 0.6 corresponds to the inverse root of 3, the equations are quite similar except for the factor β, because the AISC reference is to the electrode. In addition, according to the AISC, the strength of the weld can be increased according to the loading angle. As mentioned earlier, transverse actions lead to a bigger strength at the expense of ductility. The increase in resistance is

$$(1 + 0.5 \sin^{1.5} \vartheta)$$

which means there is an additional 50% in strength for a 90° angle.

3.9.2 Tack Welding (Intermittent Fillet Welds)

Refer to Figure 3.21 for design considerations for tack welding (intermittent fillet welds).

3.9.3 Eccentricity

In the case of eccentricity, it is necessary to evaluate the additional forces stressing the welds. If the eccentricity adds extra bending moment (perpendicular to the plane of the welds) over a double-fillet weld, for example, this could be evaluated considering the top part, above the neutral axis, working in tension and the other in compression, with a lever arm equal to the distance of the centers of the two groups. A less conservative and more accurate alternative is to calculate the plastic (or elastic) module of the weld, using it to get stresses. Chapter 4 will present some numerical examples.

As for eccentric bolted joints, the AISC manual introduces an elastic method and an instantaneous center-of-rotation method to evaluate welds with eccentric loads. See Ref. [10] for further details.

3.9.4 Fillet Weld Groups

For a group of welds, the stresses can be calculated similarly to a beam, whereas the composed section is obtained by turning the weld resisting sections on one side. Otherwise, the actions (bending, shear, and so on) can be assigned to different groups appropriately, similar to what is done with beams. For example, for a welded section at the base of a cantilever, the shear can be assigned to the welds

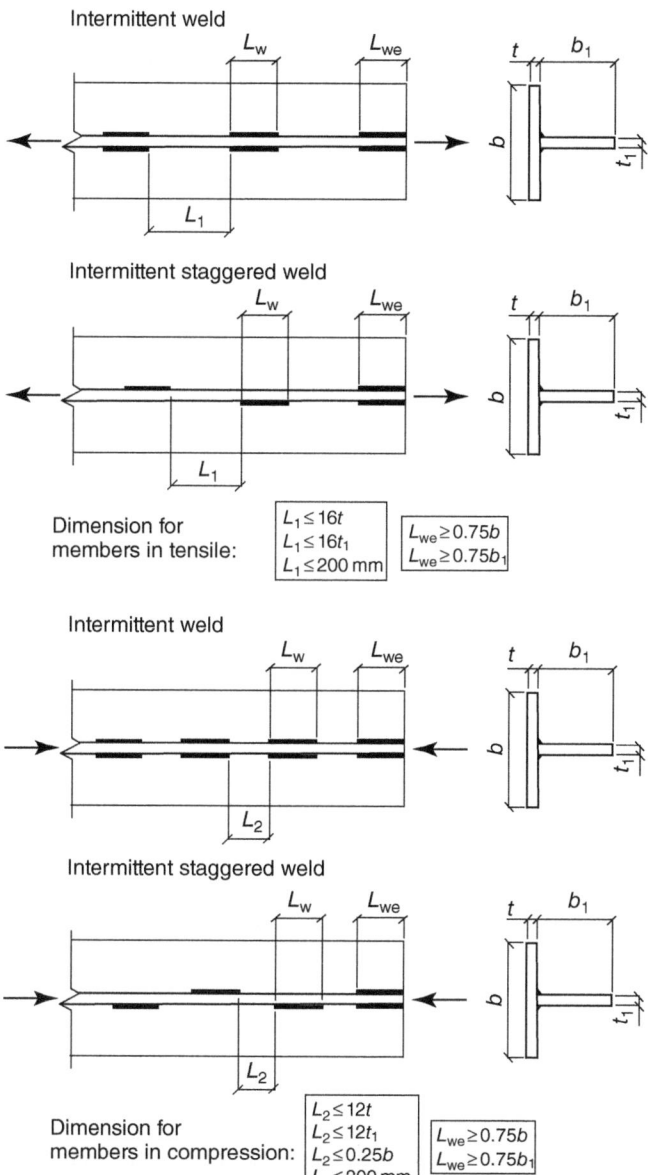

Figure 3.21 Instructions for intermittent fillet welds. Source: Adapted from EC 3.

of the web while the bending moment is assigned to the welds of the flanges. For a comprehensive theoretical approach please refer to [17], which also deals with situations of torsion or complex stress, and see Chapter 4 for some practical examples.

A good rule to apply is based on the fact that groups of fillets have approximately equal throat thicknesses in order to guarantee a similar global stiffness and ensure that the forces are distributed evenly.

3.9.5 Welding Methods

Here we mention the most commonly used welding methods in metalwork:

- Gas metal arc welding (GMAW, or flux-cored arc welding (FCAW) if the wire electrode contains flux in the center, using American Welding Society (AWS) terminology), commonly used in workshops especially for small- to medium-sized fillet welds
- Shielded-metal arc welding (SMAW according to AWS), which is the common method for field welds
- Submerged arc welding (SAW), usually made by automatic or semiautomatic machinery, also for welds of relevant size.

For information on preparations and beveling of welds, see Ref. [19], where the instructions are an abbreviated form of the more typical cases in [10].

3.9.6 Inspections

We will now take a quick closer look at the various methods of inspecting welds.

3.9.6.1 Visual Testing
Visual testing is the most economical system, yet it is important that it be done effectively. It would be desirable to also inspect the bevels before welding.

3.9.6.2 Penetrant Testing
A liquid, usually red, is sprayed on the welds and left to seep into the cracks and imperfections. Then the excess red liquid is removed and another contrast or detector liquid is sprayed. This second mixture is usually white so it is clearly visible when the red liquid has penetrated into cracks.

It is only possible to identify the cracks open on the surface. It may also be useful to check the extent of the defects that appear when conducting the first visual inspection or to verify that a correction of previous defect has been effectively carried out.

3.9.6.3 Magnetic Particle Testing
By using particle magnetization, it is possible to identify cracks up to a depth of about 2–3 mm (0.1 in.) because those cracks, when present, distort the magnetic field. The magnetic "dust" identified with this method gives not only the position but also the shape and size of the cracks.

3.9.6.4 Radiographic Testing
Radiography (the concept is intuitive) has poor applicability in tube-shaped joints or similar situations with variations in thickness and irregular shape. It is also a process that becomes expensive due to the difficulty in protecting the surrounding environment from X-rays if not performed in the laboratory, and therefore it has been largely supplanted by ultrasonic testing when detailed inspections are required.

3.9.6.5 Ultrasonic Testing

Ultrasonic testing identifies cracks (even internal ones) with an electromagnetic system that reflects the signal, similarly to radar/sonar. The system is faster and less expensive than radiography but is not recommended (that is, the skill and experience of the operator becomes the main factor so results may not be solid and consistent) for situations such as small thicknesses (less than about 8 mm, 5/16 in.), HSS (hollow structural steel), fillet welds, or austenitic steels.

3.10 T-stub, Prying Action

The concept of *T-stub*, typical of the EC, is necessary to analyze many kinds of connections, such as end plates, with tension and/or bending moment. The bending moment can be divided into a couple of forces (generally using as lever arm the depth of the profile), which engage the T-stub in axial tension, as shown in Figure 3.22 with Eurocode symbols. There is also the T-stub in compression (to be used in column base plates, as discussed in Chapter 4) but what is treated here and is normally considered a "T-stub" is the T-stub in tension.

A plate with two bolts working in tension and an additional perpendicular plate is shaped like a T. The former plate could be a "real" plate or the flange of a column.

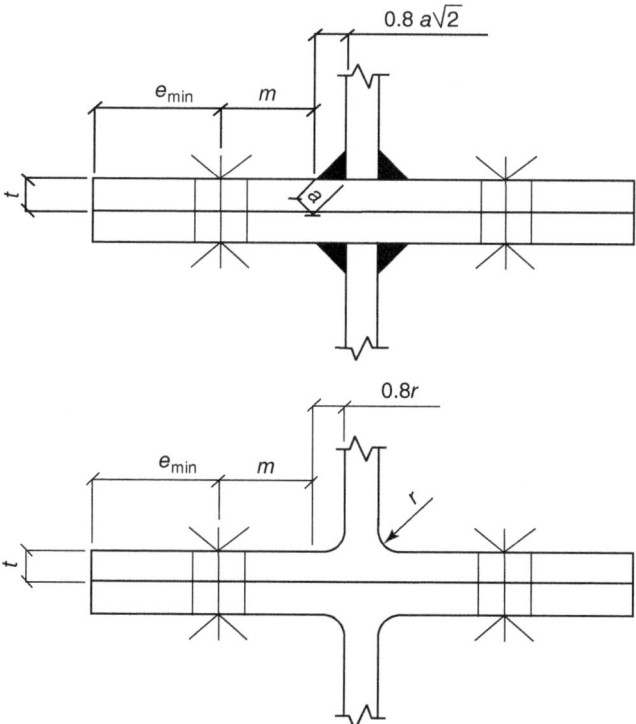

Figure 3.22 EC symbols for two T-stub cases – how to calculate *m*.

The latter perpendicular plate could be the web or a flange of a member (beam or column) or a stiffener.

3.10.1 T-stub with Prying Action

The tension acting on the system will apply bending to the plate.

There are three possible collapse mechanisms for a T-stub:

1. The plate yields completely, which means that a mechanism will be created after plastic hinges develop at the flange–web junction and near the plate section at the bolts.
2. The bolts collapse after a plastic hinge develops at the flange–web junction (or the equivalent if there are plates).
3. Only the bolts fail, which means that the prying action is negligible: the bolts share the load and the plate (rigid and strong) will work in bending as shown in Figure 3.23.

This means that the design of a resisting system can occur in two opposite ways:

(a) Minimizing the action upon bolts at the expense of the plate (a thick plate will make the prying action negligible); this approach corresponds to the collapse mode in Case 3 of Figure 3.23.
(b) Minimizing the plate thickness at the expense of larger bolts since they will have to share the load and the prying action arising at the edges of the plate; this corresponds to Case 1.

Case (mode) 2 in Figure 3.23 corresponds to an intermediate solution.

To evaluate this important phenomenon quantitatively we need to review the list of "basic" equations (for m and n see Figure 3.23).

In Case 1 the design resistance of the T-stub is given by

$$F_{T,1,Rd} = \frac{4M_{pl,1,Rd}}{m}$$

where $M_{pl,1,Rd}$ is the plastic modulus of the plate, equal to

$$M_{pl,1,Rd} = \frac{1}{4} \sum \ell_{eff,1} t^2 \frac{f_y}{\gamma_{M0}}$$

The term $\sum \ell_{eff}$ and more generally ℓ_{eff} appear here for the first time. The latter represents the effective length of the T-stub while the sum, according to EC, coincides with ℓ_{eff} when analyzing a single row of bolts, while it is the sum of the various contributions when a group of bolts is taken into account.

Note that the above is the easier approach in the EC [1] because, considering the distribution of the load given the bolt head (or the washer), some modified formulas can be used and they allow an increased resistance, that is,

$$F_{T,1,Rd} = \frac{(8n - 2e_w)M_{pl,1,Rd}}{2mn - e_w(m+n)}$$

Figure 3.23 Mechanisms of T-stub failure.

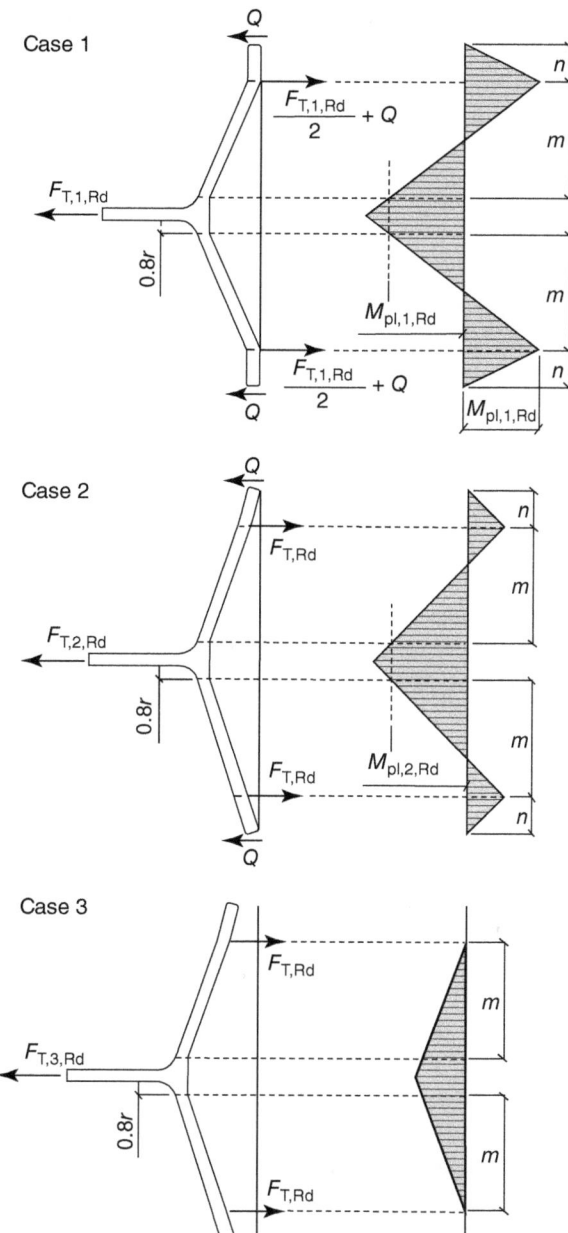

where e_w represents one-fourth of the diameter of the washer (or the bolt head or the nut if there is no washer). See Chapter 6 for some reference values for washers, heads, and nuts in this regard.

In Case 3 instead we have

$$F_{T,3,Rd} = \sum F_{t,Rd}$$

Attention must be paid to the difference between the subscripts T and t, with the first indicating the T-stub while the second indicates the bolt (which, combined with the other symbols, refer to their respective tensile strengths). The last equation tells us that the T-stub resistance is calculated as the sum of the tensile strengths of the bolts.

As already mentioned, Case 2 fits in an intermediate mode with the formula

$$F_{T,2,Rd} = \frac{2M_{pl,2,Rd} + n \sum F_{t,Rd}}{m+n}$$

The definition of $M_{pl,2,Rd}$ is similar to $M_{pl,1,Rd}$ with $\ell_{eff,2}$ instead of $\ell_{eff,1}$.

Now, to make the formulas truly operational, it is essential to identify ℓ_{eff}. This will be seen for each case in the following sections.

We should remember that the final strength is the smallest of the three values and, therefore, they are all to be evaluated unless the exact failure mode has been located in advance (by minimum thickness, see Section 3.10.2, or other considerations).

Given a certain bolt diameter, if the engineer wants to maximize the resistance, he or she may, with respect to any other spacing and tolerance issues:

- Use a greater plate thickness.
- Decrease the distance of the bolt from the center, that is, b' in Figure 3.24.
- Increase the distance of the bolt from the fulcrum of the lever, namely, the external part of the plate.

3.10.2 Possible Simplified Approach According to AISC

Depending on the stiffness of the plate, that is, its thickness, the prying action becomes negligible. In this respect there is an AISC formula for the minimum thickness of the plate (a similar one can also be derived from simple considerations for the moment resistance) above which prying is in fact irrelevant:

$$t_{min} = \sqrt{\frac{4.44 T b'}{p f_u}}$$

where T represents half of the (factored) force loading the pair of bolts, f_u the material ultimate resistance, and p the effective length for the pair of bolts (equal to the distance between the bolts in the plane perpendicular to the picture but

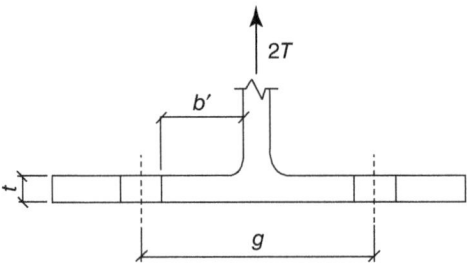

Figure 3.24 Negligible prying action – symbols for minimum-thickness formula.

not exceeding g in Figure 3.24 as a stress distribution at 45° without overlapping areas must be considered). Note the presence of f_u instead of the yield strength, although it is indeed a limit state corresponding to yield and not to failure: the reason is a better consistency found with laboratory tests because there is considerable hardening.

The simplified strategy (not recommended except for special cases) might therefore be to choose a plate that meets the minimum-thickness requirement and to check the bolts only for resistance in tension (that is, mode 3).

3.10.3 Backing Plates

Case 1 can also benefit from the contribution of "backing plates" (reinforcing bolted plates as shown in Figure 3.25).

Figure 3.25 Backing plates.

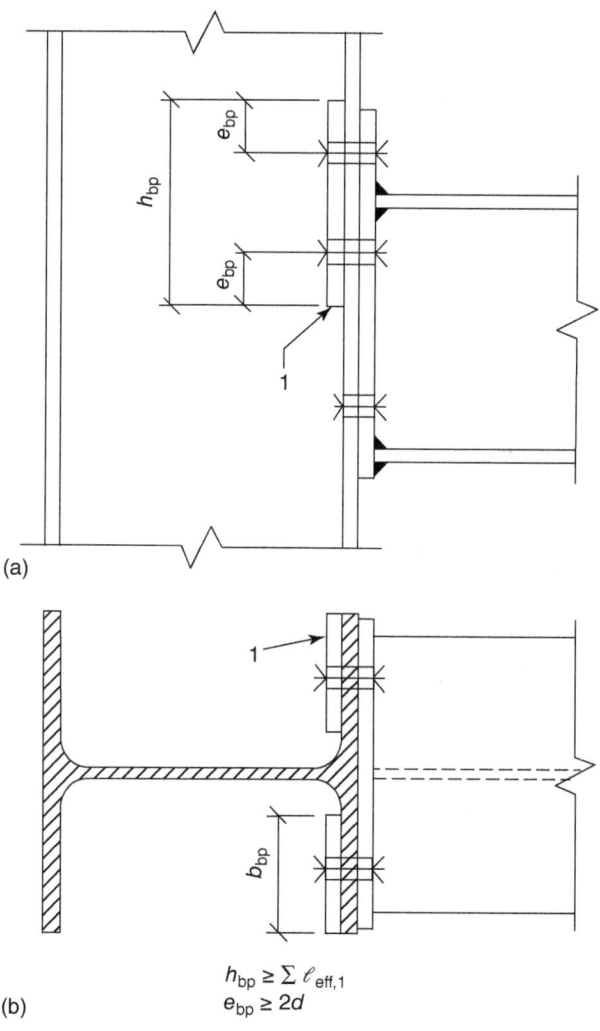

$$h_{bp} \geq \Sigma \, \ell_{eff,1}$$
$$e_{bp} \geq 2d$$

In this case, the formula for the T-stub resistance with respect to Case 1 becomes

$$F_{T,1,Rd} = \frac{4M_{pl,1,Rd} + 2M_{bp,Rd}}{m}$$

with (t_{bp} is the backing plate thickness)

$$M_{bp,Rd} = \frac{1}{4}\sum \ell_{eff,1} t_{bp}^2 \frac{f_{y,bp}}{\gamma_{M0}}$$

3.10.4 Length Limit for Prying Forces and T-stub without Prying

Eurocode defines a length limit of the bolts to assess the presence or absence of prying. If the elongation of the bolt (or the anchor bolt) is greater than the length limit, we can consider that there are no prying forces, whereas if it is below the length limit, the prying forces must be taken into consideration.

It is important to remember that in the event that there is no prying action, $F_{T,2,Rd}$ is no longer to be considered and $F_{T,1,Rd}$ is halved and becomes (with change of name in subscript in EC)

$$F_{T,1-2,Rd} = \frac{2M_{pl,1,Rd}}{m}$$

The length limit is defined as

$$L_b^* = \frac{7m^2 n A_s n_b}{\sum \ell_{eff,1} t^3}$$

where some of the symbols have already been defined, A_s is the bolt net area, and n is the minimum between $1.25m$ and e but for the extended part of an end plate becomes $n = \min(1.25m_x, e_x)$. In [1] the term $7m^2 n$ is simplified to $8.8m^3$, assuming $n = 1.25m$. Also note the presence of n_b (correction of the original [1] that was not providing this term), indicating the number of bolt rows (assumption of two bolts per row).

The length limit L_b^* is to be compared with the elongation length L_b, equal for a bolt to the sum of the connected thicknesses, washers, and half the depth of the nut and the bolt head. For an anchor bolt, the elongation length is equal to eight times the diameter summed with the base plate thickness, the mortar thickness, the washer, and half the height of the nut.

In reality, the formula is not of immediate practical application because some values ($\sum \ell_{eff,1}$, m, and n) can be different depending on the accounted bolt row (and, in a beam–column connection, also depending on the side that is considered, which also varies t), which means the engineer does not know what to insert in the formula (an average? the value of the external bolts?).

Checking the length is not usually necessary in beam–column joints since [1] says, in a note relating to Table 6.2, to take the prying action for granted. For the length limit in other cases (e.g. base plates) that do not have specific guidelines, the recommended approach is to verify both assumptions and then choose the most conservative. However, for base plates, the latest EC instructions (errata of [1]) recommend considering the prying action for the check of the anchor bolts but not when verifying the base plate thickness. Numerically this means using $F_{T,1-2,Rd}$ (no prying) instead of $F_{T,1,Rd}$ but at the same time calculating $F_{T,2,Rd}$ (prying), which might govern the design.

This illustrates how the prying effect calculation routine might therefore be beneficial to a plate with relatively modest thickness as the verification method accounts for the creation of two plastic hinges to form a mechanism.

3.10.5 T-stub Design Procedure for Various "Components" According to Eurocode

The EC approach described further for the calculation of effective lengths is difficult to apply in some parts because the topic is too subtle and complex to illustrate in a standard. Despite becoming officially a standard in 2005, some formulas are still of dubious application when compared to [18], which is complete and well-illustrated, dedicated in particular to end-plate connections between beams and columns. A significant excerpt of [18] is therefore illustrated after this section.

3.10.5.1 Column Flange

The values of m and e_{min} for use in formulas are given in Figure 3.26 (from [1]). In the absence of stiffeners, it is necessary to calculate the values in Table 3.11 to define ℓ_{eff}.

Then the following can be calculated:

$$\ell_{eff,1} = \min(\ell_{eff,nc}, \ell_{eff,cp})$$

$$\ell_{eff,2} = \ell_{eff,nc}$$

Note that the above two formulas for the calculation of $\ell_{eff,1}$ and $\ell_{eff,2}$ are valid not only for unstiffened column flanges but also for the other cases dealing with stiffened flanges and end plates.

The term "end" bolt row in the EC tables points to a bolt row near the free edge (of the plate or of the column flange); see also Figure 3.27.

Assuming there are stiffeners, the calculation of $\ell_{eff,cp}$ and $\ell_{eff,nc}$ is as per Table 3.12. Note how the values for "other inner row" and "other end row" correspond to those of Table 3.11.

To give a value to the α parameter in the table, refer to Figure 3.28. In [18] the value 4.45 is given as a minimum and 2π as a maximum. For EC, since indications are not provided but the line is drawn for $\alpha = 8$, and 8 is recommended as a maximum. An analytical method exists in [18] for obtaining α yet it is not recommended since there are fourth-degree polynomials with coefficients of six decimal places.

68 | *3 Limit States for Connection Components*

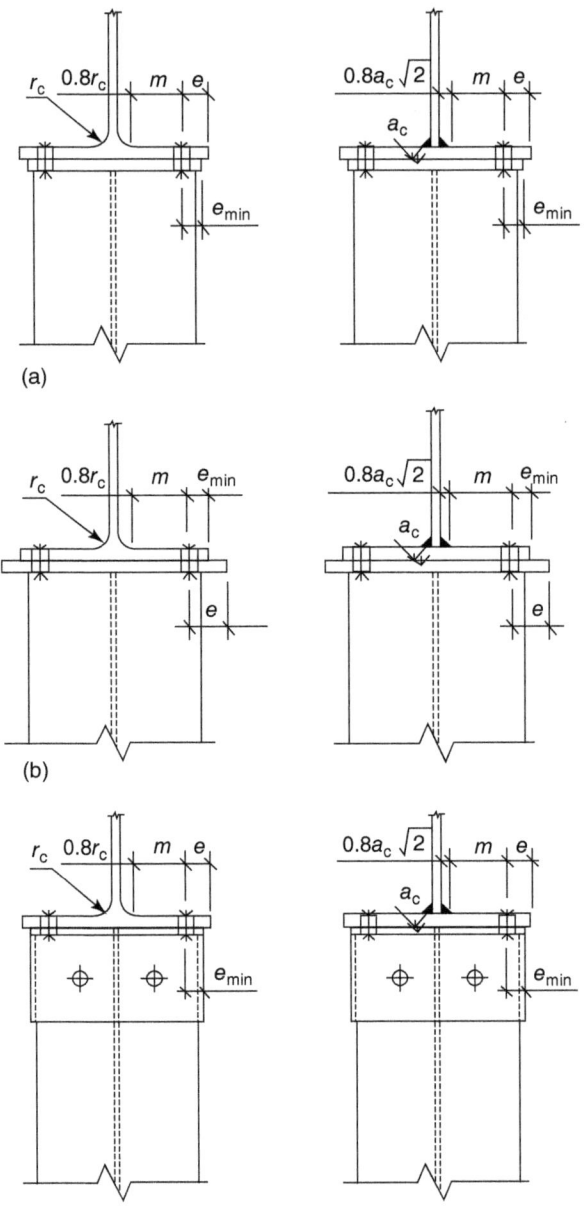

Figure 3.26 Symbol representations from EC 1993-1-8. (a) End plate narrower than column flange, (b) end plate wider than column flange, and (c) angle flange cleats.

Table 3.11 Values for the calculation of ℓ_{eff} for an unstiffened flange column.

Bolt row location	Bolt row considered individually		Bolt row considered as part of a group	
	Circular patterns $\ell_{eff,cp}$	Noncircular patterns $\ell_{eff,nc}$	Circular patterns $\ell_{eff,cp}$	Noncircular patterns $\ell_{eff,nc}$
Inner row	$2\pi m$	$4m + 1.25e$	$2p$	p
End row	$\min \begin{cases} 2\pi m \\ \pi m + 2e_1 \end{cases}$	$\min \begin{cases} 4m + 1.25e \\ 2m + 0.625e + e_1 \end{cases}$	$\min \begin{cases} \pi m + p \\ 2e_1 + p \end{cases}$	$\min \begin{cases} 2m + 0.625e + 0.5p \\ e_1 + 0.5p \end{cases}$

Note: e_1 is the distance of the bolt axis from the outer edge perpendicular to e.

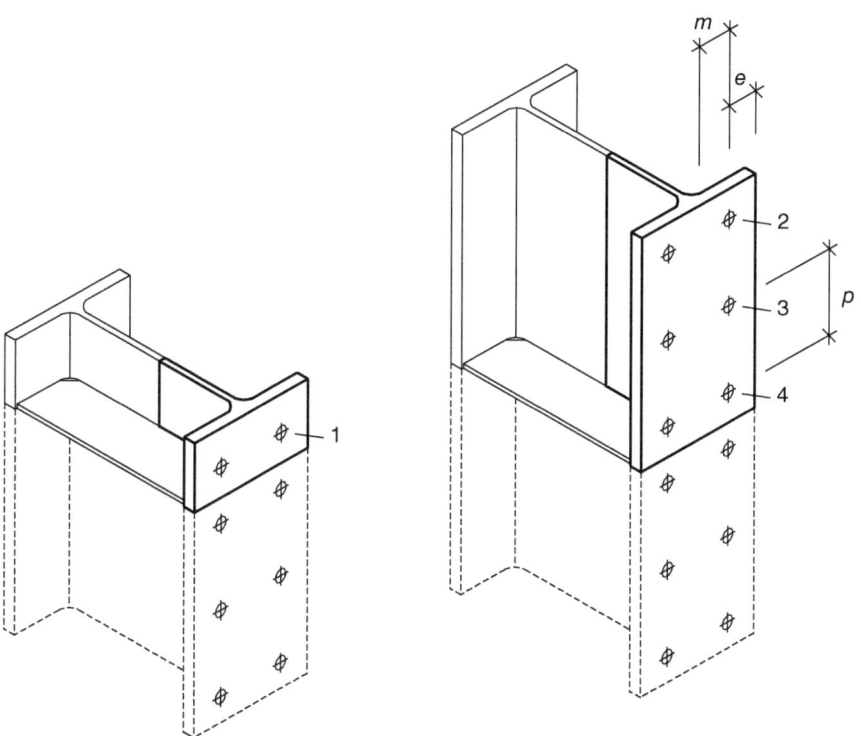

Figure 3.27 Examples of "position of a bolt row," as in [1]: (1) external row near a stiffener; (2) external row; (3) internal row; and (4) internal row near a stiffener.

Table 3.12 Values for calculating ℓ_{eff} for a stiffened column flange.

Bolt row location	Bolt row considered individually		Bolt row considered as part of a group	
	Circular patterns $\ell_{eff,cp}$	Noncircular patterns $\ell_{eff,nc}$	Circular patterns $\ell_{eff,cp}$	Noncircular patterns $\ell_{eff,nc}$
End row near stiffener	$\min \begin{cases} 2\pi m \\ \pi m + 2e_1 \end{cases}$	$e_1 + \alpha m + -(2m + 0.625e)$	—	—
Other end row	$\min \begin{cases} 2\pi m \\ \pi m + 2e_1 \end{cases}$	$\min \begin{cases} 4m + 1.25e \\ 2m + 0.625e + e_1 \end{cases}$	$\min \begin{cases} \pi m + p \\ 2e_1 + p \end{cases}$	$\min \begin{cases} 2m + 0.625e + 0.5p \\ e_1 + 0.5p \end{cases}$
Inner row near stiffener	$2\pi m$	αm	$\pi m + p$	$0.5p + \alpha m + -(2m + 0.625e)$
Other inner row	$2\pi m$	$4m + 1.25e$	$2p$	p

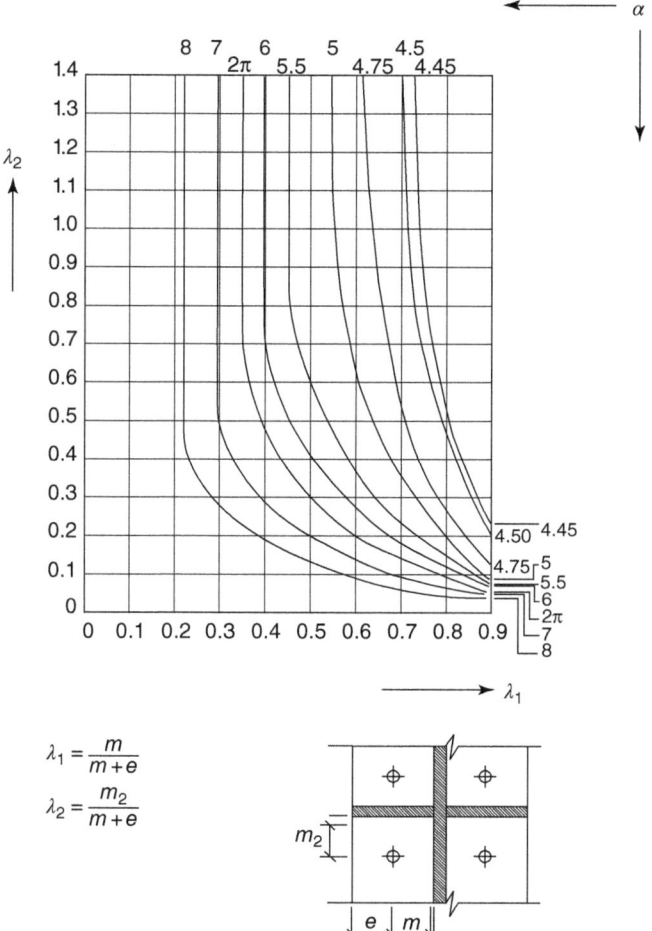

Figure 3.28 For estimating α in Tables 3.12 and 3.13. Source: From Ref. [1].

3.10.5.2 End Plate
Use Figure 3.29 and Table 3.13 to calculate ℓ_{eff} in the case of an end plate. For e_{min} refer to Figure 3.26.

3.10.5.3 Angle Flange Cleat
In this situation, $\ell_{eff} = 0.5 b_a$ as shown in Figure 3.30. For other parameters, refer to Figure 3.30.

3.10.6 T-stub Design Procedure for Various "Components" According to the "Green Book"

As already mentioned, the approach of [18] (also known as the "green book" for its cover color) is in the wake of the BS but is quite similar to EC and arguably better

72 | *3 Limit States for Connection Components*

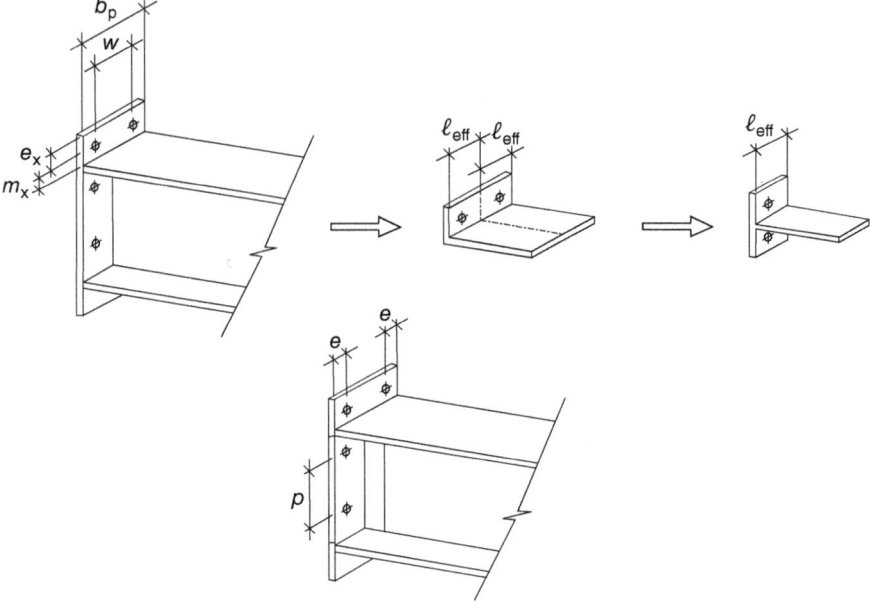

Figure 3.29 For calculating ℓ_{eff} for an end plate. Source: From EC 1993-1-8.

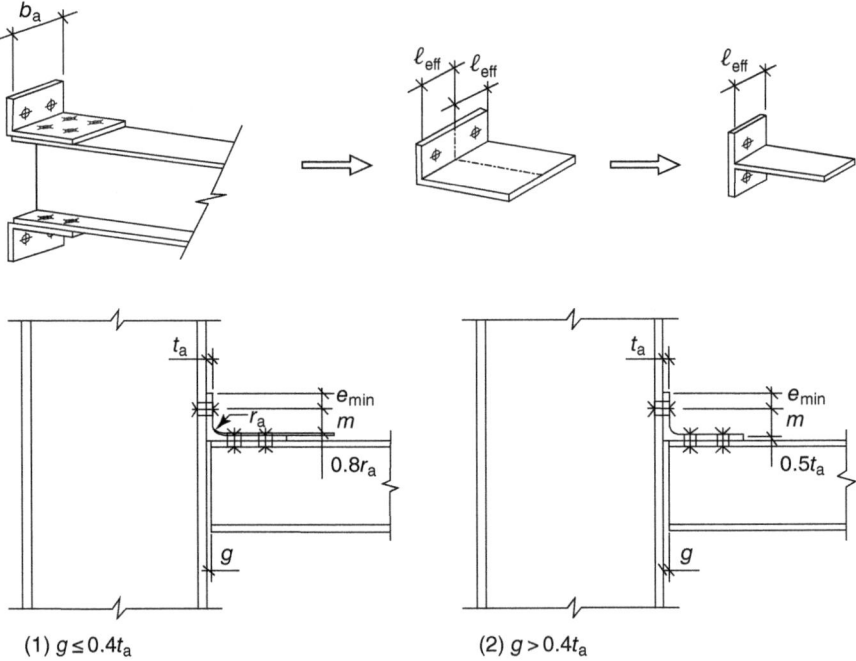

Figure 3.30 Angle flange cleat design parameters.

Table 3.13 Values for calculating ℓ_{eff} for an end plate.

Bolt row location	Bolt row considered individually		Bolt row considered as part of a group	
	Circular patterns $\ell_{eff,cp}$	Noncircular patterns $\ell_{eff,nc}$	Circular patterns $\ell_{eff,cp}$	Noncircular patterns $\ell_{eff,nc}$
Row outside tension flange of beam	$\min \begin{cases} 2\pi m_x \\ \pi m_x + w \\ \pi m_x + 2e \end{cases}$	$\min \begin{cases} 4m_x + 1.25e_x \\ e + 2m_x + 0.625e_x \\ 0.5b_p \\ 0.5w + 2m_x + 0.625e_x \end{cases}$	—	—
First row below tension flange	$2\pi m$	ffm	$\pi m + p$	$0.5p + am$ $-(2m + 0.625e)$
Other inner row	$2\pi m$	$4m + 1.25e$	$2p$	p
Other inner row	$2\pi m$	$4m + 1.25e$	$\pi m + p$	$2m + 0.625e + 0.5p$

explained in different parts. First there is no division between "circular" and "non-circular" patterns (EC aims to give a global approach also including the cases that do not have prying action) but a dedicated discussion that is simplified omitting some formulas that rarely constitute a real limit in the case of beam-to-column connections (where it is assumed the prying is always present).

Moreover, especially when it comes to defining the effective lengths within groups, the guidelines provided by EC show deficiencies and, in the writer's opinion, the engineer should use what is explained in this section as well as in [18].

Note that ℓ_{eff} values are for pairs of bolts and may not coincide with the actual lengths represented by the drawings in Tables 3.14 and 3.15.

3.10.6.1 ℓ_{eff} for Equivalent T-stubs for Bolt Row Acting Alone
See Tables 3.14 and 3.15.

Table 3.14 Formulas for evaluating ℓ_{eff} in a pair of bolts separated by a web in a column flange or end plate.

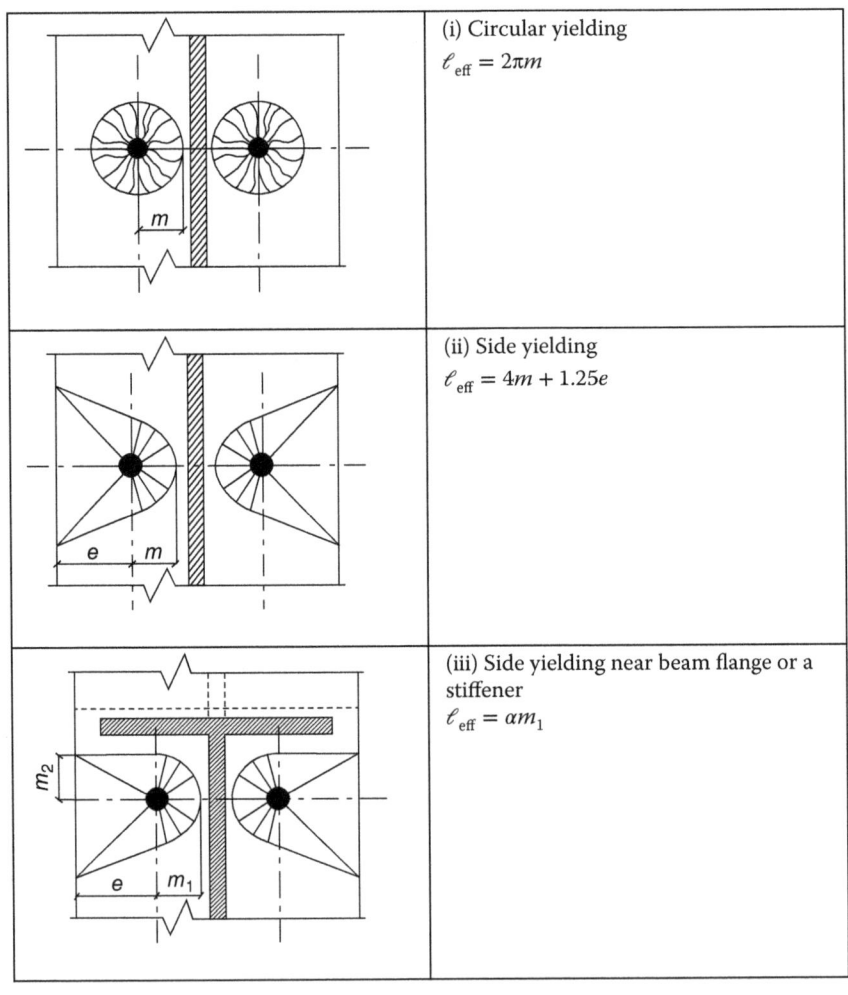

	(i) Circular yielding $\ell_{\text{eff}} = 2\pi m$
	(ii) Side yielding $\ell_{\text{eff}} = 4m + 1.25e$
	(iii) Side yielding near beam flange or a stiffener $\ell_{\text{eff}} = \alpha m_1$

3.10 T-stub, Prying Action

Table 3.14 (Continued)

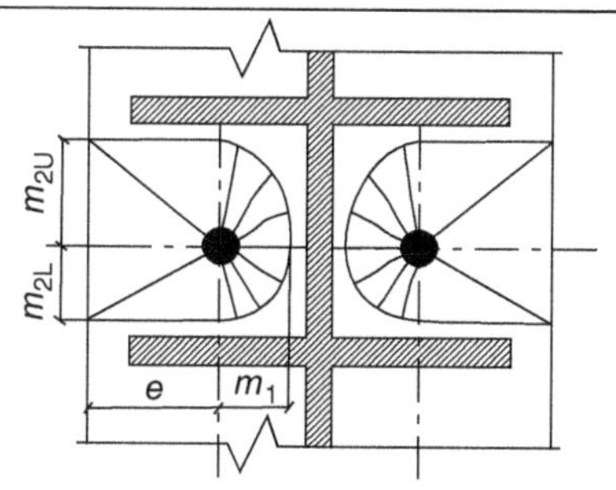	(iv) Side yielding between two stiffeners $\ell_{\text{eff}} = \alpha m_1 + \alpha' m_1 - (4m_1 + 1.25e)$ (α is calculated using m_{2U} and α' using m_{2L})
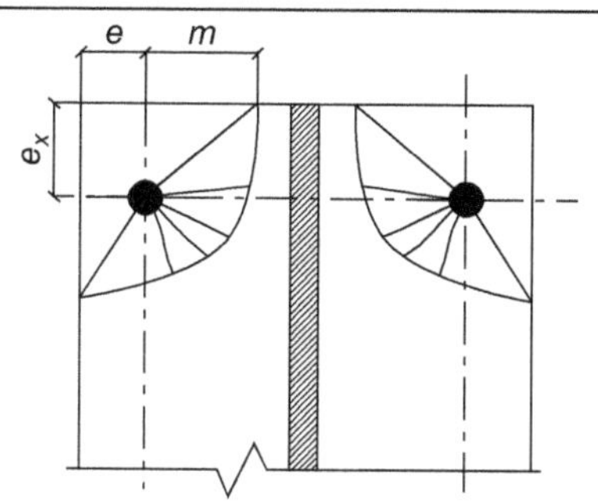	(v) Corner yielding $\ell_{\text{eff}} = 2m + 0.625e + e_x$
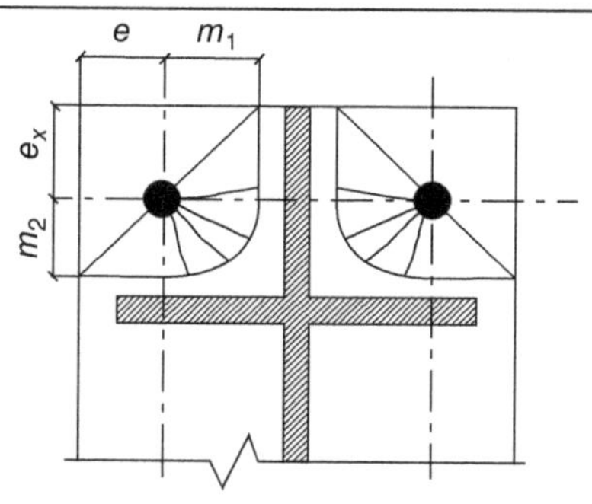	(vi) Corner yielding near a stiffener $\ell_{\text{eff}} = \alpha m_1 - (2m_1 + 0.625e) + e_x$

Source: From Ref. [18].

Table 3.15 Formulas for evaluating ℓ_{eff} for a bolt row in a plate extension.

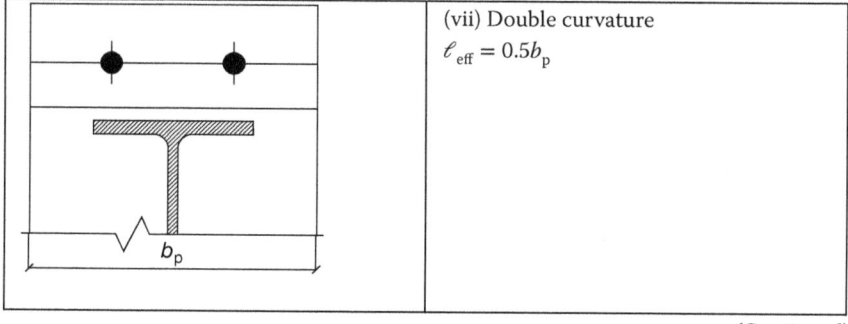	(vii) Double curvature $\ell_{\text{eff}} = 0.5 b_p$

(Continued)

Table 3.15 (Continued)

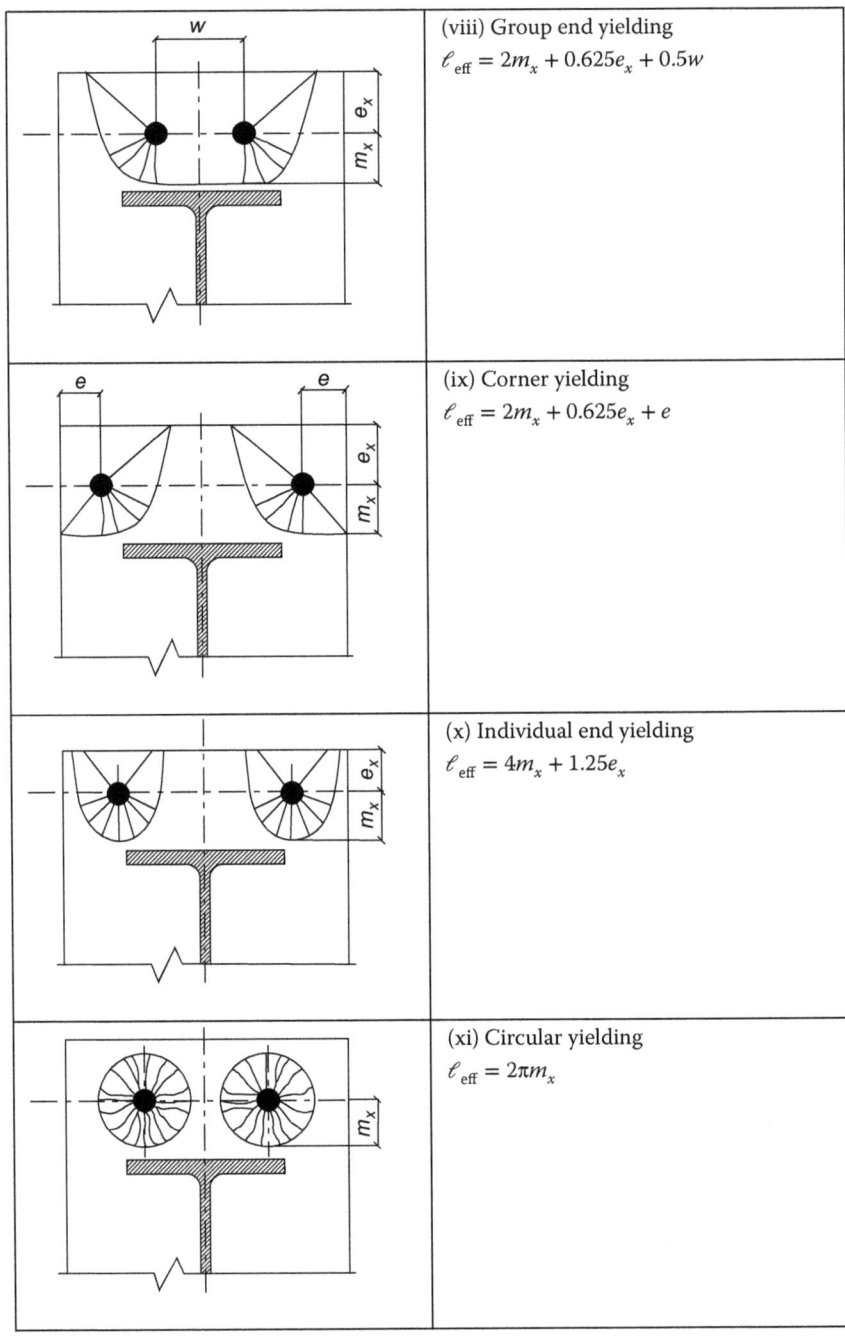

	(viii) Group end yielding $\ell_{\text{eff}} = 2m_x + 0.625e_x + 0.5w$
	(ix) Corner yielding $\ell_{\text{eff}} = 2m_x + 0.625e_x + e$
	(x) Individual end yielding $\ell_{\text{eff}} = 4m_x + 1.25e_x$
	(xi) Circular yielding $\ell_{\text{eff}} = 2\pi m_x$

Source: From Ref. [18].

3.10.6.2 ℓ_{eff} to Consider for a Bolt Row Acting Alone
See Table 3.16.

Table 3.16 ℓ_{eff} to be taken into consideration for a bolt row acting alone.

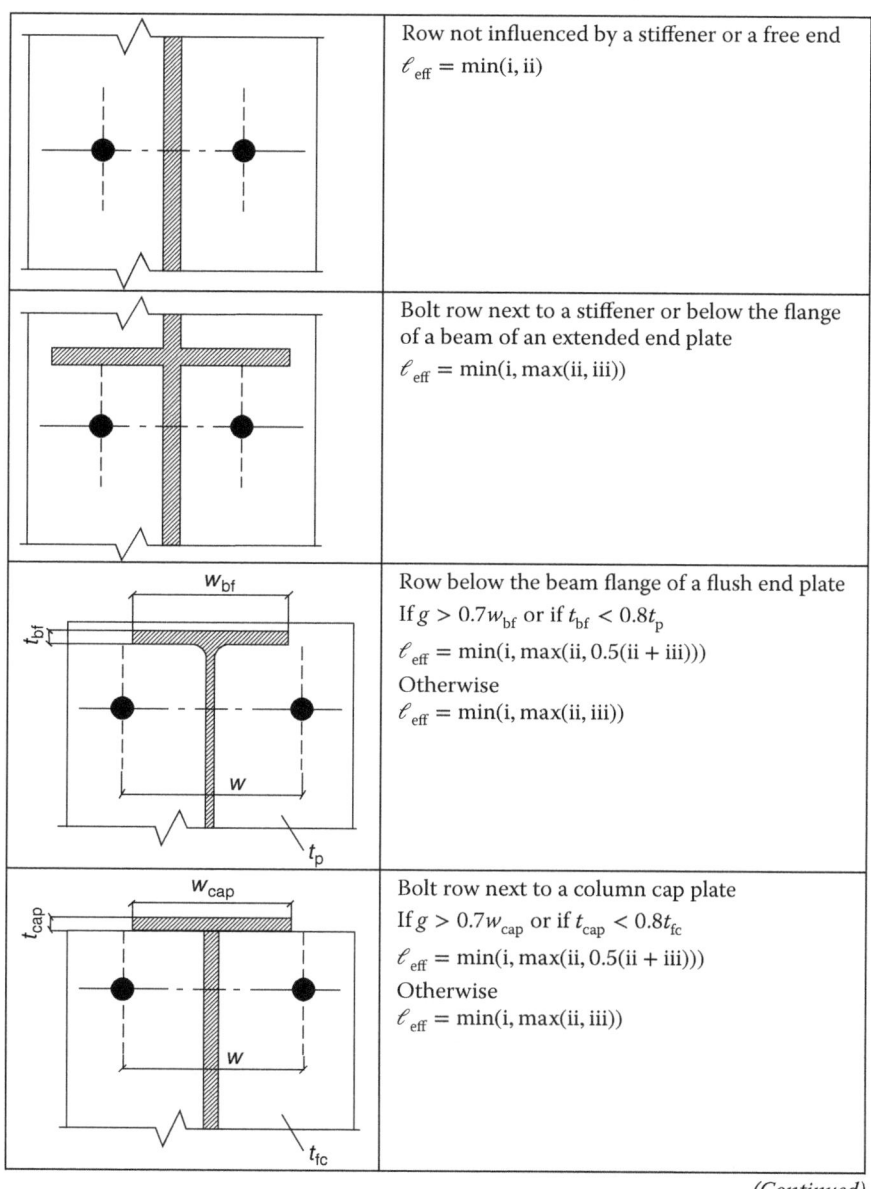

	Row not influenced by a stiffener or a free end $\ell_{eff} = \min(i, ii)$
	Bolt row next to a stiffener or below the flange of a beam of an extended end plate $\ell_{eff} = \min(i, \max(ii, iii))$
	Row below the beam flange of a flush end plate If $g > 0.7w_{bf}$ or if $t_{bf} < 0.8t_p$ $\ell_{eff} = \min(i, \max(ii, 0.5(ii + iii)))$ Otherwise $\ell_{eff} = \min(i, \max(ii, iii))$
	Bolt row next to a column cap plate If $g > 0.7w_{cap}$ or if $t_{cap} < 0.8t_{fc}$ $\ell_{eff} = \min(i, \max(ii, 0.5(ii + iii)))$ Otherwise $\ell_{eff} = \min(i, \max(ii, iii))$

(Continued)

Table 3.16 (Continued)

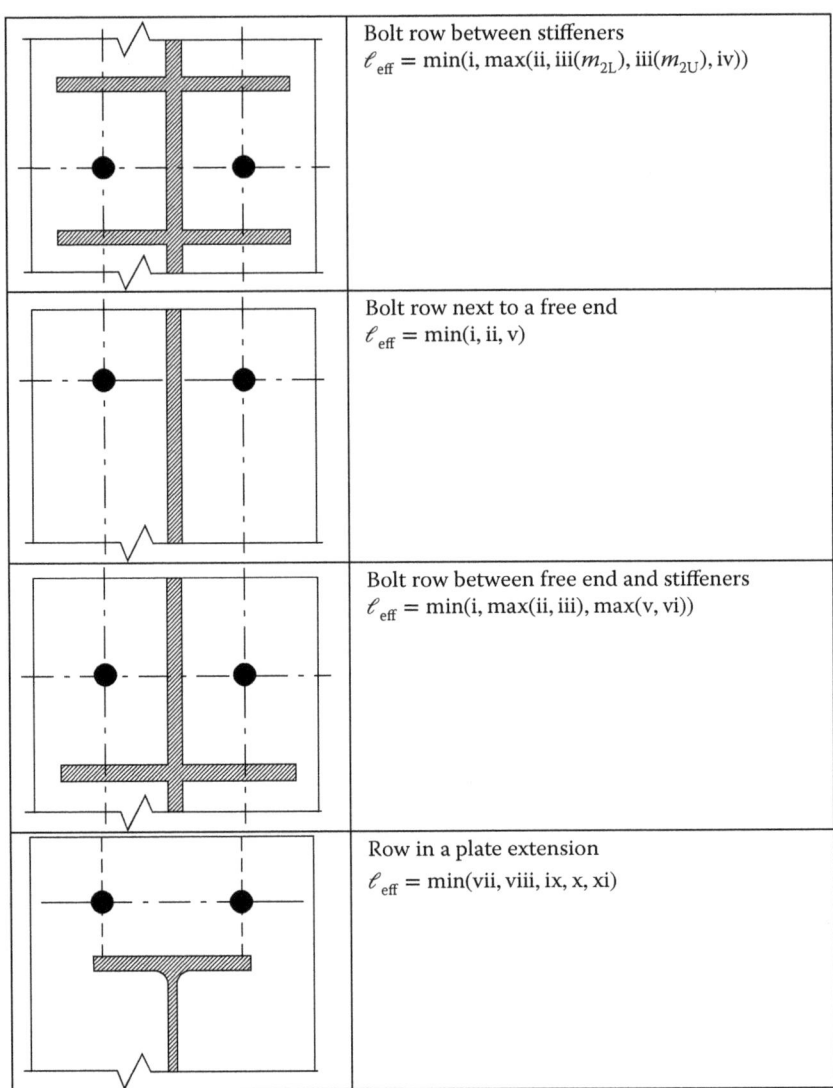

	Bolt row between stiffeners $\ell_{eff} = \min(i, \max(ii, iii(m_{2L}), iii(m_{2U}), iv))$
	Bolt row next to a free end $\ell_{eff} = \min(i, ii, v)$
	Bolt row between free end and stiffeners $\ell_{eff} = \min(i, \max(ii, iii), \max(v, vi))$
	Row in a plate extension $\ell_{eff} = \min(vii, viii, ix, x, xi)$

Source: From Ref. [18].

3.10.6.3 ℓ_{eff} to Consider for Bolt Rows Acting in Group
See Table 3.17.

Table 3.17 ℓ_{eff} to be taken into consideration for a bolt row acting in a group.

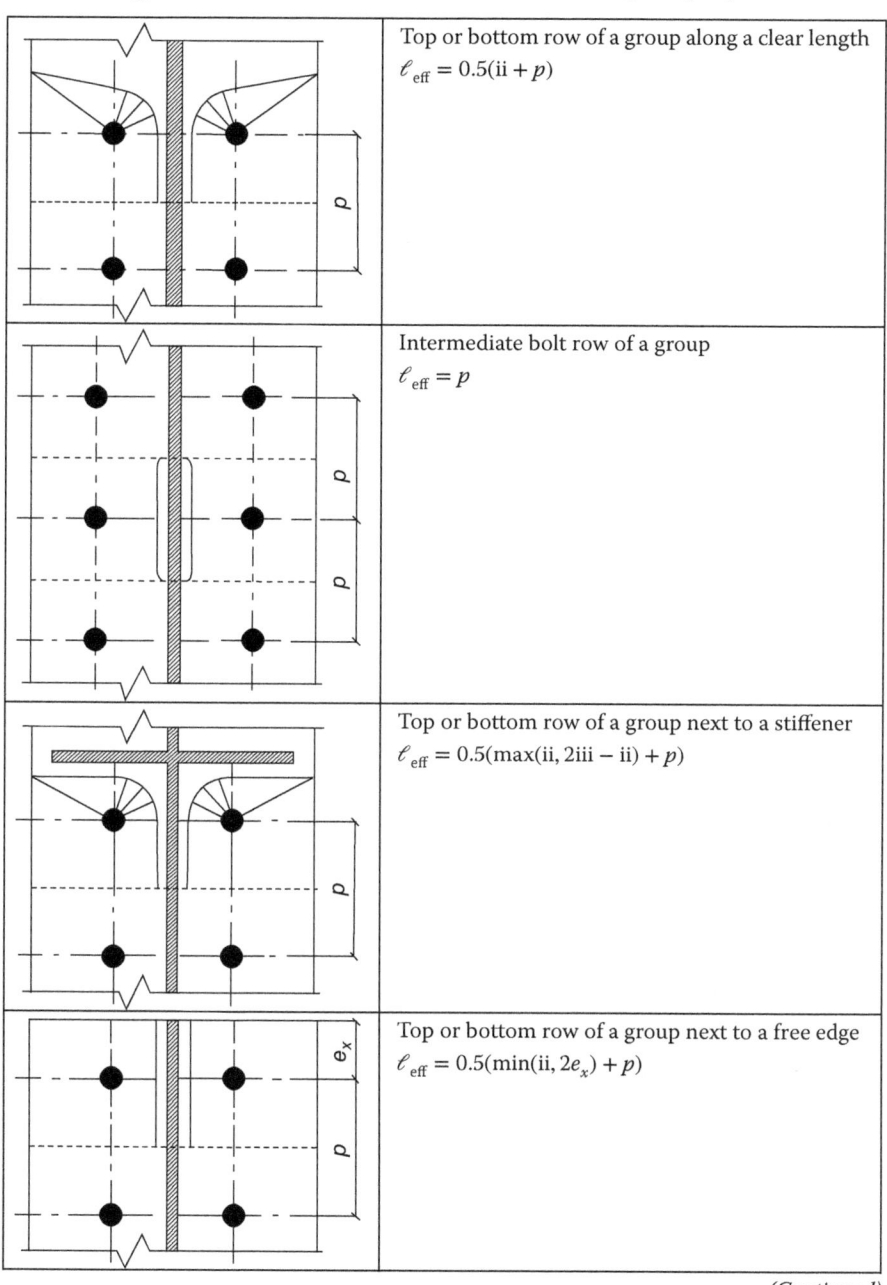

	Top or bottom row of a group along a clear length $\ell_{eff} = 0.5(ii + p)$
	Intermediate bolt row of a group $\ell_{eff} = p$
	Top or bottom row of a group next to a stiffener $\ell_{eff} = 0.5(\max(ii, 2iii - ii) + p)$
	Top or bottom row of a group next to a free edge $\ell_{eff} = 0.5(\min(ii, 2e_x) + p)$

(Continued)

Table 3.17 (Continued)

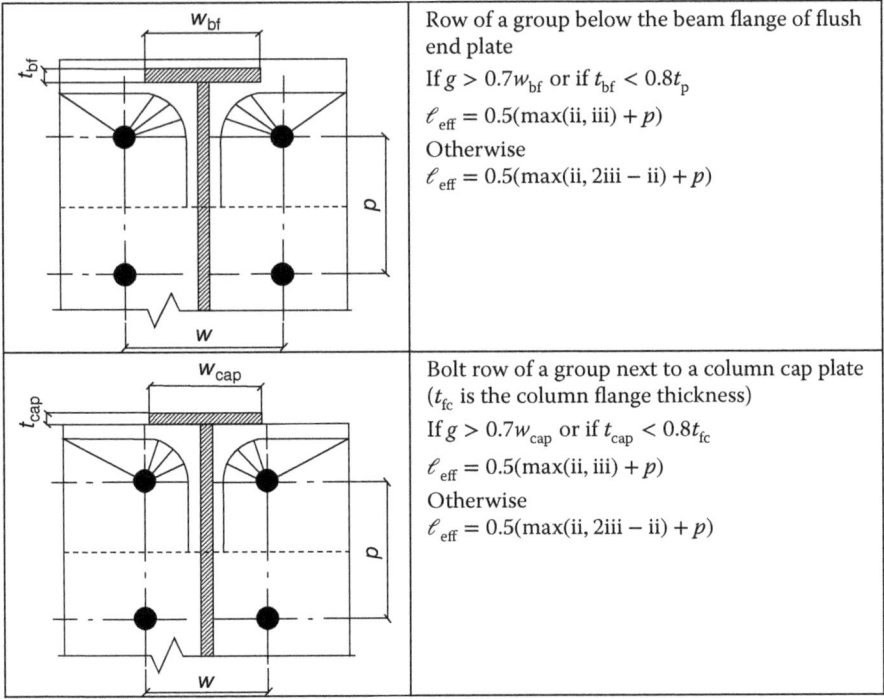

(diagram with w_{bf}, t_{bf}, w, p)	Row of a group below the beam flange of flush end plate If $g > 0.7w_{bf}$ or if $t_{bf} < 0.8t_p$ $\ell_{eff} = 0.5(\max(ii, iii) + p)$ Otherwise $\ell_{eff} = 0.5(\max(ii, 2iii - ii) + p)$
(diagram with w_{cap}, t_{cap}, w, p)	Bolt row of a group next to a column cap plate (t_{fc} is the column flange thickness) If $g > 0.7w_{cap}$ or if $t_{cap} < 0.8t_{fc}$ $\ell_{eff} = 0.5(\max(ii, iii) + p)$ Otherwise $\ell_{eff} = 0.5(\max(ii, 2iii - ii) + p)$

Source: From Ref. [18].

3.10.6.4 Examples of ℓ_{eff} for Bolts in a Group
See Table 3.18.

Table 3.18 Typical examples from [18].

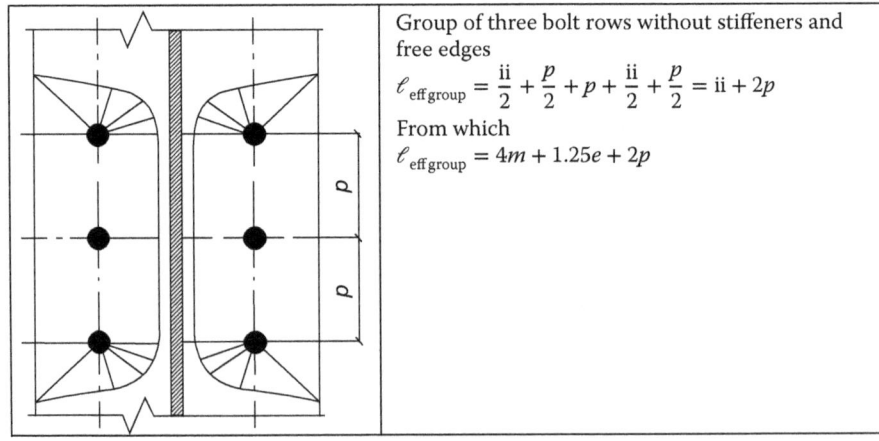

(diagram of three bolt rows)	Group of three bolt rows without stiffeners and free edges $\ell_{eff\,group} = \dfrac{ii}{2} + \dfrac{p}{2} + p + \dfrac{ii}{2} + \dfrac{p}{2} = ii + 2p$ From which $\ell_{eff\,group} = 4m + 1.25e + 2p$

Table 3.18 (Continued)

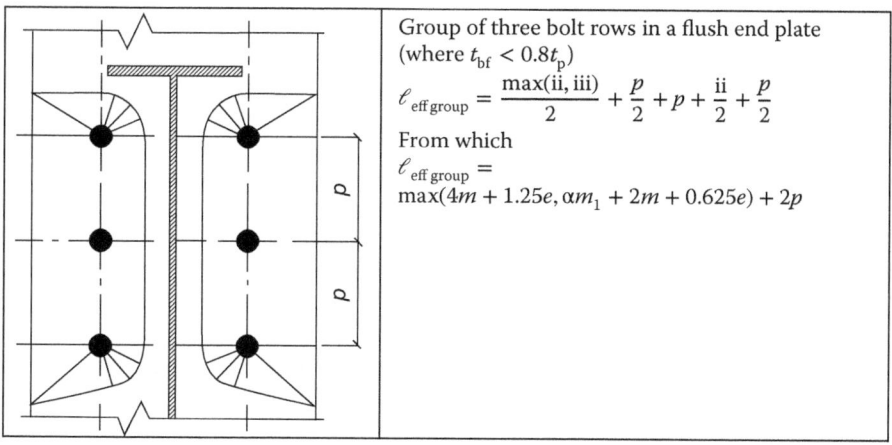

Group of three bolt rows in a flush end plate (where $t_{bf} < 0.8t_p$)

$$\ell_{\text{eff group}} = \frac{\max(\text{ii, iii})}{2} + \frac{p}{2} + p + \frac{\text{ii}}{2} + \frac{p}{2}$$

From which

$$\ell_{\text{eff group}} = \max(4m + 1.25e, \alpha m_1 + 2m + 0.625e) + 2p$$

3.10.7 T-stub for Bolts Outside the Beam Flanges

In [1] the bolts are assumed to be at a distance (gauge) not greater than the width of the beam flange. If the engineer desires to perform a calculation without implementing this hypothesis, that is, with the bolts in the corner (which is common to base plates), he or she may refer to the examples in [20], also well represented in [21].

3.10.8 Stiffness Coefficient

According to [1], the stiffness coefficient for the flange of a column in bending because of a bolt row in tension is calculated as

$$k_4 = \frac{0.9\ell_{\text{eff}} t_{fc}^3}{m^3}$$

where ℓ_{eff} must be taken as the lowest possible value among the effective lengths alone or in a group for the considered bolt row and t_{fc} represents the column flange thickness.

For an end plate in the same situation, the equation becomes

$$k_5 = \frac{0.9\ell_{\text{eff}} t_p^3}{m^3}$$

with the same indication for ℓ_{eff}, while t_p represents the plate thickness and m to be taken as m_x for an extended end plate.

Lastly, for an angle flange cleat,

$$k_6 = \frac{0.9\ell_{\text{eff}} t_a^3}{m^3}$$

3.11 Punching

For axially stressed bolts, the engineer has to verify that the design action is less than the punching resistance of the connected part or plate, which can be calculated for a single bolt (according to EC provisions) as

$$B_{p,Rd} = \frac{0.6\pi t_p d_m f_u}{\gamma_{M2}}$$

where d_m is the mean of the dimensions across points and across flats of the bolt head or the nut, whichever is smaller (it is recommended to take the minimum; if needed, check the dimension data reported in Chapter 6) while the other symbols have already been defined.

Essentially, the punching shear can become critical with small thicknesses; therefore, the engineer should pay attention to this limit state, especially in connections with cold-formed profiles.

3.12 Equivalent Systems

A method slightly outdated but effective and sometimes used as an alternative to the more "modern" T-stub is the method of equivalent systems illustrated by [17]. An end plate in bending (therefore in tension in the critical zone) is calculated as a system of equivalent beams and, equating the displacements ($f_1 = f_2$ with symbology in Figure 3.31), the force acting on each "beam" is found ($F_1 + F_2 = F_{TOT}$). From this the thickness of the plate can be verified. Also, F_{TOT} must be lower than the force transmittable by the bolt. The width of the equivalent beams (the height is equal to the thickness of the plate) is obtained by distributing the force at 45° plus the width of the bolt head (possibly the washer). The method evaluates the thickness in the elastic range and implies a behavior without prying but it is considered conservative.

3.13 Web Panel Shear

The resistance of the column web to shear is an important limit state to check (it is often necessary to reinforce the panel area) when there are moment connections, that is, end plates with remarkable bending moments. This detail is also called *web panel* and is seismically critical when it is part of the lateral load resisting system since it is heavily stressed (likely inelastically) during earthquakes.

The EC formula for calculating the (plastic) resistance is

$$V_{wp,Rd} = \frac{0.9 f_{y,wc} A_{vc}}{\sqrt{3}\gamma_{M0}}$$

The symbolism recalls the column shear area (A_{vc}) and the web yield strength ($f_{y,wc}$) of the web (for a rolled profile obviously equal to that of the column in general). The ratio between the column depth and the web thickness must be equal to or less than 69ε ($\varepsilon = \sqrt{235/f_y}, f_y$ in N mm^{-2}).

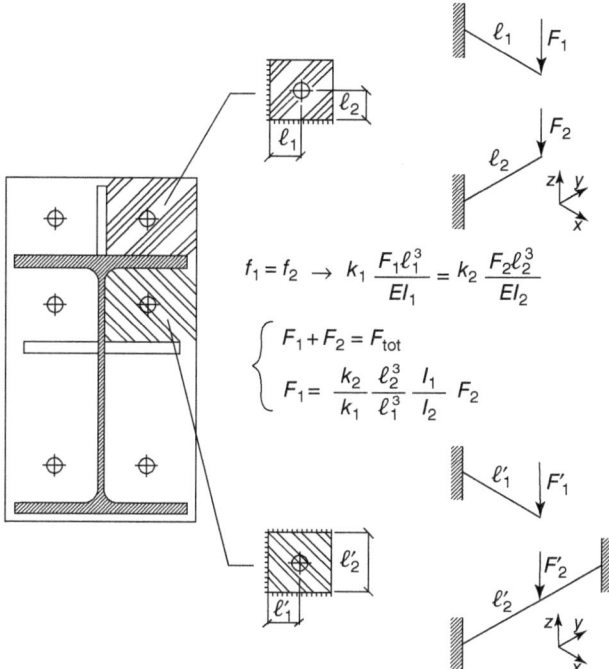

Figure 3.31 Equivalent system example.

If there are horizontal continuity plates (above and below the web at distance d_s), the value of $V_{wp,Rd}$ may be increased, as per EC, by

$$V_{wp,add,Rd} = \min\left(\frac{2M_{pl,fc,Rd} + 2M_{pl,st,Rd}}{d_s}, \frac{4M_{pl,fc,Rd}}{d_s}\right)$$

where $M_{pl,fc,Rd}$ is the design plastic bending resistance of the column flange and $M_{pl,st,Rd}$ is the stiffener resistance.

The reinforcement of the web where this is not yet verified can be performed in two ways:

1. With diagonal plates that connect the flanges transversely (transverse diagonal stiffeners); these plates can take shape as /, \ (also called N stiffeners; see Figure 4.67) or X or K or other (as the Morris stiffener; see again Figure 4.67)
2. With reinforcement plates welded to the web (also called supplementary web plates in the United Kingdom or doublers in the United States).

Both cases can easily create issues: in the first case, it is necessary to avoid interference with the bolts of the portal and the transverse plates can make it difficult for connections that arrive on the weak axis of the column. In the second case, the plate welded to the web can create problems if the column is galvanized (read the discussion in Section 6.14).

For a calculation method of transverse stiffeners see Section 4.12 on rigid end plates.

For supplementary web plates, Ref. [1] recommends the following:

- Same material as the column.
- Width so as to arrive at least at the root radius.
- Length beyond the flanges of the beam in order to completely cover the diffusion zone of the tension and compression forces (about 60° from the beam flanges should be considered).
- A thickness t_s that is not less than the column web and that can ensure the plate width is not less than $40\varepsilon t_s$.
- If plates are added on both sides of the web, only one considered to increase the shear area.

The AISC [10] formula of resistance ($\Phi R_n = \Phi 0.6 f_y d_c t_w$, with t_w the web thickness, d_c the column depth, and $\Phi = 0.9$) is also interesting because it takes into account a reduction coefficient due to the axial interaction acting upon the column. This coefficient must be considered only when compression engages the column to a value greater than 40% of the column's pure compressive yield strength, that is, without considering buckling issues. The reduction coefficient is equal to 1.4 minus the ratio between the design compression force and the design strength without considering any instability and only considering the yielding of the material.

3.13.1 Stiffness Coefficient

According to EC, the stiffness is infinite when the web is reinforced by diagonal stiffeners; otherwise (unless there is a more difficult situation to evaluate like beams with very different depths on both sides of the column)

$$k_1 = \frac{0.38 A_{vc}}{\beta z}$$

where β is the transformation parameter (see the next section) and z the lever arm (see Section 4.12.2 for a discussion of rigid end plates).

3.14 Web in Transverse Compression

The design resistance of the web of a column (this is the most common situation, where the formulas can also be applied to beams that, e.g. have a haunch, as later illustrated) in transverse compression can be evaluated according to EC as

$$F_{c,wc,Rd} = \min\left(\frac{1}{\gamma_{M0}}, \frac{\rho}{\gamma_{M1}}\right) \omega k_{wc} b_{eff,c,wc} t_{wc} f_{y,wc}$$

where t_{wc} and $f_{y,wc}$ are the thickness and yield of the column web (the subscript wc stands for "web, column") and ω is an interaction coefficient with the shear and can be calculated after having defined a couple of additional coefficients:

$$\omega_1 = \frac{1}{\sqrt{1 + 1.3\left(\frac{b_{eff,c,wc} t_{wc}}{A_{vc}}\right)^2}}$$

Table 3.19 ω as a function of β.

$0 \leq \beta \leq 0.5$	$\omega = 1$
$0.5 < \beta < 1$	$\omega = \omega_1 + 2(1 - \beta)(1 - \omega_1)$
$\beta = 1$	$\omega = \omega_1$
$1 < \beta < 2$	$\omega = \omega_1 + (\beta - 1)(\omega_2 - \omega_1)$
$\beta = 2$	$\omega = \omega_2$

and

$$\omega_2 = \frac{1}{\sqrt{1 + 5.2\left(\frac{b_{\text{eff,c,wc}} t_{\text{wc}}}{A_{\text{vc}}}\right)^2}}$$

where A_{vc} is the shear area of the column while β is the transformation parameter (see Section 3.14.1) needed to get ω from Table 3.19 and $b_{\text{eff,c,wc}}$ is the column web in compression effective width, which can be evaluated, depending on the cases (see Figure 3.32 for the symbols):

$$b_{\text{eff,c,wc}} = t_{\text{fb}} + 2\sqrt{2}a_p + 5(t_{\text{fc}} + s) + s_p$$

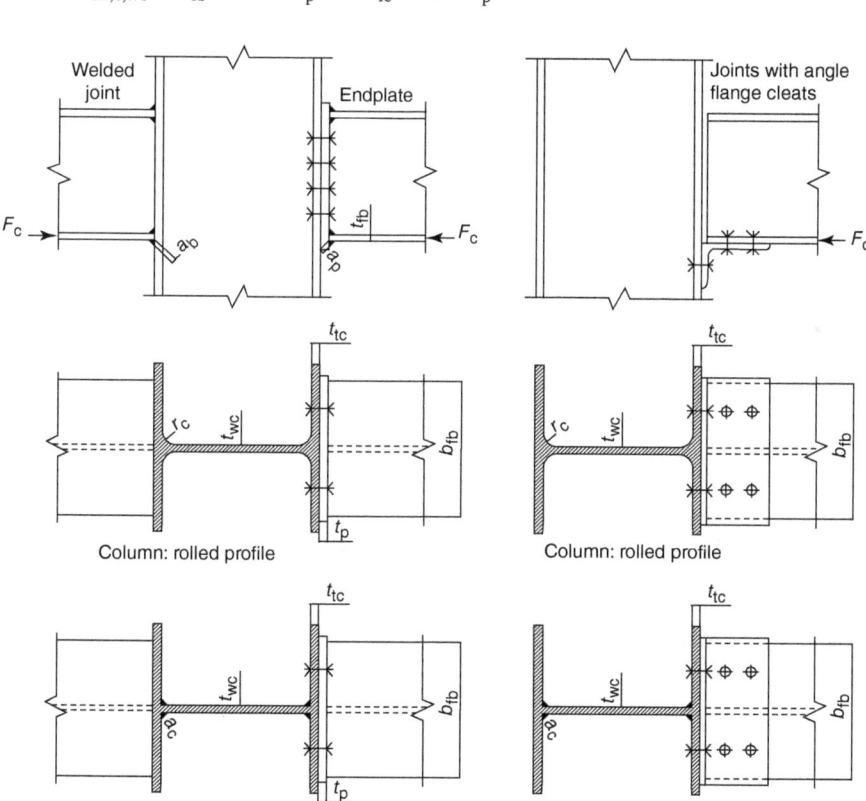

Figure 3.32 Definitions of symbols. Source: From Ref. [1].

for a bolted end plate,

$$b_{\text{eff,c,wc}} = t_{\text{fb}} + 2\sqrt{2}a_b + 5(t_{\text{fc}} + s)$$

for a welded joint, and

$$b_{\text{eff,c,wc}} = 2t_a + 0.6r_a + 5(t_{\text{fc}} + s)$$

for a bolted connection with angle flange cleats.

In the above formulas, $s = r_c$ for rolled profiles but $s = a_c\sqrt{2}$ for welded profiles; the value for s_p is included between t_p and $2t_p$ (the latter when the extension of the end plate below the flange is enough to allow a complete distribution of the force at 45°); also consider that the term $5(t_{\text{fc}} + s)$ comes from considering a force distribution at almost 70° on each side and might need to be reduced in the rare cases where the column is not long enough (the moment should be negative) to allow the full distribution.

The reduction factor ρ for plate bucking is 1 when $\overline{\lambda}_p i \leq 0.72$; otherwise

$$\rho = \frac{\overline{\lambda}_p - 0.2}{\overline{\lambda}_p^2}$$

where the plate slenderness is given as

$$\overline{\lambda}_p = 0.932\sqrt{\frac{b_{\text{eff,c,wc}} d_{\text{wc}} f_{\text{y,wc}}}{E t_{\text{wc}}^2}}$$

with $d_{\text{wc}} = h_c - 2(t_{\text{fc}} + r_c)$ for rolled sections but $d_{\text{wc}} = h_c - 2(t_{\text{fc}} + a_c\sqrt{2})$ for welded profiles.

The coefficient k_{wc} is 1 when the maximum web longitudinal stress (given by axial load and moment) $\sigma_{\text{com,Ed}}$ is less than $0.7f_{\text{y,wc}}$, but when $\sigma_{\text{com,Ed}}$ is over 70% of the yield strength, $k_{\text{wc}} = 1.7 - \sigma_{\text{com,Ed}}/f_{\text{y,wc}}$. Generally, k_{wc} is 1 since the bending moment does not impact the web heavily, the 70% of axial utilization being very high if measured upon taking into account only the yield strength.

The resistance in compression and tension (see Section 3.15) can be increased with dedicated transverse stiffeners or reinforcing plates welded to the web. In this second case, [1] allows to consider $1.5t_{\text{wc}}$ as the maximum value to use in calculations when only one plate is welded and $2t_{\text{wc}}$ when there are two plates (one each side).

3.14.1 Transformation Parameter β

The parameter β accounts for the interaction with shear when determining the resistance and the rotation capacity of beam-to-column connections.

The following equations are provided in [1] for calculating the exact value (the symbols are given in Table 3.20):

$$\beta_1 = \left|1 - \frac{M_{j,b2,\text{Ed}}}{M_{j,b1,\text{Ed}}}\right| \leq 2$$

Table 3.20 For obtaining approximate values of β in the most common cases.

Joint configuration	Case	β
(single-sided joint with $M_{b1,Ed}$)		$\beta_1 \approx 1$
(double-sided joint with $M_{b2,Ed}$ and $M_{b1,Ed}$)	$M_{j,b1,Ed} = M_{j,b2,Ed}$	$\beta_1 = \beta_2 = 0$
	$M_{j,b1,Ed} = -M_{j,b2,Ed}$	$\beta_1 = \beta_2 = 2$
	$\dfrac{M_{j,b1,Ed}}{M_{j,b2,Ed}} > 0$	$\beta_1 \approx 1$; $\beta_2 \approx 1$
	$\dfrac{M_{j,b1,Ed}}{M_{j,b2,Ed}} < 0$	$\beta_1 \approx 2$; $\beta_2 \approx 2$

Source: Adapted from Ref. [1].

$$\beta_2 = \left|1 - \frac{M_{j,b1,Ed}}{M_{j,b2,Ed}}\right| \leq 2$$

It is probably easier for the most common situations to get an approximate (though more conservative) value from Table 3.20.

3.14.2 Formulas for Other Local Buckling Limit States

The AISC and BS codes distinguish more distinct limit states about web compression (e.g. the AISC terms *crippling*, *sidesway*, and *compression buckling*); refer to [10, 18] for detailed formulas. For example, what is called *sidesway* in AISC is called *column sway* in EC, but EC does not provide quantitative formulas but only some general advice to prevent it by constructional restraints.

3.14.3 Stiffness Coefficient

In instances where the web is not stiffened for compression (the coefficient value is infinite when there are stiffeners), Ref. [1] provides the following expression to evaluate the rotational stiffness:

$$k_2 = \frac{0.7 b_{\text{eff,c,wc}} t_{\text{wc}}}{d_c}$$

where d_c is the column depth.

3.14.4 T-stub in Compression

This limit state is only treated by EC and it applies to base plates. See Section 4.4 for more information.

3.15 Web in Transverse Tension

For the transverse resistance in tension of a column web, Ref. [1] indicates

$$F_{\text{t,wc,Rd}} = \frac{\omega b_{\text{eff,t,wc}} t_{\text{wc}} f_{y,\text{wc}}}{\gamma_{M0}}$$

where the symbols are as already defined. Even ω can be calculated in the same way, replacing $b_{\text{eff,c,wc}}$ with $b_{\text{eff,t,wc}}$. In welded connections $b_{\text{eff,t,wc}}$ is obtained as in transverse compression (again, see Section 3.14) while in bolted connections it is taken as ℓ_{eff} of the equivalent T-stub representing the column flange.

As for compression, it is possible to stiffen the column and the limitations are similar to those given in Section 3.14 (for additional guidance about welds, see Ref. [1]).

It is easier to remember and to apply (since choosing ℓ_{eff} can create doubts) the BS formula (refer to [18]), which does not consider ω and takes $b_{\text{eff,t,wc}}$ supposing a 60° diffusion of the force into the column web (see Figure 3.33).

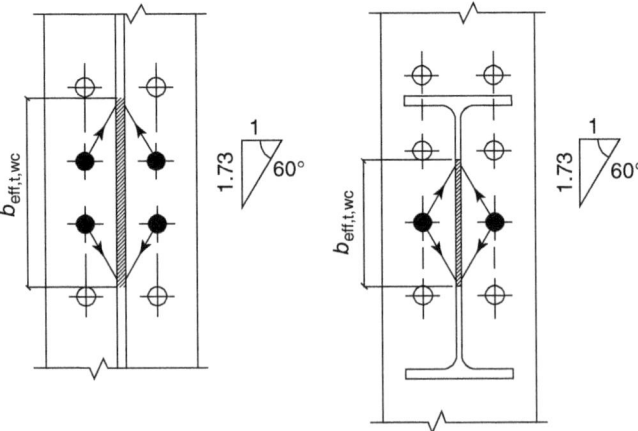

Figure 3.33 Calculation according to BS. Source: From Ref. [18].

3.15.1 Stiffness Coefficient

The equation for determining the stiffness coefficient is the same as the one used for transverse compression, with $b_{eff,t,wc}$ instead of $b_{eff,c,wc}$. In [1] k_3 is called stiffness coefficient.

3.16 Flange and Web in Compression

The following formula gives the design resistance to local compression of the flange and web of a beam (the beam is the standard situation but it can also be applied to a column in a base plate):

$$F_{c,fb,Rd} = \frac{M_{c,Rd}}{h - t_{fb}}$$

where $M_{c,Rd}$ represents the design moment that the beam can resist and t_{fb} and h are the flange thickness and the depth of the section, respectively.

If the beam is reinforced with a part cut from another beam or with a pair of plates (one as a web, the other as a flange) as in Figure 3.34, creating what is called a "haunch," the web must be checked as explained in Section 3.14. The haunch must be, according to [1], of the same material as the beam, with thicknesses that are not less than the ones in the profile (valid for web and flanges) and with an angle (from the horizontal) less than 45°.

3.17 Beam Web in Tension

The design resistance of a beam web in tension as stressed, for example, by a bolted end plate can be evaluated as

$$F_{t,wb,Rd} = \frac{b_{eff,t,wb} t_{wb} f_{y,wb}}{\gamma_{M0}}$$

where $b_{eff,t,wc}$ can be taken as ℓ_{eff} of the equivalent T-stub representing the end plate in bending.

Figure 3.34 Haunch.

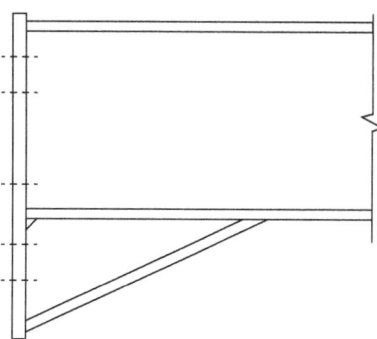

3.18 Plate Resistance

The plate is a crucial element in practically every connection and, therefore, it is fundamental to evaluate its stress correctly.

Calculating the resistance can have as a reference the yielding or the ultimate resistance of the material, depending on the limit state (and, partially, on the reference standard).

3.18.1 Material Properties

Table 3.21 provides, in simplified form (for complete values see the standards), selected yield and ultimate resistance values for typical materials of plates and profiles used in fabrication.

3.18.2 Tension

The most recent provisions are oriented to give, for tension design, the minimum value between the ultimate resistance of the material considering the net area and the yielding considering the gross area (bolt holes are not considered). In other words, the check must include the verification of both limit states.

As already indicated elsewhere in this book, it is actually preferable, if design forces are exceeded, the plate yields instead of other fragile limit states happening, since yielding allows a redistribution of forces. This means that, on the one hand, it is necessary to check that the limit state is not overtaken and, on the other, that the yielding governs the design: if a ductile limit state steps in before other fragile limit states, it can avoid, for unexpected large forces such as seismic actions, nonductile collapses such as weld failure (especially if transversely loaded), shear rupture of bolts, or even the collapse by tension of the net area.

As just discussed, EC recommends taking the tension design resistance as the minimum between

$$N_{pl,Rd} = \frac{A f_y}{\gamma_{M0}}$$

and

$$N_{u,Rd} = \frac{0.9 A_{net} f_u}{\gamma_{M2}}$$

The symbols here are defined as before, with the net area defined as the nominal area less the holes for bolts (if the bolts are staggered see further discussion).

Eurocode requires considering the net area in the first expression if bolts are designed as slip resistant at the ultimate limit state (class C). In EC symbols

$$N_{net,Rd} = \frac{A_{net} f_y}{\gamma_{M0}}$$

3.18 Plate Resistance

Table 3.21 Material properties in MPa (N mm^{-2}); "thk" stands for thickness and "lim." for limit.

Material	thk ≤ lim. 1			thk ≤ lim. 2			thk ≤ lim. 3			
	f_y	f_u	lim. 1 (mm)	f_y	f_u	lim. 2 (mm)	f_y	f_u	lim. 3 (mm)	β_w (EC)
EN 10025-2004										
S235	235	360	16	225	360	40	215	360	100	0.8
S275	275	410	16	265	410	40	235	410	100	0.85
S355	355	470	16	345	470	40	315	470	100	0.9
S275 N/NL	275	370	16	265	370	40	235	370	100	0.85
S355 N/NL	355	470	16	345	470	40	315	470	100	0.9
S420 N/NL	420	520	16	400	520	40	360	520	100	1.0
S460 N/NL	460	540	16	440	540	40	400	540	100	1.0
S450	450	550	16	430	550	40	380	550	100	1.0
S275 M/ML	275	370	16	265	370	40	255	360	100	0.85
S355 M/ML	355	470	16	345	470	40	335	450	100	0.9
S420 M/ML	420	520	16	400	520	40	390	520	100	1.0
S460 M/ML	460	540	16	440	540	40	430	540	100	1.0
S235 W	235	360	16	225	360	40	215	360	100	0.8
S355 W	355	470	16	345	470	40	315	470	100	0.9
S460 Q/QL/QL1	440	550	50	400	550	100				
DIN 18800-1, 2008										α_w (DIN)
S235	240	360	40	215	360	100				0.95
S235	275	410	40	255	410	80				0.85
S355	360	470	40	335	470	80				0.8
S450	440	550	40	410	550	80				0.7
AISC 14th Ed.										
A36	248	400	203	221	400	∞				
A529 Gr.50	345	448	∞							
A529 Gr.55	379	483	∞							
A572 Gr.42	290	414	∞							
A572 Gr.50	345	448	∞							
A572 Gr.55	379	483	∞							
A572 Gr.60	414	517	∞							
A572 Gr.65	448	552	∞							
A913 50	345	414	∞							
A913 60	414	517	∞							
A913 65	448	552	∞							
A913 70	483	621	∞							

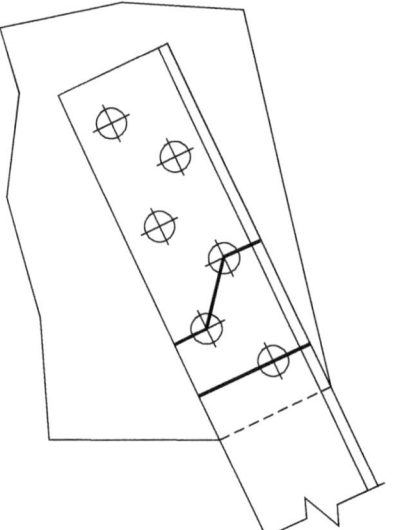

Figure 3.35 Staggered bolts.

3.18.2.1 Staggered Bolts

To increase the net area, it might be convenient to use a staggered pattern, as shown in Figure 3.35.

The net area is in fact the minimum of the depicted areas: the engineer must evaluate, even along a zigzag path, which is the smallest area, that is, the "classical" perpendicular or the one obtained by combining all the bolt holes diagonally. It might be a good optimization, respecting the minimum distances if possible, to make the two areas equal.

Using staggered bolts can be exploited in fin plates/shear tabs (Section 4.6) or similarly when the net area is an issue; it is mostly helpful in splices (Section 4.13) or in bracing connections involving angles, as in Figure 3.35.

3.18.3 Compression

According to [11, 22], the net area must not be considered to check compression loads if bolts are inserted and the clearances are normal; therefore,

$$N_{c,Rd} = \frac{A f_y}{\gamma_{M0}}$$

For sections in class 4 of EC, the effective area must be taken instead of A.

3.18.4 Shear

The general formula for checking shear (A_v represents the net area) is

$$V_{c,Rd} = \frac{A_v f_y}{\sqrt{3}\gamma_{M0}}$$

Some codes (e.g. the AISC) use 0.6 instead of $1/\sqrt{3}$.

3.18.5 Bending

Even if AISC regulations mentions not to consider bending in connection parts such as plates and angles in the typical beam-to-column connections, it is usually good practice to check it (even for the weak axis if there are those kinds of actions) to guarantee balance and consistency. When performing the check, the designer has to choose between using the elastic or plastic modulus. The choice can be according to the slenderness class even if the elastic is more conservative (once any local buckling issue is avoided).

From a numerical point of view the check means that the design resistance (modulus by yielding including necessary safety coefficients) is larger than the design bending action.

3.18.6 Design for Combined Forces

Again, the AISC says explicitly that it is not necessary to combine actions when checking connection plates. As in bending, let us take note of this, but the general conservative advice is, as before, to combine forces in the verifications.

The formulas to use vary depending on the standards used, and the engineer is invited to refer directly to the design norm.

3.18.7 Whitmore Section

When the plate area is much larger than the connected element, it is necessary to calculate the *Whitmore section*, that is, the effective area that can be utilized in checking the connection. R. E. Whitmore was an American researcher who, in the 1950s, studied the problem. He found that actions spread with a 30° angle from the beginning of the connection (i.e. the first column of bolts or the start of the weld) to its end, as in Figures 3.36–3.38. As in Figure 3.38, the effective section can "expand" into connected elements, but not beyond a side that is not connected. It must be noted that there have been proposals (Ref. [23] is one of these) to extend the load transmission angle to 45°.

The Whitmore section must therefore be large enough to take the load. For a full-strength design, the Whitmore section should be bigger than the connected element.

3.19 Reduced Section of Connected Profiles

One important limit state to bear in mind in design is that a profile connecting to a joint might have a reduced section to be considered.

Typical examples are fin plates (shear tabs) or double angles where the secondary beam is notched.

Notching one flange (or both) and partially the web might create problems since the bending, shear, and axial capacity in particular are impacted. This check does not always govern but, when it does, the following actions can be taken:

- Some local reinforcing, as an additional plate welded over the web or a plate welded perpendicularly on the top of the web to re-create a flange in a

94 | *3 Limit States for Connection Components*

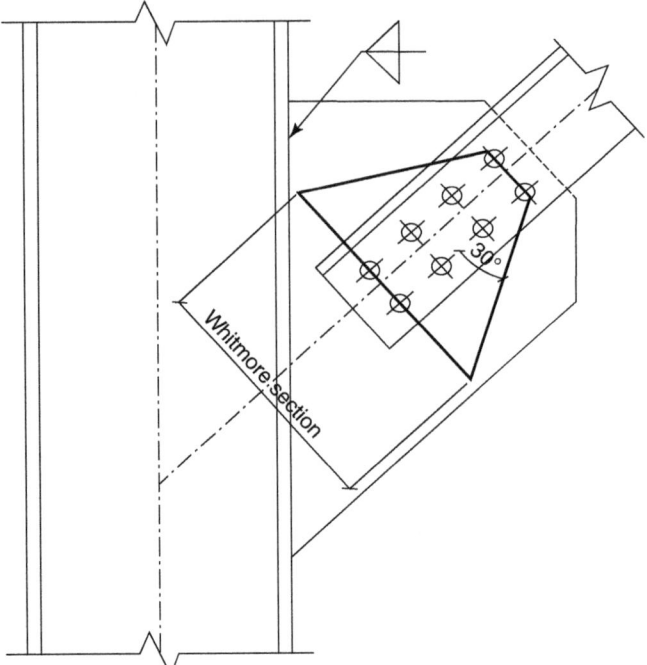

Figure 3.36 Bolted joint.

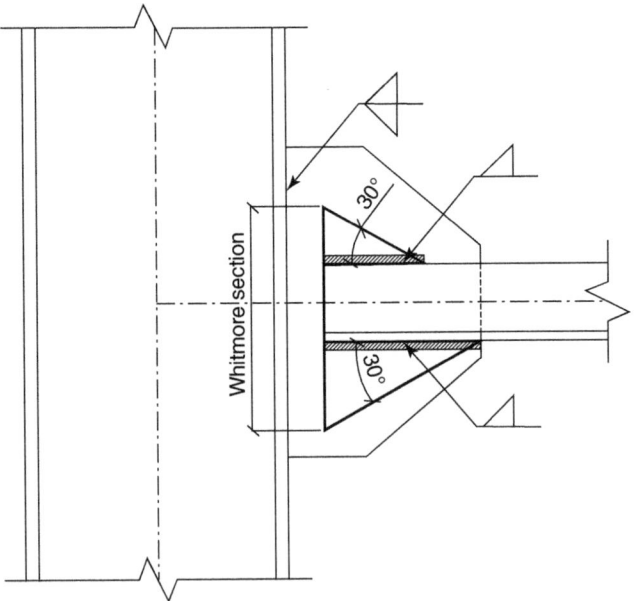

Figure 3.37 Welded joint.

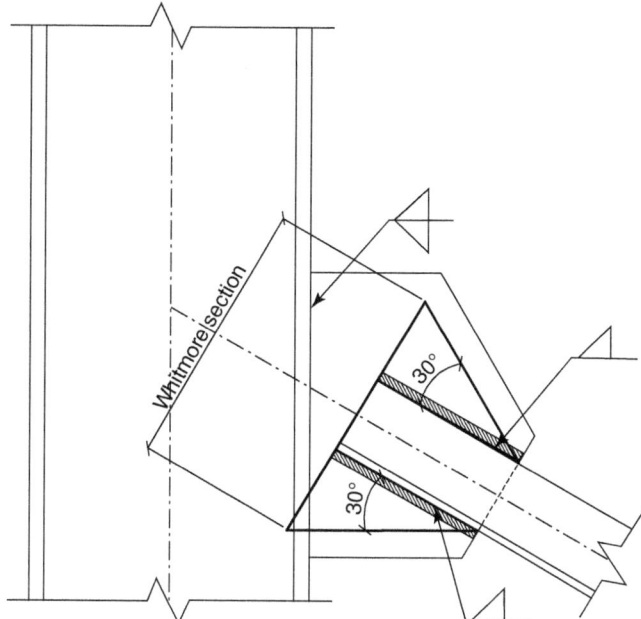

Figure 3.38 The section can expand into a connected element.

position lower than the original flange, a solution known as "false flange" (see Figure 4.42).
- A change in geometry, in order to make the notch smaller and consequently reduce the bending moment acting on the reduced section (the more the connection extends far from the connection axis, the bigger is the bending moment for ideal pin connections).
- A change in the type of connection if no viable solutions are geometrically available.

Similar situations can arise in connections of braces or trusses made with angles (or, similarly, with channels) that are connected only on one leg (or one flange). In those cases the effective section is not the nominal but it must be reduced.

3.19.1 Shear Lag

The term *shear lag* refers to profiles in tension that are connected by bolts or welds only on some elements of the profile section, that is, an angle only connected on one leg. Such a configuration implies that a reduction factor must be applied when calculating the design resistance of the profile. It is necessary to underline that the shear lag term relates to the AISC definition since, according to EC, the shear lag term refers more precisely to a reduction of the nominal area for welded profiles with very slender plates [24]. This does not mean that EC does not consider this situation of a profile only connected on some parts; rather it means simply that it is not called shear lag and is introduced, for example, for L connected only on one leg as a special case of the area reduction to apply

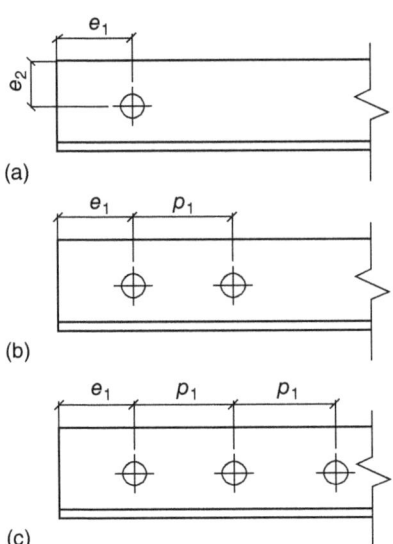

Figure 3.39 Angles connected on one leg only. Source: From Ref. [1].

when bolt holes are designed. Having several holes helps in "collecting" the profile resources and better exploits the parts that are not connected because the area of load transfer is wider and the loads are spread more uniformly.

Design equations for the tension resistance of an angle connected as in Figure 3.39 are given in [1]. In the case of only one bolt, the equation is

$$N_{u,Rd} = \frac{2t(e_2 - 0.5d_0)f_u}{\gamma_{M2}}$$

with t the angle thickness and d_0 the hole diameter. If there are more bolts on the same row, the formula becomes

$$N_{u,Rd} = \frac{\beta A_{net} f_u}{\gamma_{M2}}$$

where β must be taken as in Table 3.22, choosing β_2 if there are two columns of bolts and β_3 if there are three or more. For intermediate values of p_1, the β value is obtainable by linear interpolation.

The formulas are, once again, for angles connected on only one leg and must also be applied for large angles with two or more rows of bolts when they are only on one leg. The number of "columns" of bolts (perpendicular to load direction) will be the key variable to choose the correct value.

Table 3.22 β value for shear lag formula.

Pitch p_1	$\leq 2.5d_0$	$\geq 5d_0$
β_2 for two bolts	0.4	0.7
β_3 for three or more bolts	0.5	0.7

Eurocode [1] clarifies the conditions that are to be met for using the so-called lug angle, that is, an additional angle in the connection that greatly helps the resistance:

- The bolts connecting the lug angle to the brace must be designed for 1.4 times the force in the outstand of the angle member; if a channel is connected, 1.1 times the force in the channel flange must be considered.
- The bolts connecting the lug angle to the gusset plate must be designed for 1.2 times the force in the brace outstand (valid for both angles and channels); there must be at least two bolts.
- The connection must have (Figure 3.40) a lug angle that runs from the end of the brace to a point beyond the supporting plate.

The AISC provisions have a more general approach for determining the shear lag that is quite interesting because it is largely applicable. The formula is simply using a reduction coefficient U for the net area. Generally speaking (plates and connections with hollow sections are not included because additional considerations apply) the coefficient is defined as

$$U = 1 - \frac{x}{l}$$

where x is the distance of the connected part of the member from the center of gravity (CG) while l is the connection length, that is, the distance between the first and last bolts in the force direction for bolted connections (see Figure 3.42 for welded joints). Figure 3.41 shows some cases for calculating x which allow applying the reduction coefficient also for H- and I-shaped profiles connected only to the web or only to the flanges.

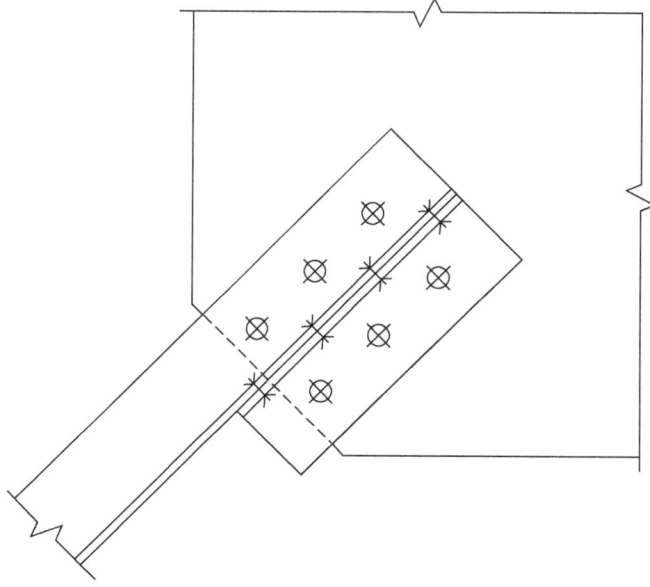

Figure 3.40 Lug angle connection.

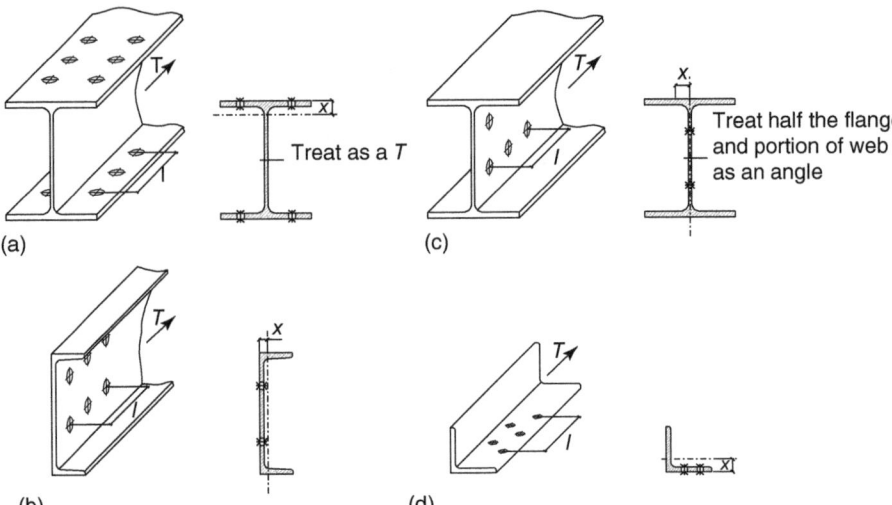

Figure 3.41 Instructions to evaluate x and l in bolted joints. Source: From Ref. [10].

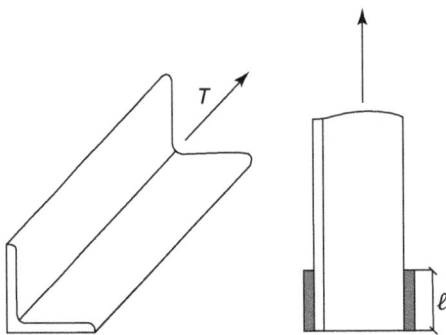

Figure 3.42 Evaluation of l in welds. Source: From Ref. [10].

AISC instructions for connecting tube members with a plate welded in the middle of the member are also quite useful (typical brace connection). The weld length to connect the plate must not be less than the tube diameter or the width (dimension parallel to the inserted plate) if the tube is rectangular.

If the tube is round and $l \geq 1.3D$, a value of $U = 1$ can be taken. For intermediate cases (to apply in the equation previously seen for the general case), if it is a round hollow section,

$$x = \frac{D}{\pi}$$

while, if it is rectangular (B is the dimension perpendicular to the connected plate and H the parallel one; see Figure 3.43),

$$x = \frac{B^2 + 2BH}{4(B + H)}$$

It must be noted that in both cases there is a discontinuity for l that is slightly less than $1.3D$ ($U \sim 0.75$) and l equal to $1.3D$ ($U = 1$).

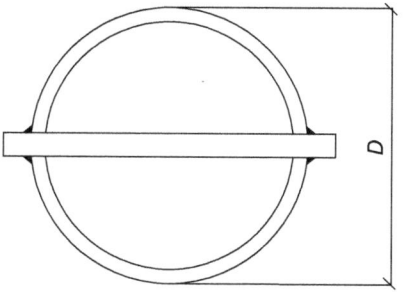 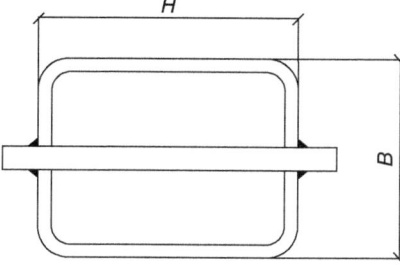

Figure 3.43 Literal symbols. Source: From Ref. [10].

Figure 3.44 Detail to be considered when evaluating the net section in hollow sections. Source: From Ref. [10].

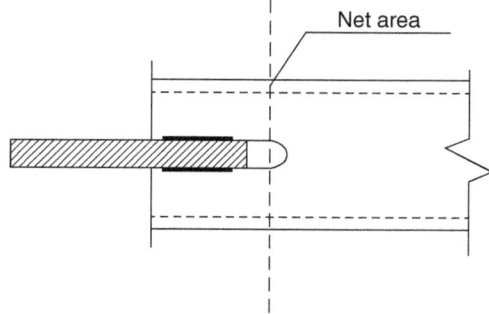

Attention must be paid when evaluating the critical area of the tube due to the slot created for inserting the plate: Figure 3.44 shows the potential problem unless special care is given during fabrication.

3.20 Local Capacity

The connections can locally stress the elements in various patterns to evaluate case by case. If there are dubious situations, a simplified hand check for strength (and buckling) should be done. Eventually, some local reinforcement can be added.

If, say, there is a shear tab (or any similar situation), it is necessary to verify locally the web of the main member (from [13], where A_v does not have the factor 2 as multiplier but it is checked against half of the design shear), taking a resistant area equal to (see Figure 3.45 for symbols)

$$A_v = 2 \times 0.9 \, lt_w$$

There are many possible nonstandard situations where the designer must carefully evaluate his or her design using approximate but conservative methods.

Also, the *tie force* could be considered as a local capacity check, see Section 3.22 for an introduction and the relevant sections of Chapter 4 or [13] directly for numeric approaches case by case.

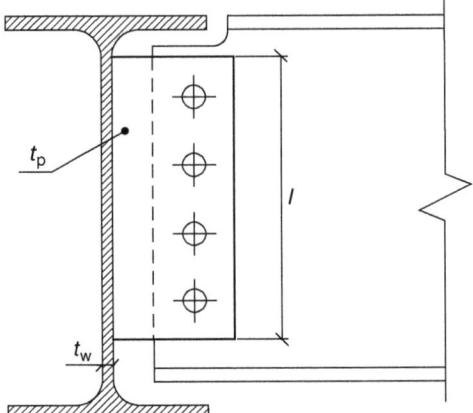

Figure 3.45 Symbols used in formulas.

3.21 Buckling of Connecting Plates

The engineer must carefully assess all the possible limit states in a connection, including the tricky local buckling phenomena. For brace connections, for example, the gusset plate can easily be critically affected by instability if the brace works in compression.

3.21.1 Gusset Plate Buckling

When a plate is in compression, the buckling resistance can be evaluated considering it as a column, where the critical axis is obviously the weak one involving the thickness.

As a general reference to calculate the unbraced length of a plate, the distance between the CG of the bolt group (conservative, otherwise it is more usual to take the near end of the bolted group) and the CG of the weld could be taken, depending on the kind of connections, that might have two bolt groups or be fully welded. A gusset plate for a brace connection is a typical example, with the additional problem that the joint can be in a corner and can have various characteristics, as the examples in Figure 3.46 show. Several research papers dedicated to gusset plates in compression (e.g. Refs. [23, 25–28]) discuss methods to evaluate the correct unbraced length and the effective length factors. The results are not univocal. With the symbols l_1, l_2, and l_3 defined in Figure 3.47 and calculating l_{avg} as the average of the three, Table 3.19 provides a synthesis from the sources and sums up the coefficients to apply. The values are quite different and it is up to the engineer whether to choose the most conservative or the best performing one. It must be recalled that the mathematical definition of the limit between a compact and a noncompact plate (the effective length factors vary according to this; see Table 3.23) is given in [25]. The limit thickness t_β is evaluated as

$$t_\beta = 1.5\sqrt{\frac{f_y c^3}{E l_1}}$$

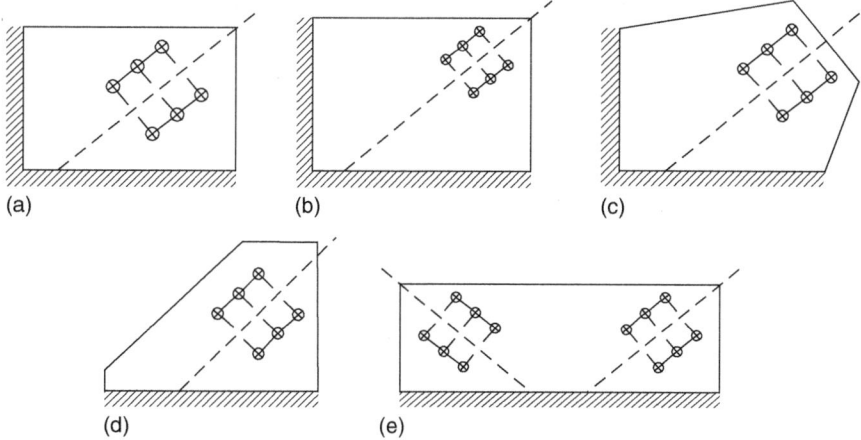

Figure 3.46 Possible configurations for brace connections. (a) Compact, (b) noncompact, (c) extended, (d) on one sinde, (e) chevron type. Source: From Ref. [25].

Figure 3.47 Calculation of l_{avg} and c.

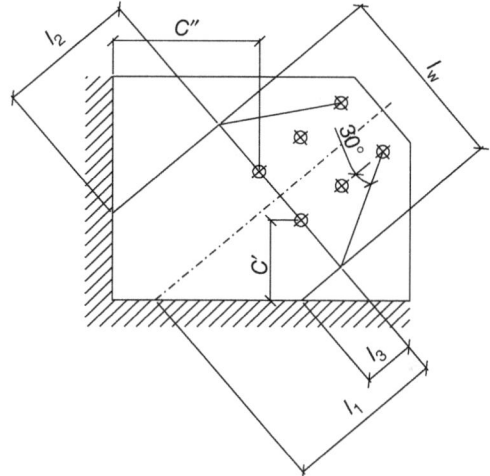

where c is the minimum value between c' and c'' in Figure 3.47. Thicker plates classify as compact.

In [27] it is recommended to use stiffeners in the various configurations in order to have effective length factors of 0.65 instead of 2. Some examples are given in Figure 3.48. The one in Figure 3.48c might actually look excessive and therefore should be used advisedly.

3.21.2 Fin Plate (Shear Tab) Buckling

The British Standard [14] provides formulas to check the buckling of a plate welded to the primary member and bolted to the secondary member. The instructions are given in [13] but they are as well reported and referenced in [16], which means that this check can be considered complementary and valid

Table 3.23 Synthesis of instructions for evaluating buckling in gusset plates.

Configuration	Effective length factor	Reference length
Compact in a corner	0[a], 0.5[b], 0.65[c]	l_{avg}
Noncompact in a corner	0.5[b], 0.65[c], 1.0[a]	l_{avg}
Extended in a corner	0.6[a], 2.0[d]	l_1
Single	0.7[a], 1.2[b], 2.0[d]	l_1
Chevron	0.75[a], 2.0[d]	l_1

a) From Ref. [25].
b) From Ref. [29].
c) From Ref. [23].
d) From Ref. [27].

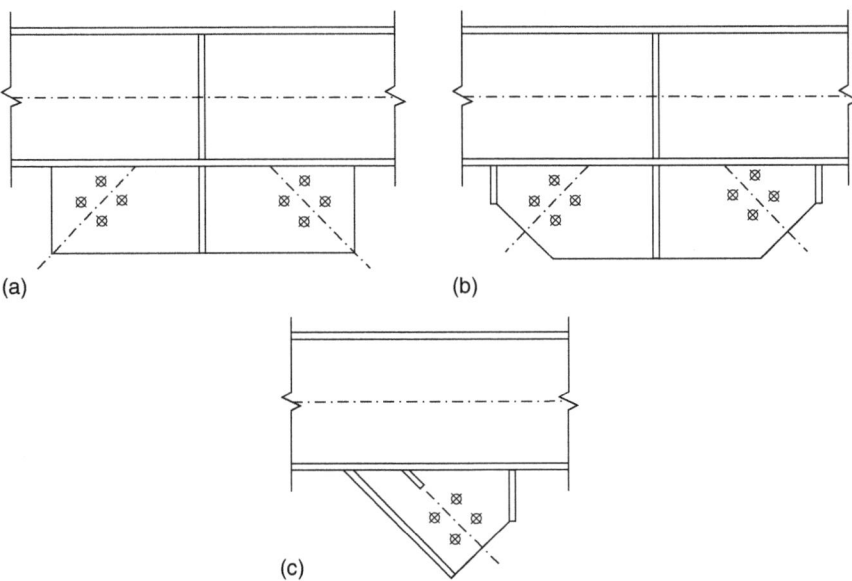

Figure 3.48 Stiffeners to apply according to [27].

also for an analysis according to the Eurocode, not only for a design based on the British code.

The method consists in first calculating a slenderness coefficient for the plate defined as

$$\lambda_{LT} = 2.8 \sqrt{\frac{z_p h_p}{1.5 \, t_p^2}}$$

where h_p and t_p are the plate depth and thickness and z_p is the distance between the side of the plate in contact with the primary member (i.e. the welded area) and the first column of bolts (see Figure 3.49).

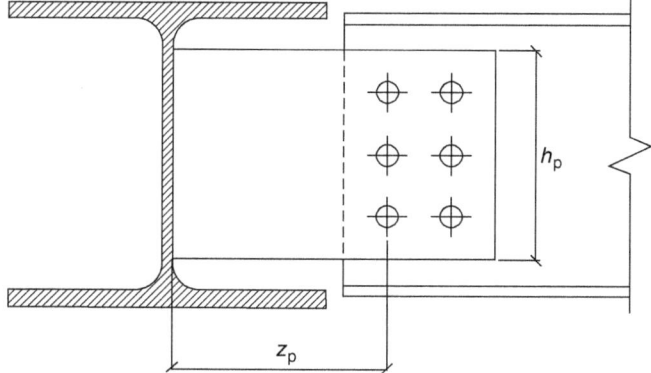

Figure 3.49 Symbols for fin plate buckling.

Using Table 3.24 (Table 17 in [14]) with λ_{LT} value and the plate nominal yield, an f_{pLT} value (p_b in the BS) is obtained and it can be used for evaluating the plate flexural torsional design resistance through the equation (adapted for EC)

$$M_{Rd,buckling} = \frac{W_{el} f_{pLT}}{m_{LT} \gamma_{M1}}$$

The coefficient m_{LT} is usually taken as 0.6; it actually varies from 0.44 to 1 according to Table 18 of [14], with the value equal to 1, valid for constant moment, making the check as conservative as possible. The final value should not be bigger than the design elastic resistance equal to $W_{el} f_y / \gamma_{M0}$.

The resulting resistance should be greater than the bending moment in the plate (more precisely, Refs. [13, 16] evaluate the bending moment as the shear multiplied by z_p).

Note that, if

$$z_p \leq t_p / 0.15$$

the check becomes negligible and can be omitted. Welding the plate to the top flange or to a stiffener, if allowed by the geometry, can also avoid the possible problem: this is effective in two ways since the plate is better restrained and the bending moment decreases.

3.22 Structural Integrity (and Tie Force)

After the World Trade Center collapse in 2001, the need for an "additional" structural resistance for some structures became a sensitive issue.

Terrorist attacks might indeed generate unexpected load conditions (Figure 3.50), and therefore structures should be capable of absorbing some "extra" loads, potentially extreme in value and unknown in direction.

A famous incident that was the first (occurring in 1968, well before 2001) to stimulate the international engineering community to study this issue was the English "Ronan Point" case: a 22-story building in precast concrete near London

Table 3.24 Values of f_{pLT} used to check plate buckling.

λ_{LT}	Steel yield (N mm^{-2})														
	235	245	255	265	275	315	325	335	345	355	400	410	430	440	460
25	235	245	255	265	275	315	325	335	345	355	400	410	430	440	460
30	235	245	255	265	275	315	325	335	345	355	390	397	412	419	434
35	235	245	255	265	272	300	307	314	321	328	358	365	378	385	398
40	224	231	237	244	250	276	282	288	295	301	328	334	346	352	364
45	206	212	218	224	230	253	259	265	270	276	300	306	316	321	332
50	190	196	201	207	212	233	238	243	248	253	275	279	288	293	302
55	175	180	185	190	195	214	219	223	2227	232	251	255	263	269	281
60	162	167	171	176	180	197	201	205	209	212	237	242	253	258	269
65	150	154	158	162	166	183	188	194	199	204	227	232	242	247	256
70	139	142	146	150	155	177	182	187	192	196	217	222	230	234	242
75	130	135	140	145	151	170	175	179	184	188	207	210	218	221	228
80	126	131	136	141	146	163	168	172	176	179	196	199	205	208	214
85	122	127	131	136	140	156	160	164	167	171	185	187	190	192	195
90	118	123	127	131	135	149	152	156	159	162	170	172	175	176	179
95	114	118	122	125	129	142	144	146	148	150	157	158	161	162	164
100	110	113	117	120	123	132	134	136	137	139	145	146	148	149	151
105	106	109	112	115	117	123	125	126	128	129	134	135	137	138	140
110	101	104	106	107	109	115	116	117	119	120	124	125	127	128	129
115	96	97	99	101	102	107	108	109	110	111	115	116	118	118	120
120	90	91	93	94	96	100	101	102	103	104	107	108	109	110	111
125	85	86	87	89	90	94	95	96	96	97	100	101	102	103	104
130	80	81	82	83	84	88	89	90	90	91	94	94	95	96	97
135	75	76	77	78	79	83	83	84	85	85	88	88	89	90	90
140	71	72	73	74	75	78	78	79	80	80	82	83	84	84	85
145	67	68	69	70	71	73	74	74	75	75	77	78	79	79	80
150	64	64	65	66	67	69	70	70	71	71	73	73	74	74	75
155	60	61	62	62	63	65	66	66	67	67	69	69	70	70	71
160	57	58	59	59	60	62	62	63	63	63	65	65	66	66	67
165	54	55	56	56	57	59	59	59	60	60	61	62	62	62	63
170	52	52	53	53	54	56	56	56	57	57	58	58	59	59	60
175	49	50	50	51	51	53	53	53	54	54	55	55	56	56	56
180	47	47	48	48	49	50	51	51	51	51	52	53	53	53	54
185	45	45	46	46	46	48	48	48	49	49	50	50	50	51	51
190	43	43	43	44	44	46	46	46	46	47	48	48	48	48	48
195	41	41	42	42	42	43	44	44	44	44	45	45	46	46	46
200	39	39	40	40	40	42	42	42	42	42	43	43	44	44	44
210	36	36	37	37	37	38	38	38	39	39	39	40	40	40	40
220	33	33	34	34	34	35	35	35	35	36	36	36	37	37	37
230	31	31	31	31	31	32	32	33	33	33	33	33	34	34	34
240	28	29	29	29	29	30	30	30	30	30	31	31	31	31	31
250	26	27	27	27	27	28	28	28	28	28	29	29	29	29	29
λ_{L0}	37.1	36.3	35.6	35	34.3	32.1	31.6	31.1	30.6	30.2	28.4	28.1	27.4	27.1	26.5

Note: λ_{L0} is reported as in the standard [14] but it is not necessary in calculations.

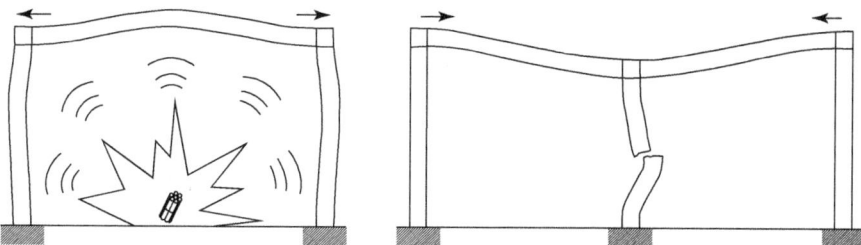

Figure 3.50 Situations where structural integrity comes into play.

collapsed from top to bottom on a corner of the building because of an explosion in a kitchen on the 18th floor. Surprisingly, the explosion was not that violent, as testified by the fact that a woman in the same room as the explosion was only wounded. However, the event showed that the structural system was very weak in resisting unexpected loads, quite small in value when compared to the resistance of the building to the gravity loads.

In the United States, a special committee is currently studying this kind of problem and the proposal is, for buildings that are tall or in special categories (hospitals or sensible targets), to impose some additional requirements that can help guaranteeing a good *structural integrity*.

Among the considerations that the committee is listing, some interesting and meaningful ones are as follows:

- Checking that column connections can bear the same load in tension as they can bear in compression.
- Guaranteeing that simple shear connections can also resist an axial load (not simultaneous with the shear) equal to the shear (or two-thirds of it, depending on the verification method).

The "lesson" that the structural engineer can learn is to design connections that can take loads in directions different from the one strictly coming from the design model. Those forces are also known as *tying forces*. In Chapter 4, suggestions to reach an acceptable structural integrity are given for many kinds of joints.

3.23 Ductility

Ductility is not a real limit state but it is an additional source (supply) of resisting capacity that is good to have. Being a "supply," it is not essential and might even be superfluous; at the same time, though, it can suddenly become extremely useful and can consistently increase the collapse limit of a structural system.

The reader is referred to Section 2.4 and the considerations about the different kinds of connections in Chapter 4, where there is advice to provide better ductility.

3.24 Plate Lamellar Tearing

Plates can have imperfections due to the lamination process that can become a structural concern in some conditions.

The lamellar tearing phenomenon can become critical in thick plates that are loaded perpendicular to the lamination plane, for example, with T, corner, and cruciform junctions with relevant welds. The examples in Figure 3.51 help visualize the problem: the material will tear because the contraction of the weld deposit will amplify any lamellar inclusion.

Eurocode instructions are based on defining, depending on the fabrication details (see Figures 3.52 and 3.53), a Z design value (Z_{Ed}) obtained summing all the contributions Z_a, Z_b, Z_c, Z_d, and Z_e.

Once this is done, a minimum required value ($Z_{Rd} \geq Z_{Ed}$) is specified for the supplied material, as per EN 10164, for example, Z15, Z25, or Z35.

One note in the German version of [30] is interesting since it suggests forgetting the prescription when $Z \leq 10$. According to [17], the potential trouble is in fact a design consideration only when the plate is over 40 mm thick, even though some defects have been recorded with thickness around 25–30 mm.

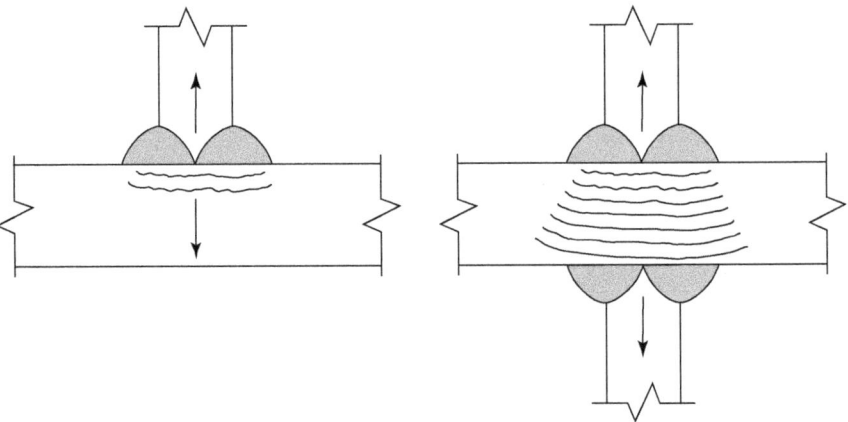

Figure 3.51 Lamellar tears.

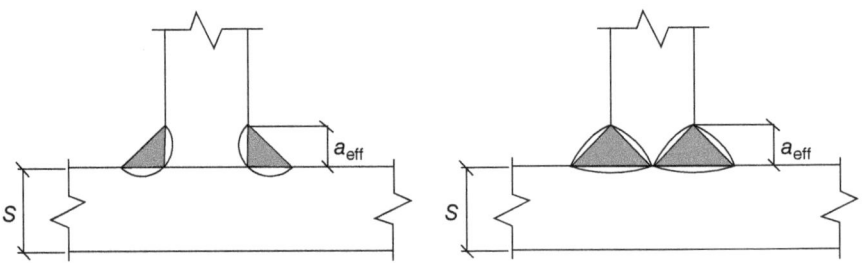

Figure 3.52 How to calculate a_{eff} in Figure 3.53.

3.24 Plate Lamellar Tearing

		Criteria affecting the target value of Z_{Ed}	
(a) Weld depth relevant for straining from metal shrinkage		Effective weld depth "a_{eff}" (see pic. 3.52) = throat thickness "a" of fillet welds	
	$Z_a = 0$	$a_{eff} \leq 7$ mm	$a = 5$ mm
	$Z_a = 3$	$7 < a_{eff} \leq 10$ mm	$a = 7$ mm
	$Z_a = 6$	$10 < a_{eff} \leq 20$ mm	$a = 14$ mm
	$Z_a = 9$	$20 < a_{eff} \leq 30$ mm	$a = 21$ mm
	$Z_a = 12$	$30 < a_{eff} \leq 40$ mm	$a = 28$ mm
	$Z_a = 15$	$40 < a_{eff} \leq 50$ mm	$a = 35$ mm
	$Z_a = 15$	50 mm $< a_{eff}$	$a > 35$ mm
(b) Shape and position of welds in T- and cruciform- and corner-connections	$Z_b = -25$		
	$Z_b = -10$	Corner joints	
	$Z_b = -5$	Single-run fillet welds $Z = 0$ or fillet welds with $Z > 1$ with buttering with low-strength weld material	
	$Z_b = 0$	Multirun fillet welds	
	$Z_b = 3$	With appropriate welding sequence to reduce the shrinkage affects. Partial-and full-penetration weld. Case 1, Case 2, Case 1, Case 2	
	$Z_b = 5$	Partial-and full-penetration weld	
	$Z_b = 8$	Corner joints	
(c) Effect of material thickness s on restraint to shrinkage	$Z_c = 2$	$s \leq 10$ mm	
	$Z_c = 4$	$10 < s \leq 20$ mm	
	$Z_c = 6$	$20 < s \leq 30$ mm	
	$Z_c = 8$	$30 < s \leq 40$ mm	
	$Z_c = 10$	$40 < s \leq 50$ mm	
	$Z_c = 12$	$50 < s \leq 60$ mm	
	$Z_c = 15$	$60 < s \leq 70$ mm	
	$Z_c = 15$	70 mm $< s$	
(d) Remote restraint of shrinkage after welding by other portions of the structure	$Z_d = 0$	Free shrinkage possible (e.g. T-joints)	
	$Z_d = 3$	Free shrinkage restricted (e.g. diaphragms in box girders)	
	$Z_d = 5$	Free shrinkage not possible (e.g. stringers in orthotropic deck plates)	
(e) Influence of preheating	$Z_e = 0$	Without preheating	
	$Z_e = -8$	Preheating ≥ 100 °C	

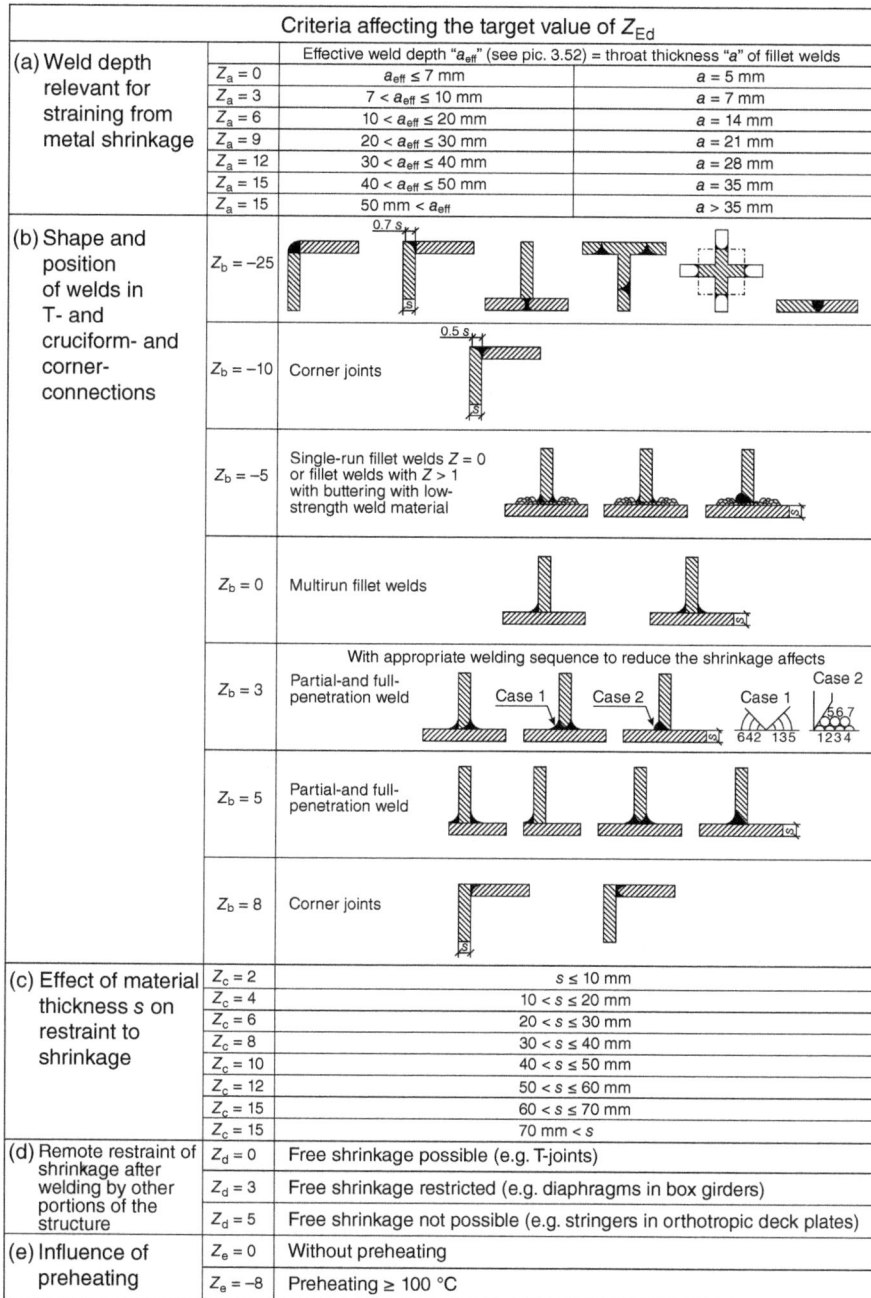

Figure 3.53 Criteria affecting the target value of Z_{Ed}. Source: Table from Ref. [30].

At the end of the day, the engineer has a couple of approaches to solve the problem when it is really a design concern:

- To prescribe the necessary fabrication details since they represent the best prevention as it is clear from the table in Figure 3.53 (preheating, bevels, number of runs are all beneficial).
- To specify a high minimum Z (which likely translates into plates that are ultrasonically controlled since they guarantee defects within a certain tolerance).

AISC addresses the problem by reminding about good detailing practices.

3.25 Other Limit States in Connections with Sheets and Cold-formed Steel Sections

When the connection is made by self-drilling or self-threading screws that fix cold-formed steel sections and/or sheets, other limit states should be considered, as illustrated in Figure 3.54 (for more details, please refer to [31]).

Since experimental testing is recommended in those cases, it may be best to ask the material supplier for instructions and design tables that indicate where to choose the details of the connections.

3.26 Fatigue

Fatigue assessment is not a common practice for standard steel structures (i.e. we are not talking about special structures such as amusement rides), but it must be evaluated in particular cases where loads change frequently and consistently in value and direction. This includes:

- Bridges
- Platforms with machinery
- High towers or stacks (the wind itself can generate fatigue)
- Crane runways.

Details about fatigue checks are not within the scope of this book and the reader is referred to the applicable standards (which might follow different approaches) and to specific texts, for example, [32], for instructions about applying the EC method [33]. Generally speaking, the fatigue assessment (which must be done on serviceability combinations that are created on purpose) is based on the following fundamental variables:

- Algebraic difference between maximum and minimum stress (possible reductions on compression according to some approaches)
- Number of cycles
- Type of details
- Possible load concentrations.

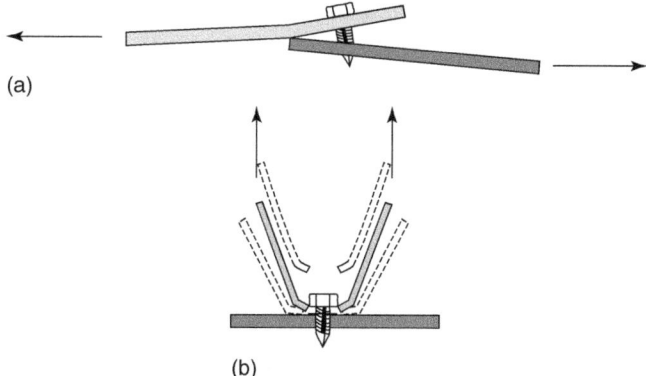

Figure 3.54 Limit states. Source: From Ref. [31].

It is true that the welding details should be carefully studied to increase the fatigue resistance and it is also well known that high-resistance bolts provide good fatigue behavior if they do not routinely work also in compression, that is, if compression in a cycle does not go beyond the bolt pretension (to be remembered as among the benefits of pretensioning).

3.27 Limit States of Other Materials in the Connection

It is common for steel connections to also have concrete involved in the joint, for example, in base plates. Other materials are wood in some roof connections and aluminum.

Dealing with the limit states of those materials is not within the scope of the book, but the engineer is encouraged to carefully consider them.

References

1 CEN (2005). *Eurocode 3: design of steel structures – Part 1–8: Design of joints*, EN 1993-1-8: 2005. Brussels: CEN.
2 Council on Tall Buildings and Urban Habitat, Committee 43 (1993). *Semi-Rigid Connections in Steel Frames*. New York: McGraw-Hill.
3 Bjorhovde, R., Colson, A., and Zandonini, R. (1996). *Connections in Steel Structures III: Behaviour, Strength and Design*. Oxford: Pergamon.
4 Faella, C., Piluso, V., and Rizzano, G. (2000). *Structural Steel Semi-Rigid Connections – Theory, Design and Software*. Boca Raton, FL: CRC Press.
5 Kulak, G.L., Fisher, J.W., and Struik, J.H.A. (2001). *Guide to Design Criteria for Bolted and Riveted Joints*, 2e. Chicago, IL: RSCS and American Institute of Steel Construction.
6 Australian Building Codes Board (1998). *Australian standard – steel structures*, AS 4100: 1998. Sydney: Standards Australia.

7 Bureau of Indian Standards (BIS) (2007). *General construction in steel – code of practice*, 3rd rev., IS 800: 2007. New Delhi, India: BIS.
8 CEN (2008). *Execution of steel structures and aluminium structures – Part 2: Technical requirements for the execution of steel structures*, EN 1090-2:2008. Brussels: CEN.
9 Research Council on Structural Connections (RCSC), Committee A.1 (2009). *Specification for Structural Joints Using ASTM A325 or A490 Bolts*: RCSC www.boltcouncil.org (accessed 23 January 2018).
10 American Institute of Steen Construction (AISC) (2011). *Steel Construction Manual*, 14e. Chicago, IL: AISC.
11 Ministero delle Infrastrutture e dei Trasporti (2008). *NTC—Norme tecniche per le costruzioni*, Decreto Ministeriale (D.M.) 14 January 2008. Rome: Gazzetta Ufficiale.
12 UNI – Ente Nazionale Italiano di Unificazione (1988). *Costruzioni di Acciaio – Istruzioni per il calcolo, l'esecuzione, il collaudo e la manutenzione*, CNR-UNI 10011. Milan: UNI.
13 The Steel Construction Institute (SCI), The British Constructional Steelwork Association (BCSA) (2002). *Joints in steel construction: simple connections.* Ascot, UK: SCI and BCSA.
14 British Standards Institution (BSI) (2000). *Structural use of steelwork in building – Part 1: Code of practice for design – rolled and welded sections*, BS 5950-1: 2000. London: BSI.
15 Lesik, D.F. and Kennedy, D.J.L. (1990). Ultimate strength of fillet welded connections loaded in plane. *Can. J. Civil Eng.* 17 (1): 55–67.
16 Jaspart, J.P., Demonceau, J.F., Renkin, S., and Guillaume, M.L., ECCS Technical Committee 10, Structural Connections (2009). *European recommendations for the design of simple joints in steel structures*, Eurocode 3, Parts 1–8, ECCS Guide No. 126, Portugal: ECCS.
17 Ballio, G. and Mazzolani, F. (1983). *Theory and Design of Steel Structures*. London: Taylor & Francis.
18 The Steel Construction Institute (SCI) and The British Constructional Steelwork Association (BCSA) (1995). *Joints in Steel Construction: Moment Connections*. Ascot, UK: SCI, BCSA.
19 American Welding Society (AWS) (2010). *Structural welding code – steel*, AWS D1.1/D1.1M:2010. Miami, FL: AWS.
20 Wald, F., Bouguin, V., Sokol, Z., and Muzeau, J.P. (2000). *Effective Length of T-stub of RHS Column Base Plates*. Clermont Ferrand: Czech Technical University in Prague and University of Blaise Pascal.
21 De Simone, S., Rizzano, G., and Latour, M. (2010). *Influenza della variabilità dei materiali nella progettazione a completo ripristino di resistenza dei nodi di base in acciaio*: Università degli studi di Salerno.
22 CEN (2005). *Design of steel structures, general rules and rules for buildings*, EN 1993-1-1: 2005. Brussels: CEN.
23 Cheng, J.J.R. and Grondin, G.Y. (1999). *Recent development in the behavior of cyclically loaded gusset plate connections*. North American Steel Construction Conference Proceedings, Toronto.

24 CEN (2005). *Design of steel structures, plated structural elements*, EN 1993-1-5: 2005. Brussels: CEN.
25 Dowswell, B. (2006). Effective length factors for gusset plate buckling. *Eng. J. AISC* 43: 91–102, Quarter 2.
26 Thornton, W.A. (1984). Bracing connections for heavy constructions. *Eng. J. AISC* 21: 139–148, Quarter 3.
27 Lin, M.-L., Tsai, K.-C., Hsiao, P.-C., Tsai, C.-Y. (2005). Compressive behavior of buckling-restrained brace gusset connections. The First International Conference on Advances in Experimental Structural Engineering, Nagoya, Japan.
28 Cheng, R.J.J., Grondin, G.Y., Yam, M.C.H. (2000). *Design and behavior of gusset plate connections. Steel Connection IV Workshop*, Roanoke, VA.
29 American Institute of Steen Construction (AISC) (2001). *Manual of Steel Construction, Load and Resistance Factor Design*, 3e. Chicago, IL: AISC.
30 CEN (2005). *Material toughness and through thickness properties*, EN 1993-1-10: 2005. Brussels: CEN.
31 European Convention for Constructional Steelwork (ECCS), Technical Committee 7 (2008). *Cold-formed steel, composite structures connections in cold formed steel structures – the testing of connections with mechanical fasteners in steel sheeting and sections*, ECCS Guide No. 124. Portugal: ECCS.
32 Nussbaumer, A., Borges, L., and Davaine, L. (2011). *Fatigue Design of Steel and Composite Structures*, ECCS Eurocode Design Manuals. Portugal: ECCS, Wiley-Blackwell, Verlag Ernst & Sohn.
33 CEN (2005). *Eurocode 3: design of steel structures – Part 1–9: Fatigue*, EN 1993-1-9: 2005. Brussels: CEN.

4

Connection Types: Analysis and Calculation Examples

The types of steel connections can be vastly different: This chapter will try to provide guidance for the design of the ones most frequently used.

Engineering judgment will be essential to extend the basic concepts to the most complicated and singular connection types.

4.1 Common Symbols

See Figure 4.1.

4.1.1 Materials

f_{yp}	characteristic yield strength of the primary element (column or main beam)
f_{up}	characteristic ultimate strength of the primary element
f_{ys}	characteristic yield strength of the secondary element (beam)
f_{us}	characteristic ultimate strength of the secondary element
f_{ypl}	characteristic yield strength of the plate
f_{upl}	characteristic ultimate strength of the plate

4.1.2 Design Forces

N_{Ed}	design value of the axial force
$V_{major\,Ed}$	design value of the strong axis shear
$V_{minor\,Ed}$	design value of the weak axis shear
$M_{major\,Ed}$	design value of the strong axis bending moment
$M_{minor\,Ed}$	design value of the weak axis bending moment

4.1.3 Bolts

d	bolt diameter
d_0	hole diameter
A_{res}	net section of the bolt (bolt area through the shear plane)

Design and Analysis of Connections in Steel Structures: Fundamentals and Examples,
First Edition. Alfredo Boracchini.
© 2018 Ernst & Sohn Verlag GmbH & Co. KG. Published 2018 by Ernst & Sohn Verlag GmbH & Co. KG.

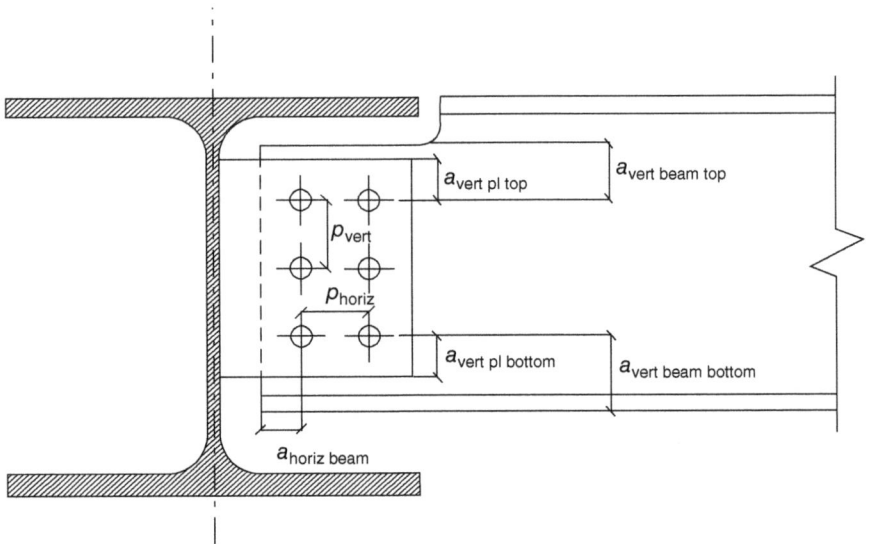

Figure 4.1 Symbols.

A	gross section of the bolt (unthreaded part)
f_{yb}	characteristic yield strength of the bolt
f_{ub}	characteristic ultimate strength of the bolt
α_v	factor to define the shear design resistance of a bolt depending on the bolt class (EC)
γ_M	partial safety factor (EC)
ϕ	resistance factor [AISC load and resistance factor design (LRFD)]
Ω	safety factor [AISC allowable stress design (ASD)]
$n_{shear\ planes}$	number of shear planes
n_{rows}	number of bolt rows (horizontal) in the joint
n_{cols}	number of bolt columns (vertical) in the joint
n_{bolts}	number of bolts in the joint
p_{horiz}	distance between columns of bolts
p_{vert}	distance between rows of bolts
$a_{vert\ beam\ top}$	vertical distance between the beam top edge and the nearest bolt (its axis)
$a_{vert\ beam\ bottom}$	vertical distance between the beam bottom edge and the nearest bolt (its axis)
$a_{horiz\ beam}$	horizontal distance between the beam edge and the nearest bolt (its axis)
$a_{horiz\ plate}$	horizontal distance between the plate edge and the nearest bolt (its axis)
$e_{bolt\ group}$	distance between the bolt group and the theoretical pin location

4.1.4 Geometric Characteristics of Plates and Profiles

t_{pl}	plate thickness
h_{pl}	plate depth (height)

h_s	secondary member depth (height)
w_s	secondary member width
t_{fs}	secondary member flange thickness
t_{ws}	secondary member web thickness
h_p	primary member depth (height)
w_p	primary member width
t_{fp}	primary member flange thickness
t_{wp}	primary member web thickness

4.2 Eccentrically Loaded Bolt Group: Eccentricity in the Plane of the Faying Surface

This section and the next do not deal with a precise type of connection; rather they explain how to deal with a "component," that is, a part of a connection that is common in several kinds of joints: a bolt group that is loaded with an eccentricity.

The EC codes do not give precise methods to calculate this kind of eccentricity and, therefore, it is convenient to look at AISC methods. Those methods can then be applied to EC design problems.

If there is an in-plane eccentricity, the shear will generate a moment in the bolt group. This means that, in addition to the "direct" shear acting on bolts, an additional shear load is necessary to balance the moment.

There are two methods that, according to the AISC manual [1], can be applied to find a solution to this kind of a problem: the elastic method and the instantaneous center-of-rotation method. The latter is more accurate but requires an iterative solution that only a dedicated software (SCS – Steel Connection Studio can do this) or, with some approximation, values in tables (see Ref. [1]) can provide. The elastic method, on the other hand, can be set up in spreadsheets but is generally more conservative since it does not account for the distribution capacity a bolted group can offer.

4.2.1 Elastic Method

Every bolt will be loaded by the "direct" shear and the additional shear that compensates eccentricity.

Defining (Figure 4.2) P as the total load (shear) and n as the number of bolts, the direct shear (T_d) for each bolt is given as

$$T_d = \frac{P}{n}$$

which can be decomposed according to the load angle (angle with the horizontal) in

$$T_{dx} = T_d \cos\theta$$
$$T_{dy} = T_d \sin\theta$$

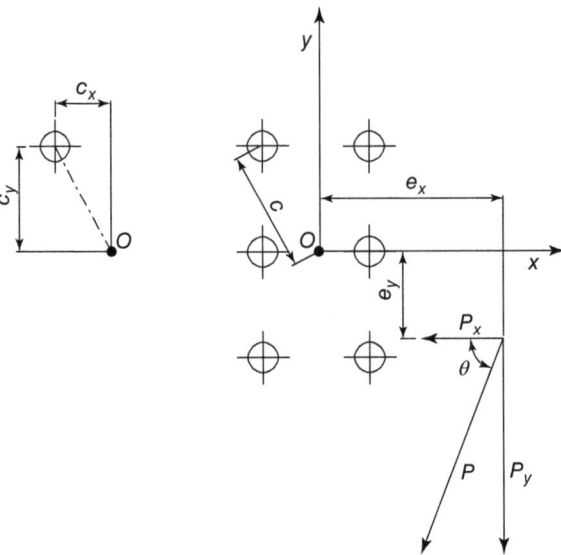

Figure 4.2 Elastic method, symbols.

The polar moment of inertia of the bolt group to its center of gravity must then be obtained. Another value that is necessary to apply the equations below is c, that is, the distance, divided in its x and y Cartesian components, of each bolt from the center of the group. The additional shear (here called T_m) Cartesian components that arise by the eccentricity are therefore

$$T_{mx} = \frac{Pec_y}{I_p}$$

$$T_{my} = \frac{Pec_x}{I_p}$$

Finally, the overall shear load is calculated, bolt by bolt, as

$$T = \sqrt{(T_{dx} + T_{mx})^2 + (T_{dy} + T_{my})^2}$$

For (at least) one bolt the algebraic signs will agree and the force will reach its maximum.

4.2.1.1 Example of Eccentricity Calculated with Elastic Method

A group of 8 cl.8.8 M16 bolts are loaded by a design shear of 160 kN with a 100 mm eccentricity. The bolt threads are in the shear plane. The bolt group is also loaded by a horizontal 28-kN force as shown in Figure 4.3. The horizontal force has no eccentricity.

"Direct" shear values are, per bolt,

$$T_{dx} = 28/8 = 3.5 \text{ kN}$$

$$T_{dy} = 160/8 = 20 \text{ kN}$$

The polar moment of inertia is (simple calculations give 109.2 mm as the distance of the outer bolts from the bolt group center of gravity and 46.1 mm as the

Figure 4.3 Geometry and loads.

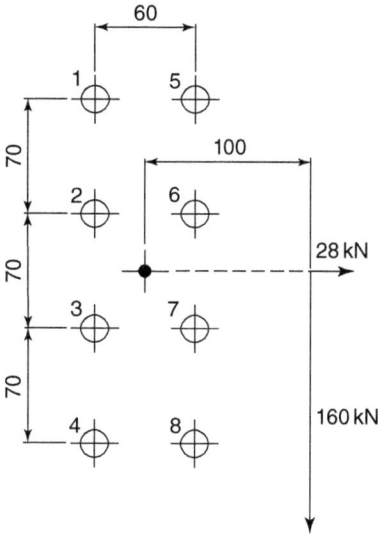

distance for the inner bolts), per area unit,

$$I_p = \sum_{i=1}^{n=8}(c^2) = (4 \times 109.2^2 + 4 \times 46.1^2) = 56\,200 \text{ mm}^4 \text{ mm}^{-2}$$

The moment given by the eccentricity can be obtained either by composing the force and finding the corresponding e' value or adding together the moments given the horizontal and vertical force, which is faster here since the horizontal force has no lever arm. The result is 160 kN × 100 mm = 16 000 kN mm, which means that the bolts labeled as 1, 4, 5, and 8 in Figure 4.4 have, in absolute value,

$$T_{mx} = 16\,000 \times 105/56\,200 = 29.9 \text{ kN}$$
$$T_{my} = 16\,000 \times 30/56\,200 = 8.5 \text{ kN}$$

which has a maximum for bolt 5 (the one where the signs of the components are the same) of

$$T = \sqrt{[(3.5 + 29.9)^2 + (20 + 8.5)^2]} = 43.9 \text{ kN}$$

(which, compared to the design resistance values of the EC, is at 73%, which is acceptable). In the same way, the forces for bolts 2, 3, 6, and 7 can be calculated but the absolute values will clearly be smaller. Final results are sketched in Figure 4.4.

Alternatively, in a faster but more conservative way, the contribution of the inner bolts could be ignored and it could be assumed that the design moment is balanced by a pair of couples resisted by the outer bolts, where the lever arm is the diagonal distance between bolts as shown in Figure 4.5.

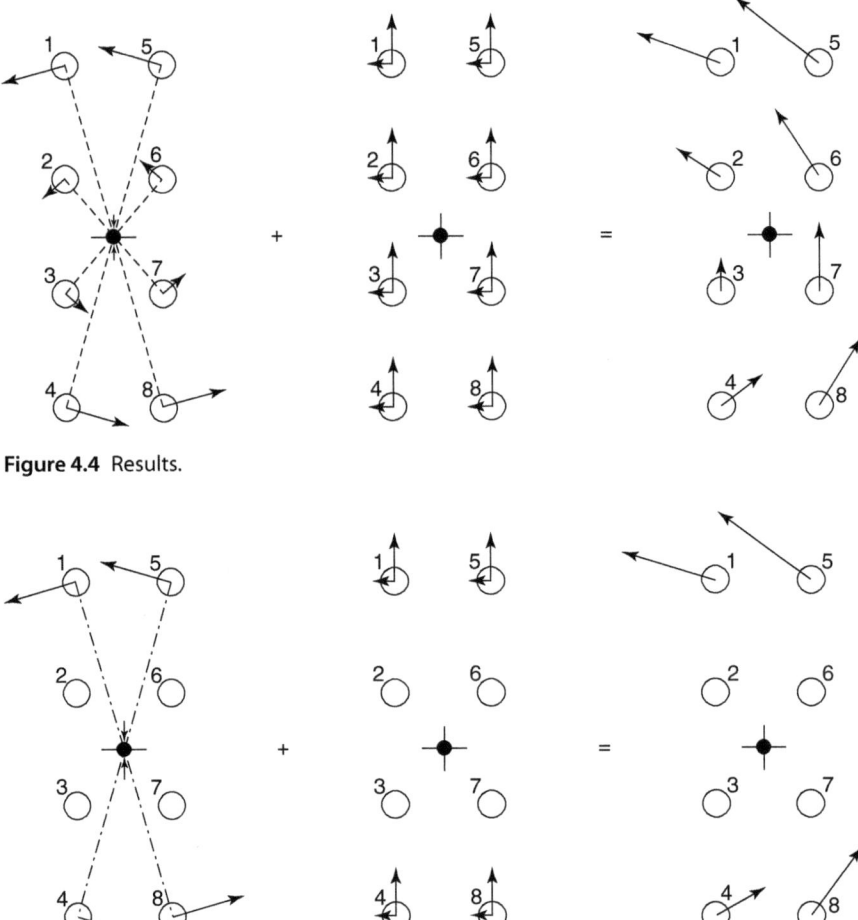

Figure 4.4 Results.

Figure 4.5 Results of the simplified computation.

Using this approach the results are

$T_m = 16\,000/(2 \times 218.4) = 36.6$ kN

$T_{mx} = 36.6 \times \cos\ 15.9° = 35.2$ kN

$T_{my} = 36.6 \times \sin\ 15.9° = 10$ kN

$T = \sqrt{[(3.5 + 35.2)^2 + (20 + 10)^2]} = 49$ kN (ok, 82% according to EC)

4.2.2 Instantaneous Center-of-Rotation Method

This method aims to locate the center of instantaneous rotation around which the bolt group rotates (Figure 4.6). The forces must therefore be perpendicular to the line drawn between each bolt and the rotation center and the sum of forces and moments must balance the applied shear (and the corresponding moment from eccentricity).

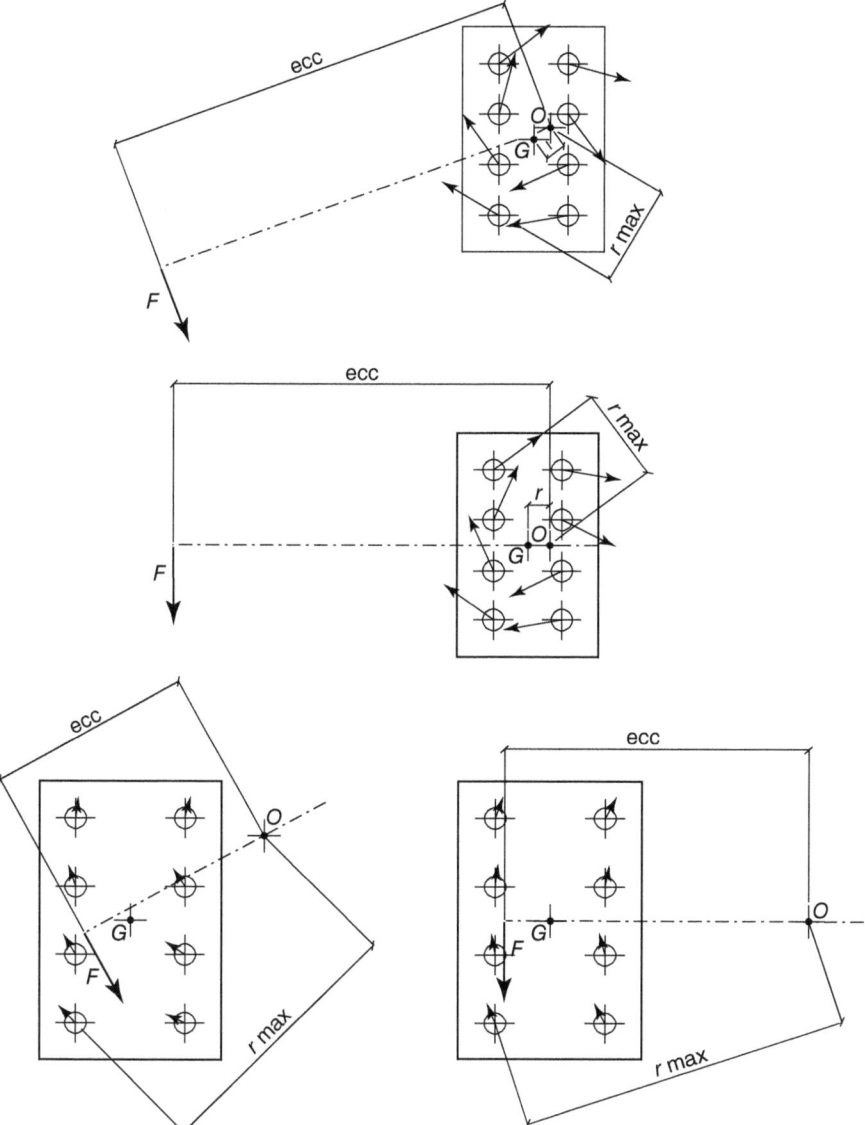

Figure 4.6 Examples of eccentric bolt groups where the centers of rotation and the balancing forces are drawn.

To collect further details about the method, which is based on a laboratory load–rotation curve of systems made by bolts and a plate, refer to [2].

4.2.2.1 Example of Eccentricity Calculated with the Instantaneous Center-of-Rotation Method

Here we will reconsider the previous example, now with the instantaneous center-of-rotation method (see Figure 4.7); but, as already mentioned, it must

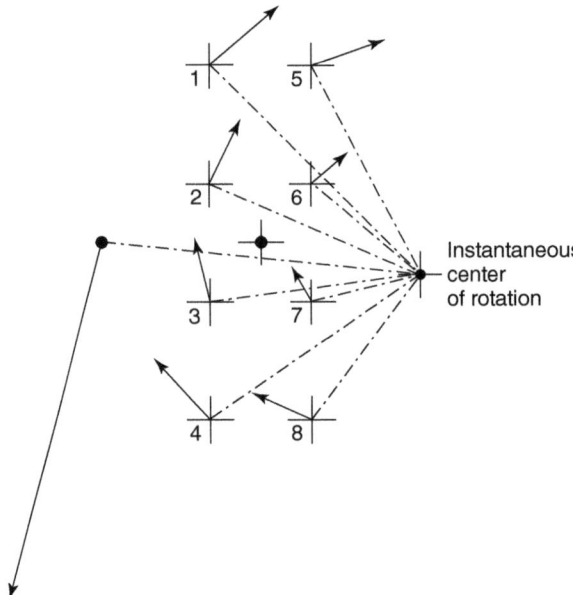

Figure 4.7 Graphical representation of the forces and the center of rotation.

be remembered that this approach requires iterative design (special software or outstanding programming skills are necessary), or special tables, as in [1].

If we look at AISC tables, we find (Ref. [1], Tables 7 and 8 considering, approximately, a 15° angle) $C \cong 4.5$. Actually, the tables have distances in inches, forcing us to continuously approximate and interpolate the values and, therefore, the procedure is not recommended when using the metric system. This C value means that an eccentrically loaded group of eight bolts has a global capacity of a group of 4.5 bolts with no eccentricity. "Translating" the result for EC, we can say that the group has a resistance of 4.5×60 kN (60 kN is the EC design resistance of one M16) = 270 kN, versus a design action of $\sqrt{(160^2 + 28^2)} = 162$ kN, meaning that the check is fine with an exploitation ratio of 60%.

The same calculation with SCS brings a resistance value for the bolt group of 277 kN and, therefore, a 58.5% design ratio. SCS gets the same exact values of the AISC tables for the same angles and distances so, the 1–2% difference is caused by the approximations in reading the tables, as already mentioned.

4.3 Eccentrically Loaded Bolt Group: Eccentricity Normal to the Plane of the Faying Surface

A "classical" elastic method to calculate forces in a bolted group where the eccentricity is perpendicular to the plane of the plate (Figure 4.9) is to define a center of compression and to scale the forces accordingly in a triangular way (Figure 4.8).

This method does not require any additional instruction, so we will look at the methods from AISC, which look easier and more immediate than EC. To apply the latter, see Section 4.12.

Figure 4.8 Triangular distribution (center of compression on the bottom).

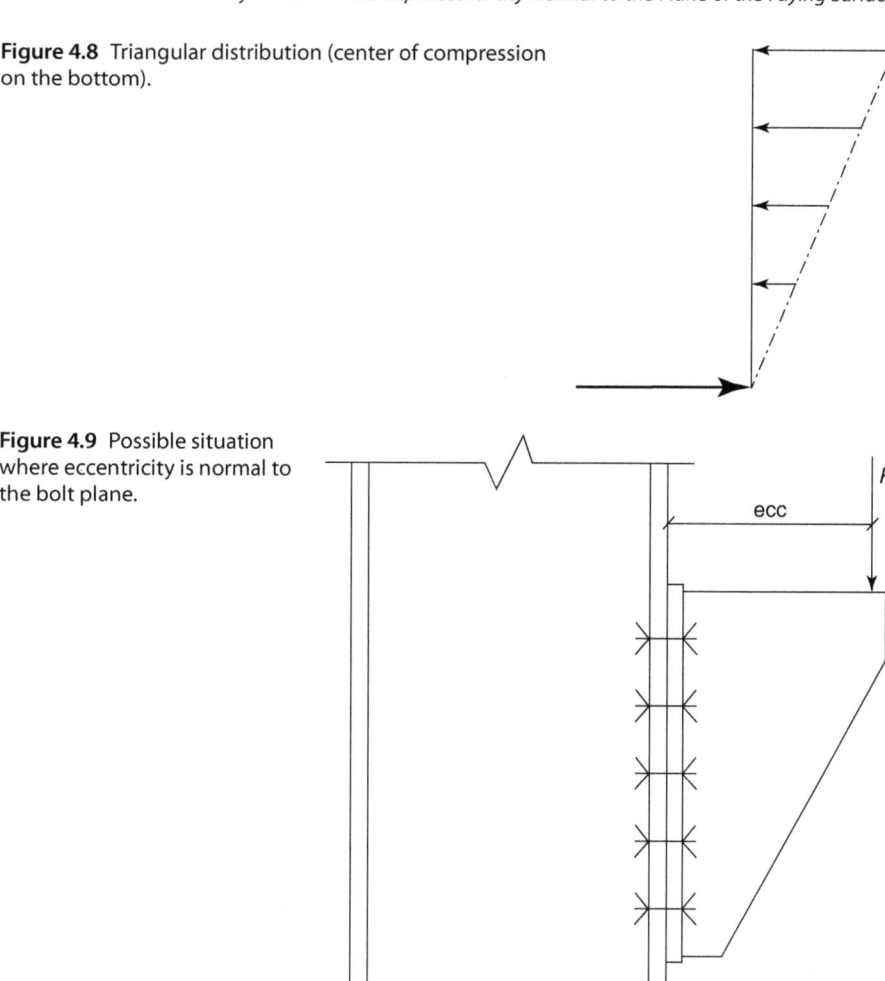

Figure 4.9 Possible situation where eccentricity is normal to the bolt plane.

AISC gives instructions on how to find forces in the bolts according to two approaches: one (neutral axis at center of gravity), which is quite easy and conservative, and the other (neutral axis not at center of gravity), which is more precise but certainly more laborious.

4.3.1 Neutral Axis at Center of Gravity

The bolt group is divided symmetrically into two parts, one working in tension and one in compression. The bolts will not work in compression: This is just a simplified method to get quick conservative results.

The bolts will work in shear and tension (in case amplified by the prying action, see Section 3.10) in the part in tension, and only in shear in the part in compression.

The method involves considering all the bolts working plastically, that is, all with the same value (no triangular scaling is needed): Even with this approach, the results are more conservative than the other method.

4.3.1.1 Example of Eccentricity Normal to Plane Calculated with Neutral Axis at Center-of-Gravity Method

A cantilever made by two welded 25-mm plates (S275) with eight M24 (cl. 8.8) that connect it to an HEB 400 column is loaded by a (factored) 400-kN load with a 300 mm eccentricity (Figure 4.10).

The bending moment given by the eccentricity is $M = 400 \times 300 = 120\,000$ kN mm. The shear per bolt is $T = 400/8 = 50$ kN. The axial force to be divided among the bolts in tension is $120\,000/240 = 500$ kN, which, divided among the four bolts, implies that the tension for each bolt is equal to $N = 500/4 = 125$ kN. The lower bolts only share the shear.

The AISC equation taken from Section 3.10.1 (as cited in the mentioned paragraph, the formula can be derived in a similar version from equilibrium conditions and is, generally speaking, valid) can be used to check if the plate thickness is enough to prevent prying action (note that here the numerator has 500 000 N instead of 500 kN):

$$t_{min} = \sqrt{\frac{4.44 \left(\frac{500\,000}{2} \frac{1}{2}\right)\left(75 - \frac{25}{2} - \frac{25}{2}\right)}{120 \times 430}} = 23 \text{ mm}$$

So, the plate is thick enough to consider the prying action as negligible.

Combining the tension and shear (e.g. according to EC in the formula below) for the four upper bolts, the result is that each bolt works (with the conservative assumption that the bolt thread is in the shear plane) as

$$\frac{50}{136} + \frac{125}{1.4 \times 203} = 0.368 + 0.44 = 0.81$$

The design by SCS gives the same solution (81%).

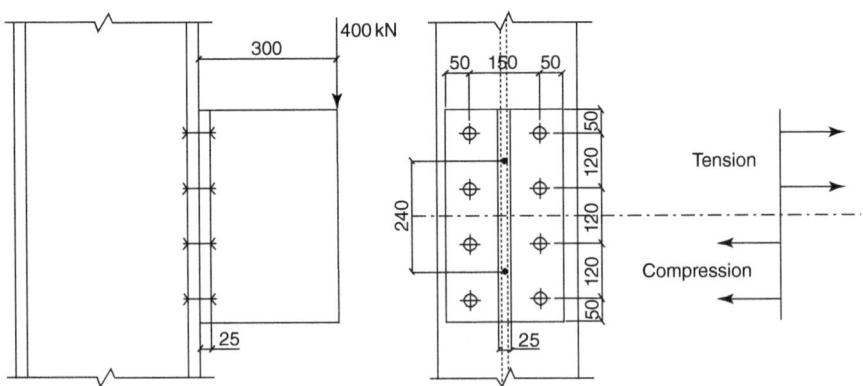

Figure 4.10 Example geometry.

4.3.2 Neutral Axis Not at Center of Gravity

As in the previous method, the shear is equally divided among the bolts. The location of the neutral axis must be calculated, taking an effective width of the flange (or of the plate) not larger than eight times the thickness (the lesser of the connected ones), which can be written as

$$b_{eff} = 8t_f \leq b_f$$

The location of the neutral axis is where the moment of the bolt areas is equal to the moment of the compression block area, that is,

$$\sum A_{b,i} y_i = b_{eff} d \frac{d}{2}$$

where A_b is the bolt gross area for AISC and d is the distance shown in Figure 4.11, not the bolt diameter. This makes the equation slightly more conservative than the same equation applied with the bolt net area, since the neutral axis shifts upward with the gross value. The same area will be used in the formulas provided further when evaluating N (tension per bolt) and I_x.

Once the neutral-axis position is localized (a first attempt can be tried, as in Figure 4.11, at one-sixth of the plate depth from the bottom), the actions on each

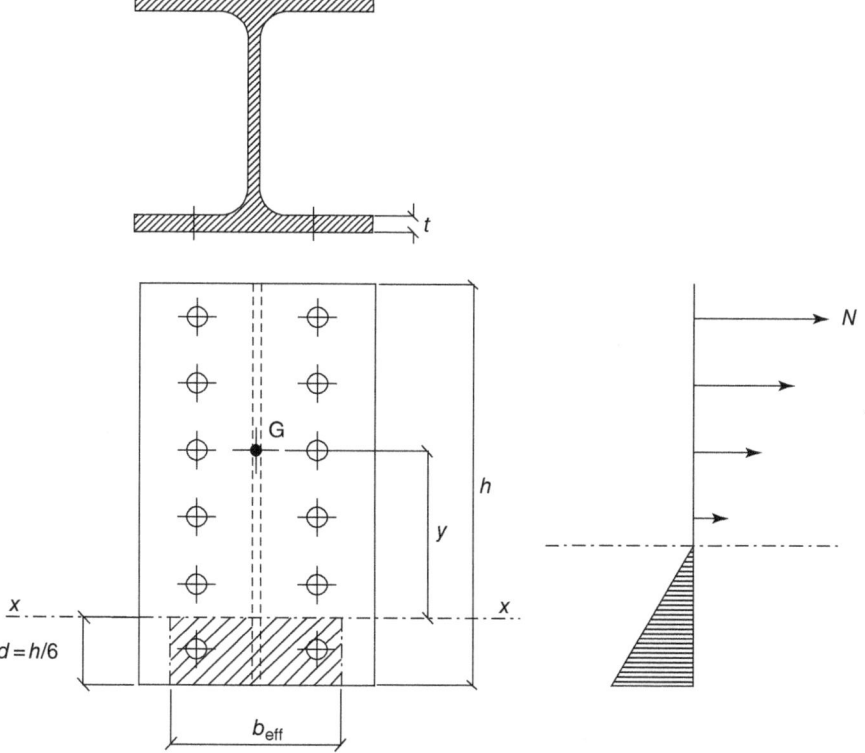

Figure 4.11 Left: initial try location; right: final location after computations. Source: From Ref. [1].

bolt can be derived (a linear distribution is assumed) from the following:

$$N = \frac{Mc}{I_x} A_b$$

where c is the bolt distance from the neutral axis, M the design moment, and I_x the combined moment of inertia (about the neutral axis) of the bolts working in tension and the compression block.

Even with this approach, as in the previous, the bolts will work in shear and tension on the top (positive moment assumption) and only in shear on the bottom.

4.3.2.1 Example of Eccentricity Normal to Plane Calculated with Neutral Axis not at Center-of-Gravity Method

To calculate eccentricity with the neutral axis not at the center of gravity, we let

$$b_{eff} = 25 \times 8 = 200 \text{ mm} < 250 \text{ mm}$$

A first attempt to find the neutral axis is tried at one-sixth of the plate depth, so $d = 80$ mm; then, through further attempts, a more refined value will be found. Alternatively, a second-degree equation can be written and solved, d being the unknown variable. Another option is using the Microsoft Excel function Goal Seek after writing formulas as a function of d.

Solving the second-degree equation ($d^2 + 27.1d - 7865 = 0$), the only positive solution is 76.2 mm, which can be conservatively rounded to 77 mm (which is quite close to the 80 mm of the one-sixth depth). Since the neutral axis is between the assumed bolt rows, we can proceed (otherwise a recalculation with the correct number of bolts in tension would have been necessary).

Let us find I_x and the tension values for the bolts, row by row:

$$I_x = 2 \times 452 \times ((410 - 77)^2 + (290 - 77)^2 + (170 - 77)^2)$$
$$+ 200 \times 77 \times (77/2)^2 = 1.72 \times 10^8 \text{ mm}^4$$

$$N_{1 \text{ bolt row 1}} = 120\,000 \times (410 - 77)/1.72 \times 10^8 \times 452 = 105 \text{ kN}$$

$$N_{1 \text{ bolt row 2}} = 120\,000 \times (290 - 77)/1.72 \times 10^8 \times 452 = 67 \text{ kN}$$

$$N_{1 \text{ bolt row 3}} = 120\,000 \times (170 - 77)/1.72 \times 10^8 \times 452 = 29 \text{ kN}$$

As verified in the previous example, the plates are thick enough to neglect prying actions and, therefore, the check of the upper bolts (the most stressed) becomes (136 kN is the design shear resistance and 203 kN is the tension resistance according to EC as in Tables 3.7 and 3.8)

$$\frac{50}{136} + \frac{105}{1.4 \times 203} = 0.368 + 0.370 = 0.74$$

The difference with the previous example is just a few percentage points but it increases with more rows of bolts.

The automatic solution by SCS confirms the results.

Let us also notice that, if the computation was made with the bolt net area instead of the nominal area, the neutral axis would have been at 69 mm from the bottom and the tension for the first row of bolts would have been about 103 kN (73% exploitation). The approach with the nominal area is confirmed to be slightly more conservative.

4.4 Base Plate with Cast Anchor Bolts

The column–foundation joint is a classical and very important connection but there are no easy and quick checks available, except for an ideal condition with no bending moment and no uplift, therefore stiffening ribs are not necessary (in other words, a thick plate can be used).

Refer to Section 1.2 for considerations about the pin or fully restrained assumption. Here, the purpose is to give practical formulas to be used in the design of the elements of the joint.

It is important to recognize that there are multiple analyses to be performed since different materials are interfaced: The designer must perform checks for the base plate as well as its ribs, if present, and for the anchor bolts, and checks for the concrete base and the soil, which, except for the contact pressure of the concrete, are not within the scope of this book.

4.4.1 Plate Thickness

The following pages will analyze how to design the thickness of the base plate according to the two main available approaches, that is, the classical AISC approach and the more recent method developed by the EC.

4.4.1.1 AISC Method

The AISC approach in designing the base plate consists in considering the plate as a grade beam (ground beam), working in the outer parts as a cantilever stressed by the concrete contact pressure. Before the EC, this was the main method used in Europe too; see, for example the approach in [3]. This behavior assumes that the pressure is uniformly distributed in the plate.

If there are no ribs/stiffeners, the cantilever lengths to be used are m, n, and $\lambda n'$. To calculate m and n refer to Figures 4.12 and 4.13.

To get n' and λ, on the other hand, use the following (where a corner yield pattern according to the *yield line method* is assumed):

$$n' = \frac{\sqrt{h_c w_c}}{4}$$

$$\lambda = \frac{2\sqrt{X}}{1 + \sqrt{1-X}} \leq 1$$

$$X = \frac{4 h_c w_c}{(h_c + w_c)^2} \frac{N_{Ed}}{N_{Rd}}$$

As is clear also from the figures, h_c and w_c represent the depth (height) and the width of the column section, respectively.

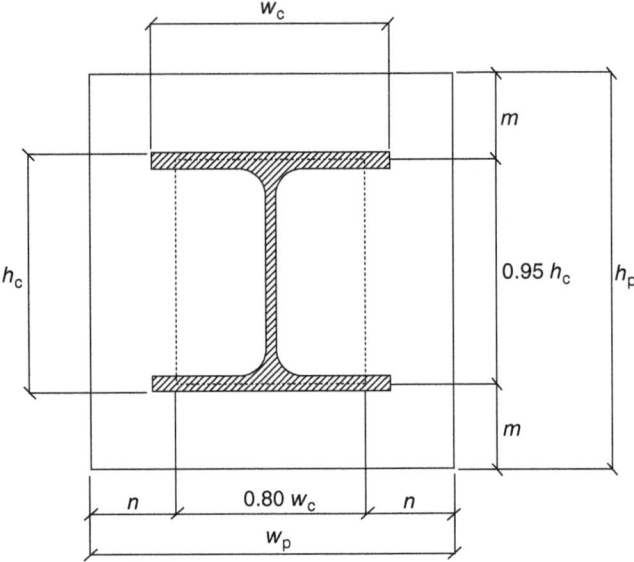

Figure 4.12 Graphical representation for reckoning *m* and *n*.

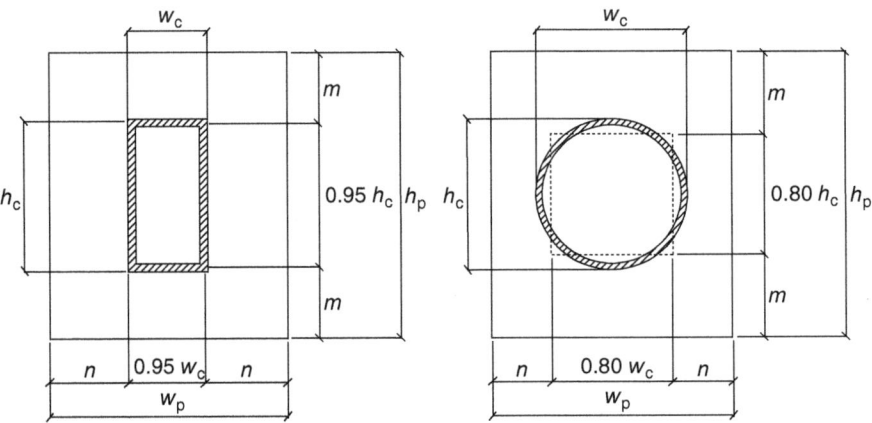

Figure 4.13 *m* and *n* for rectangular and circular hollow steel columns.

The N_{Ed}/N_{Rd} factor is the ratio between the design axial force and the design axial resistance. Therefore, N_{Ed}/N_{Rd} can be conservatively assumed, to simplify, as 1. In the same way, X and consequently λ can also be taken as 1 if the engineer wants to shorten the design procedure.

The biggest among *m*, *n*, and $\lambda n'$ will indicate the critical length to be input to compute the base plate thickness, assuming, as explained, an equivalent cantilever. This also means that similar values of *m* and *n* optimize the design.

4.4 Base Plate with Cast Anchor Bolts

Axial Load Only The thickness of the base plate can be designed, in the assumption of a plastic-type resistance of the plate and a pure axial concentric load, as

$$t = \max(m, n, \lambda n')\sqrt{\frac{2N_{Ed}}{\phi f_y h_p w_p}}$$

The equation is obtained imposing that the plate plastic resistance modulus multiplied by the yield stress reduced by resistance factor Φ (here 0.9, conceptually similar to EC partial safety factor γ_M but its reciprocal ratio) is equal to the bending moment generated by the concrete contact pressures (taken as uniformly distributed) over the previously defined cantilevers. It should be noted that if, more conservatively, the elastic modulus is considered instead of the plastic modulus, the 2 under the square root would become a 3.

Let us also point out that the shear on the base plate needs to be checked but it very rarely governs the design so engineers usually neglect this check when computing the necessary thickness t.

Axial Load and Small Eccentricity The eccentricity is defined as "small" when e (not to be confused with the e of the equivalent T-stub that has the same symbol in EC), given by the ratio between the design bending and the design axial load, is less than the critical eccentricity e_{cr}. In symbols,

$$\frac{M_{Sd}}{N_{Sd}} = e \leq e_{cr} = \frac{h_p}{2} - \frac{N_{Sd}}{2 f_{p,\max} w_p}$$

with h_p being the plate length in the relevant direction, that is, the same where the bending moment is acting (so, w_p is the width of the plate in the normal direction); $f_{p,\max}$ instead is the contact pressure between concrete and steel, see also Section 4.4.2.

The equation for the plate thickness becomes

$$t = \sqrt{\frac{2 f_p m^2}{\phi f_y}} = 1.49 m \sqrt{\frac{f_p}{f_y}}$$

when $h' \geq m$, while with $h' < m$, (to compute h', graphically represented in Figure 4.14, see once again Section 4.4.2 about the contact pressure) it is

$$t = \sqrt{\frac{4 f_p h'(m - h'/2)}{\phi f_y}} = 2.11 \sqrt{\frac{f_p h'(m - h'/2)}{f_y}}$$

If n (or $\lambda n'$) is greater than m, replace m with n (or $\lambda n'$) in the above formulas (but keep considering m when comparing the value to h' in order to decide which equation has to be applied).

Axial Load and Large Eccentricity The eccentricity is "large" when e previously defined is larger than e_{cr}.

A large-eccentricity situation brings a separation between the plate and the concrete and, therefore, the intervention of the anchor bolts in tension. The

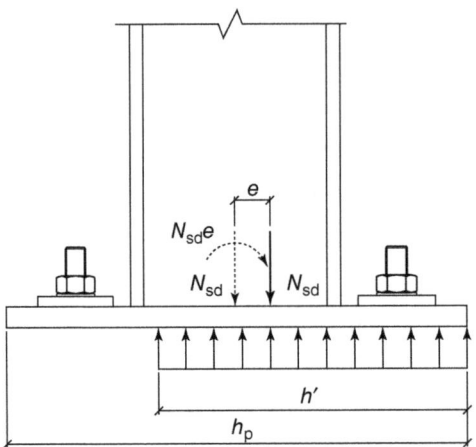

Figure 4.14 Small eccentricity.

hypothesis holds that the concrete is loaded in compression in a part of the plate (see Section 4.4.2) and the equation to use for designing t is the one used previously for small eccentricities.

The thickness must however be capable of withstanding the forces arising from the anchors in tension that load the plate with a bending moment, so it should also be checked (with the assumption that the full width of the plate can react, meaning it is inside a diffusion angle of 45° from the anchor bolts) that

$$t \geq 2.11 \sqrt{\frac{Tx}{w_p f_y}}$$

where T represents the global action on the anchors and x the distance between the anchor bolts and the critical section in bending. In the case of Figure 4.15, x is the distance from the flange center, that is,

$$x = f - \frac{h_c}{2} + \frac{t_{fl}}{2}$$

with t_{fl} indicating the column flange thickness and f the distance of the anchors from the column center.

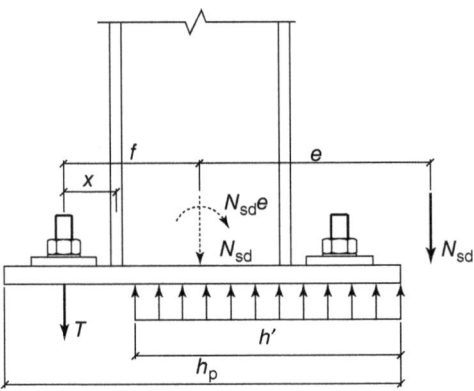

Figure 4.15 Large eccentricity.

Uplift A similar equation to the one above will be used, that is,

$$t = 2.11\sqrt{\frac{T'x'}{yf_y}},$$

where T' represents the action over the anchor bolts, x' the distance between the axial action and the critical section in bending, and y the plate width reacting in the critical section, obtainable with a force distribution at 45°. It will be up to the engineer to evaluate the possible situations and take the most unfavorable with its corresponding y value. See some possible situations sketched in Figure 4.16. It is emphasized that $T'x'$ must represent the sum of the design resistances, similar to y representing the sum of the effective resisting widths. Alternatively, $T'x'$ can be checked for a single bolt calculating the resistance with y corresponding to only that anchor.

Ribs and Stiffeners The AISC design guide [4] on base plates provides some general tips (e.g. as in [3]) on the design of base plates with ribs and stiffeners (the plate becomes like a beam and/or a plate running over multiple supports; it is

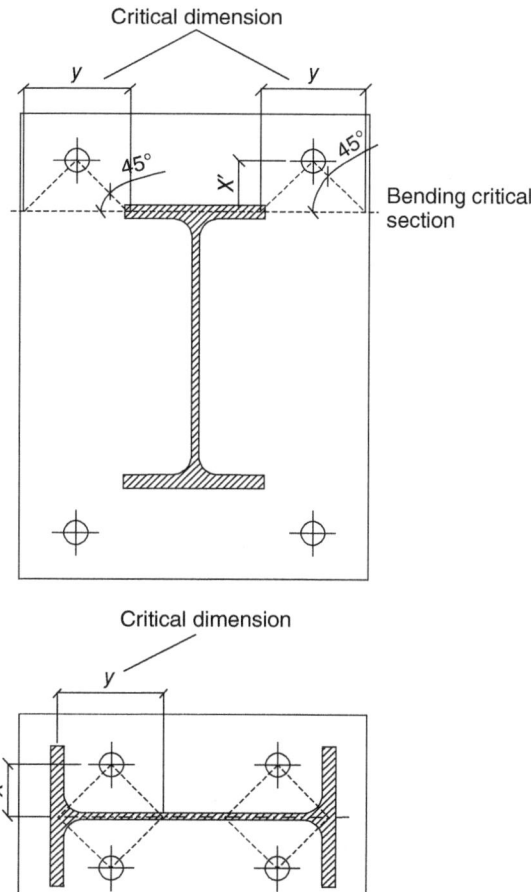

Figure 4.16 Critical dimension (width) in some example situations.

not a cantilever any more), but practical equations are not given and applying those methods is for several aspects not so immediate. This happens because in the United States it is not popular to add stiffeners to base plates with the aim of saving on the plate thickness unless the column is large and heavily loaded (even in this case it might be more common to use a double plate with filler vertical plates to distance the horizontal plates, as in Figure 4.19). The design thickness as obtained by the previous equations is therefore taken without worrying too much about saving some material with ribs, brackets, or similar, which is likely more cost-effective because there are no additional shop labor expenses.

4.4.1.2 Eurocode Method

Compared to the AISC method, the analysis according to the EC is based on a different assumption in the sense that the pressure is not considered as uniformly spread on the plate but only on the stiffest areas below the profile: near the web and the flanges.

Axial Load Only If there is only axial compression, the three T-stub regions considered are the ones shown in Figure 4.17.

To get the width of the T-stubs in compression, refer to Section 4.4.2 about the contact pressure. It has to be noticed that additional vertical plates (stiffeners) could help in spreading the base plate area that resists compression.

Axial Load and Bending Moment If there is also some bending moment, EC conservatively suggests not considering the T-stub below the web.

The procedure starts with obtaining the load eccentricity (again defined as the ratio between the bending moment and the axial load) and then comparing it with the z values of reactions as in Figure 4.18 in order to evaluate for each side (right and left) if there is tension or compression. Attention to the direction of the

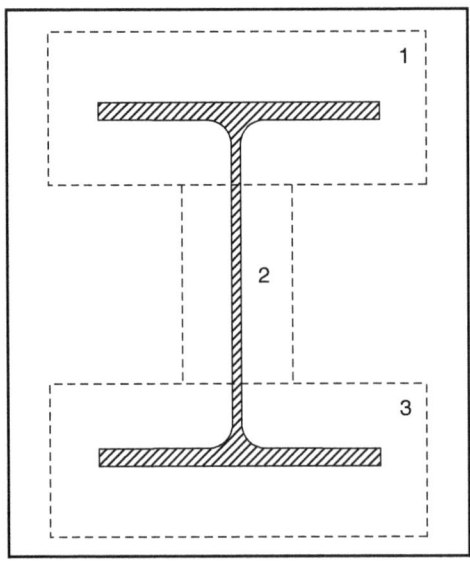

Figure 4.17 Regions of the base plate working in compression (axial load only).

4.4 Base Plate with Cast Anchor Bolts | 131

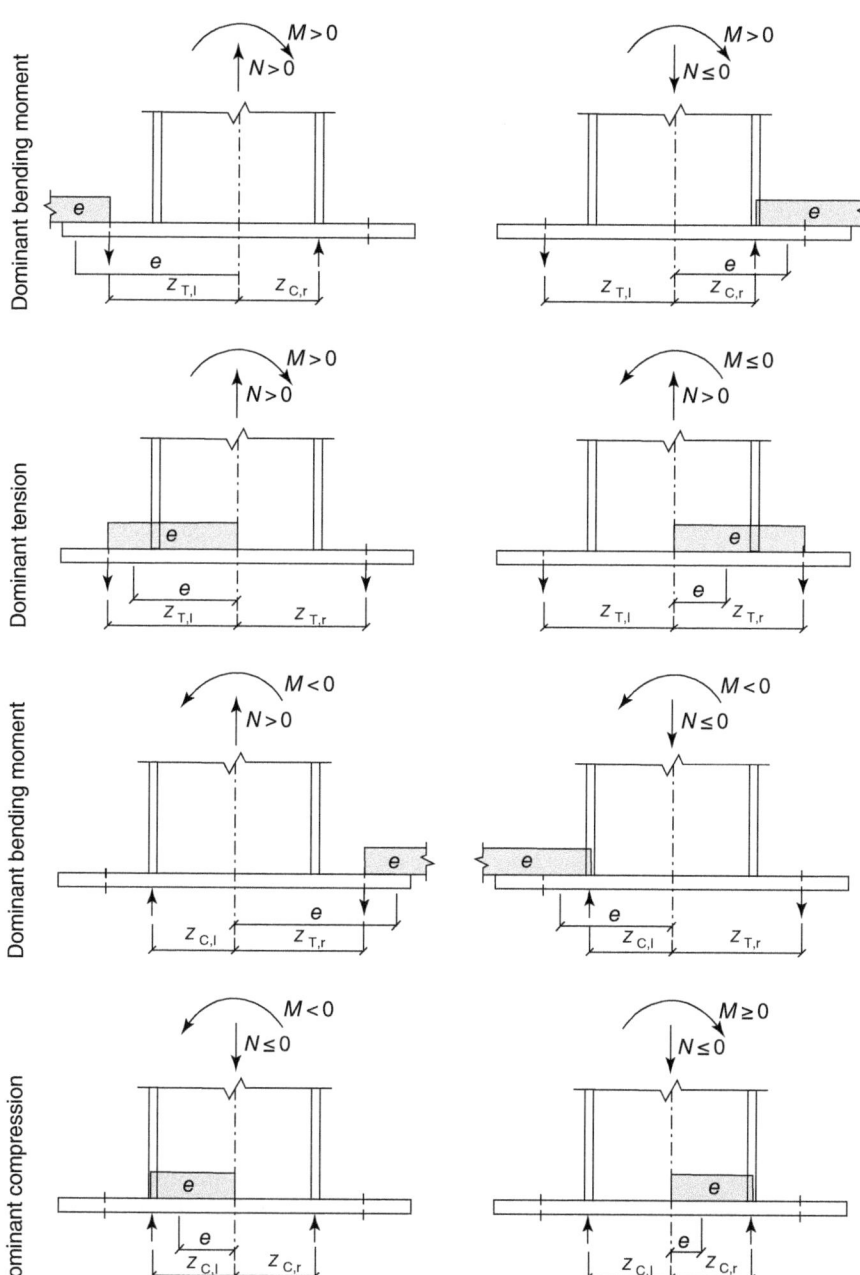

Figure 4.18 Symbols in the EC for various base plate design situations.

axial load must be paid because having uplift would imply dealing with z_T instead of z_C. The convention given by EC is to consider the bending moment as positive if clockwise and the axial load positive if in tension (therefore it is negative if the load is downward, the opposite of the AISC convention). This means that three possibilities can arise:

- Upward reaction on both sides, thus a situation of prevailing axial load (downward)
- Discordant reactions on the two sides, thus a situation of prevailing moment
- Upward reaction on both sides, thus a situation of prevailing uplift.

Once this is established, the following resisting components are evaluated:

(a) On the side(s) with prevailing upward reaction, hence the compression zone, the local compression of the column web and flange and the compression in the concrete; the lesser of the two will give $F_{C,l,Rd}$ and $F_{C,r,Rd}$.
(b) On the side(s) with prevailing downward reaction, hence the tension zone, the local tension of the column web and the plate in bending; the lesser of the two will give $F_{T,l,Rd}$ and $F_{T,r,Rd}$.

It has to be noted that [5] calls for the column web tension paragraph related to the web in transverse tension (Section 3.15) while it would seem more appropriate to reference the web tension for beams (Section 3.17) since the load is not transverse as in an end plate. After all the formulas differ for the ω coefficient only and this could be assumed as 1.

Finally, the formulas in Table 4.1 are applied to calculate the design moment resistance (the table is slightly different from the EC table since, making a comparison with [6], it is the author's opinion that the one in [5] has some inaccuracies) and the rotational stiffness (see also further on Section 4.4.6). Table 4.1 is clearer when read together with Figure 4.18. Please notice that where $M_{j,Rd}$ is the maximum result between two terms it is because the moment is negative that the design moment is the least in absolute terms. Once again, $N_{Ed} > 0$ means tension and $M_{Ed} > 0$ stands for the clockwise moment; the eccentricity e is the ratio M_{Ed}/N_{Ed}.

Ribs and Stiffeners Having ribs/stiffeners changes the pattern and ℓ_{eff} of T-stubs, which allows us to readily have formulas to evaluate the numeric impact of their usage through ℓ_{eff}. However, in the literature, there are no design examples so it is advised to proceed with caution (by the way it must be emphasized that the EC method designs thickness values that are usually lower than the AISC method also without using stiffeners). Adding ribs sometimes also eases the application of formulas or the hypotheses on top of them, in the sense that the assumption of [5] that the profile is at least as wide as the gauge between bolts is commonly not verified in base plates. Adding ribs to broaden the flanges can satisfy this hypothesis, as shown in the design example at the end of the chapter (see Figure 4.33 in Section 4.4.10).

When adding stiffeners, the plate thickness should be approximately the same as that of the flanges. The stiffener height could be about 12–13 times the thickness (in order to make it class 3 if we reference EC).

Table 4.1 Equations for calculating design moment resistance and rotation stiffness of base plates according to EC.

		$N_{Ed} > 0$ and $e > z_{T,l}$	$N_{Ed} \leq 0$ and $e \leq -z_{C,r}$
Left side in tension, right side in compression	$z = z_{T,l} + z_{C,r}$	$M_{j,Rd} = \min\left(\dfrac{F_{T,l,Rd}\,z}{\dfrac{z_{C,r}}{e}+1}, \dfrac{-F_{C,r,Rd}\,z}{\dfrac{z_{T,l}}{e}-1}\right)$ $S_j = \dfrac{Ez^2}{\mu\left(\dfrac{1}{k_{T,l}}+\dfrac{1}{k_{C,r}}\right)}\dfrac{e}{e+e_k}$ con $e_k = \dfrac{z_{C,r}k_{C,r}-z_{T,l}k_{T,l}}{k_{C,r}+k_{T,l}}$	
		$N_{Ed} > 0$ and $0 < e \leq z_{T,l}$	$N_{Ed} > 0$ and $-z_{T,r} < e \leq 0$
Left side in tension, right side in tension	$z = z_{T,l} + z_{T,r}$	$M_{j,Rd} = \min\left(\dfrac{F_{T,l,Rd}\,z}{\dfrac{z_{T,r}}{e}+1}, \dfrac{F_{T,r,Rd}\,z}{\dfrac{z_{T,l}}{e}-1}\right)$ $S_j = \dfrac{Ez^2}{\mu\left(\dfrac{1}{k_{T,l}}+\dfrac{1}{k_{T,r}}\right)}\dfrac{e}{e+e_k}$ con $e_k = \dfrac{z_{T,r}k_{T,r}-z_{T,l}k_{T,l}}{k_{T,r}+k_{T,l}}$	$M_{j,Rd} = \max\left(\dfrac{F_{T,l,Rd}\,z}{\dfrac{z_{T,r}}{e}+1}, \dfrac{F_{T,r,Rd}\,z}{\dfrac{z_{T,l}}{e}-1}\right)$

(Continued)

Table 4.1 (Continued)

		$N_{Ed} > 0$ and $e \leq -z_{T,r}$	$N_{Ed} \leq 0$ and $e > z_{C,l}$
Left side in compression, right side in tension	$z = z_{C,l} + z_{T,r}$	$M_{j,Rd} = \max\left(\dfrac{F_{T,r,Rd}\, z}{\dfrac{z_{C,l}}{e} - 1}, \dfrac{-F_{C,l,Rd}\, z}{\dfrac{z_{T,r}}{e} + 1}\right)$	
		$S_j = \dfrac{Ez^2}{\mu\left(\dfrac{1}{k_{C,l}} + \dfrac{1}{k_{T,r}}\right)} \dfrac{e}{e+e_k}$ con $e_k = \dfrac{z_{T,r}k_{T,r} - z_{C,l}k_{C,l}}{k_{C,l} + k_{T,r}}$	
		$N_{Ed} \leq 0$ and $0 < e \leq z_{C,l}$	$N_{Ed} \leq 0$ and $-z_{C,r} < e \leq 0$
Left side in compression, right side in compression	$z = z_{C,l} + z_{C,r}$	$M_{j,Rd} = \max\left(\dfrac{-F_{C,l,Rd}\, z}{\dfrac{z_{C,r}}{e} + 1}, \dfrac{-F_{C,r,Rd}\, z}{\dfrac{z_{C,l}}{e} - 1}\right)$	$M_{j,Rd} = \min\left(\dfrac{-F_{C,l,Rd}\, z}{\dfrac{z_{C,r}}{e} + 1}, \dfrac{-F_{C,r,Rd}\, z}{\dfrac{z_{C,l}}{e} - 1}\right)$
		$S_j = \dfrac{Ez^2}{\mu\left(\dfrac{1}{k_{C,l}} + \dfrac{1}{k_{C,r}}\right)} \dfrac{e}{e+e_k}$ con $e_k = \dfrac{z_{C,r}k_{C,r} - z_{C,l}k_{C,l}}{k_{C,r} + k_{C,l}}$	

Note: To calculate k values (various subscripts) see Section 4.4.6 about stiffness; for μ refer to Section 3.1.

Figure 4.19 Column base detail with a double plate to be considered for heavily loaded cases.

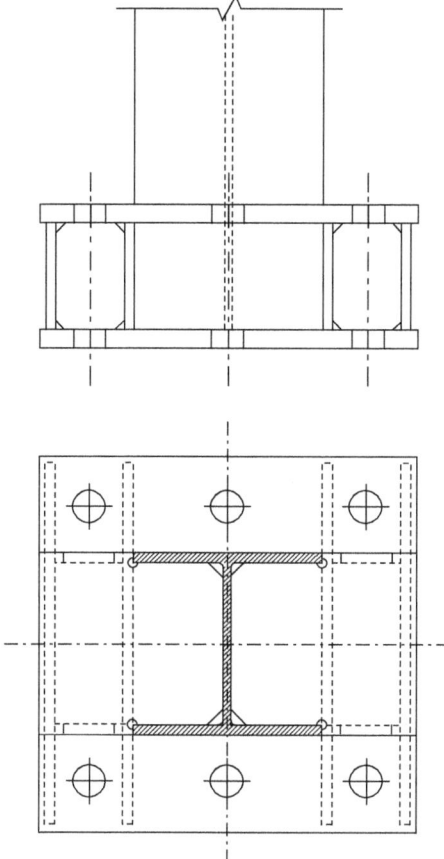

When additional plates are added locally under the anchor bolt nut, working similarly to washers, their numeric influence could be evaluated considering them as backing plates of T-stubs (by the EC method).

If columns that are big in size are also heavily loaded, a solution with a double plate as in Figure 4.19 could be considered.

A similar double plate has a section modulus that is very high.

4.4.2 Contact Pressure

The contact pressure between the base plate and the concrete is the first element to check: It is necessary to define the plate geometry and then design the other components.

4.4.2.1 AISC Method

According to the AISC, the maximum value that the pressure contact (f_p) between concrete and steel can reach is given by

$$f_{p,max} = \phi_c 0.85 f_c' \sqrt{\frac{A_f}{A_p}}$$

where the term A_f/A_p is equal to 4 maximum (hence the square root is maximum 2), f'_c represents the compression resistance of the concrete, A_p is the base plate area, and A_f is the concrete foundation relevant to the base plate similar to the EC method (see Section 4.4.2.2); ϕ_c is instead a resistance factor taken as 0.65 following [7] or 0.60 following [1].

Axial Load Only If there is only axial load, the whole base plate will react uniformly and the contact pressure value will be

$$f_p = \frac{N_{Sd}}{A_p}$$

which will be compared to the value $f_{p,max}$ seen above.

Eccentricity According to this method, the plate reaction is partial but the pressure is uniform. In the previous edition of [4], that is, [8], a triangular distribution of the contact pressure was instead assumed and, though this approach is still possible (Appendix B of [4]), it is not recommended because the calculations are a little more complicated (the results generally translate into a slightly thicker plate and slightly smaller anchor bolts).

The resisting part of the plate is w_p (width) while the depth h' can be obtained for small eccentricities as

$$h' = h_p - 2e$$

with e representing the eccentricity defined in the previous section where the thickness design was discussed.

The contact pressure can then be calculated as

$$f_p = \frac{N_{Sd}}{h' w_p}$$

If the eccentricity is large and f is defined as the distance between anchor bolts and the column center, h' is designed using

$$h' = \left(f + \frac{h_p}{2}\right) - \sqrt{\left(f + \frac{h_p}{2}\right)^2 - \frac{2N_{Sd}(e+f)}{f_p w_p}}$$

If the square root is less than zero, the geometry must be changed, usually increasing the plate dimensions.

The assumed design hypotheses mean that the pressure has a value f_p smaller than $f_{p,max}$ for small eccentricities while f_p coincides with $f_{p,max}$ if the eccentricity is large.

4.4.2.2 Eurocode Method

In the EC, the maximum design pressure contact (f_{jd}) between steel and concrete is

$$f_{jd} = \frac{\beta_j F_{Rdu}}{A_{eff}}$$

4.4 Base Plate with Cast Anchor Bolts

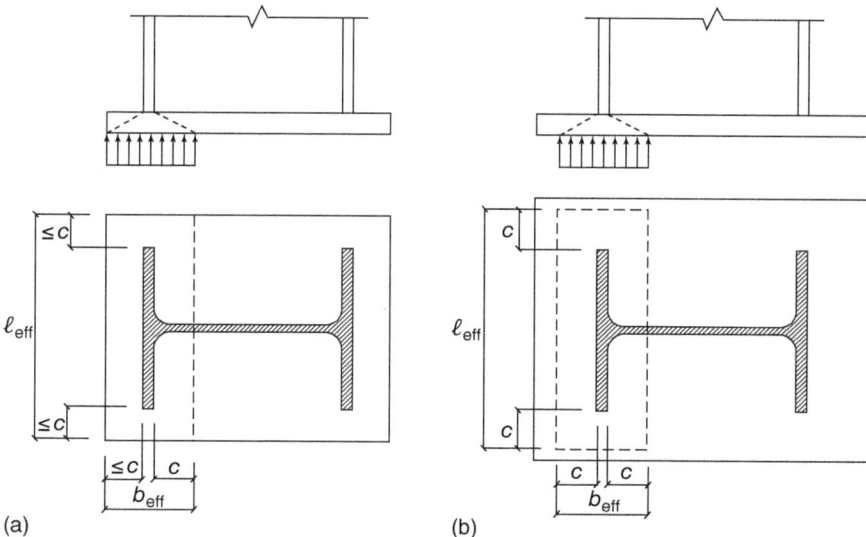

Figure 4.20 How to calculate the effective contact area according to [5]. (a) Short projection, (b) large projection.

where A_{eff} is the effective (see below) reference area, β_j is a coefficient based on the grout characteristics, and F_{Rdu} is the design resistance from EC 2 (the EC series about concrete). β_j is 0.66 in the typical situation of grouting under the plate with the characteristic grout resistance that is at least 20% of the concrete resistance and the grout thickness that is not more than 20% of the smaller base plate dimension (the lesser between the width and the length). If the grout thickness exceeds 50 mm (2 in.), then its characteristic resistance must match the concrete resistance.

To evaluate A_{eff} it is first necessary to get c (see Figure 4.20) and then b_{eff} and l_{eff} ($A_{eff} = b_{eff}\, l_{eff}$). The equation for c is

$$c = t\sqrt{\frac{f_{y,p}}{3 f_{j,d} \gamma_{M0}}}$$

where t is the base plate thickness and $f_{y,p}$ is its yield strength. Then F_{Rdu} (see Ref. [9] for details) can generally be obtained by

$$A_{c0} f_{cd} \sqrt{\frac{A_{c1}}{A_{c0}}}$$

where the term A_{c1}/A_{c0} is 9 maximum (therefore the square root is 3 maximum) and f_{cd} represents, with EC symbols, the compression design resistance of the concrete. Eurocode suggests taking a value of A_{c0} that is the same as A_{eff}, but the procedure would become complicated (to determine A_{c1}) and iterative because c and f_{jd} depend on each other.

The procedure suggested in [6] is to get the ratio $k_j = \sqrt{(A_{c1}/A_{c0})}$ (which is three maximum as just noted) taking the full base plate (conservative) in order to deduce

$$f_{jd} = \beta_j k_j f_{cd}$$

and then c, then b_{eff} and l_{eff}, up to

$$N_{Rd,\text{T-stub}} = f_{jd} A_{eff}$$

Please notice that A_{eff} is here referred to only one T-stub in compression and the total $N_{Rd,\text{bearing}}$ will be the sum of the different contributions.

With regard to the A_{c1}/A_{c0} ratio, A_{c1} is calculated as $w_f h_f$ (see symbols in Figure 4.21).

Numerically (again from [6]) w_f and h_f can be calculated as follows:

$$w_f = \min(w_p + 2w_1, 5w_p, w_p + s, 5h_f)$$
$$h_f = \min(h_p + 2h_1, 5h_p, h_p + s, 5w_f)$$

It is also noticed that [10] advises against considering resistant contact pressures that are more than $15\,\text{N}\,\text{mm}^{-2}$ if there is no inspection after the grout has

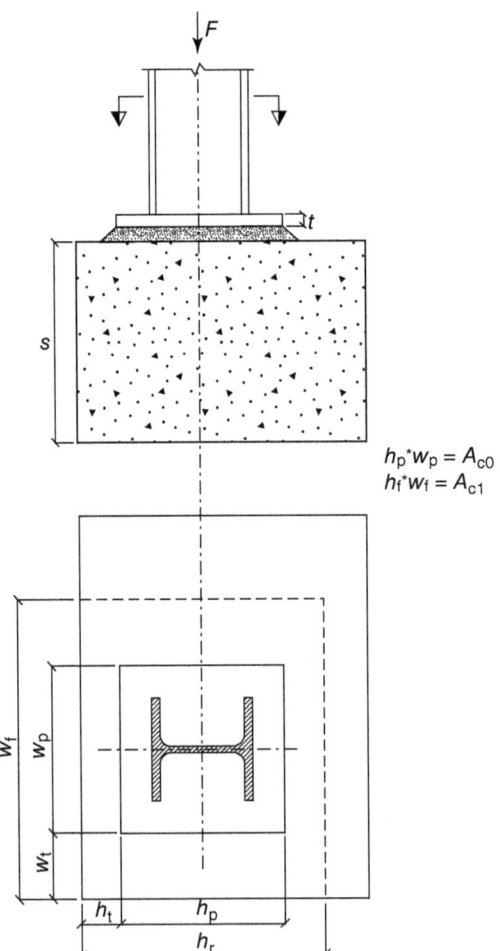

Figure 4.21 Geometric representation of symbols.

been laid: Any voids or air bubbles under the plate that are likely to originate if the workmanship does not provide the necessary care and quality may in fact invalidate the assumptions of large contact pressures guaranteed by high-resistance concrete.

4.4.3 Anchor Bolts in Tension

4.4.3.1 AISC Method

If there is uplift or large eccentricity as previously defined, anchor bolts (or at least a part of them) work in tension. Although it may be trivial it is important to remember that anchor bolts are necessary even when the theoretical design load on them is zero because anchors are also fundamental during erection so as to avoid column overturns when positioned during installation. According to US Occupational Safety and Health Administration (OSHA) regulations for construction, a base plate must have at least four anchor bolts and must be able to resist a vertical load of 300 lb (135 kg) with a 18 in. (about 450 mm) eccentricity from the outer face of the column. The only columns that do not have to follow this requirement are the ones that weigh less than 300 lbf.

The total tension (on the relevant side) over anchors when the eccentricity is large can be evaluated imposing the equation for the equilibrium of the vertical forces (see also previous paragraphs):

$$T = f_p w_p h' - N_{Sd}$$

For the bolts, Ref. [8] recommends a minimum length of at least 12 diameters inside the concrete (which becomes $17d$ for high-resistance materials). The minimum distance from the foundation edge must instead be more than the larger of 10 cm (4 in.) and $5d$ (for high-resistance anchors, 7 diameters and again 10 cm).

The possible shapes and installation systems to make the anchors capable of resisting uplift inside the concrete are several. AISC [8] suggests three types, as in Figure 4.22.

Hooked Bar The hooked bar is not recommended unless there is only little uplift or moment because the anchor might fail by straightening and pulling out of concrete (most of all anchors that are smooth since oily substances might be present).

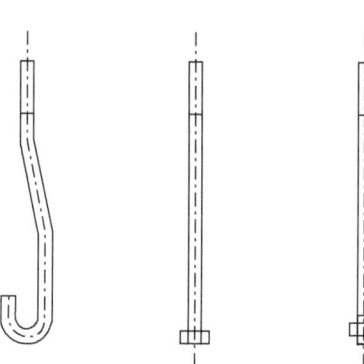

Figure 4.22 Possible solutions for realizing anchor bolts according to AISC.

When installing this kind of anchor, the formulas to use are several (the "new" ones in [4] are quite different from the old ones in [8]); the reader is referred to the original papers because the design issues are mostly related to concrete resistance, which is outside the scope of this book.

Threaded Bar (or Bolt) with Nut The AISC design guides advise welding or bolting a nut to the anchor (in any case some tack welding should follow the bolt tightening to prevent any loosening when the outside bolt is tightened). Washers in the lower nut are not necessary if following [8] when the anchor material is S235 or similar. If the resistance of the material is higher, small washers are enough to spread the concrete cone (but large washers are not recommended in [8] because they lower the edge distance).

The resistant area A_{psf} is shown in Figure 4.23 and it is a circular area on the concrete surface obtained by spreading the effect of the anchor bolt at 45° (the nut area can be neglected to simplify the math). If anchor bolts are adjacent and their cones overlap, the values must be modified as illustrated in Figure 4.24.

4.4.3.2 Eurocode Method

Eurocode also recommends hooking the anchor or, preferably, fixing a washer (or equivalent) with a nut in the lower part as in Figure 4.25 (not the nut only as just seen according to AISC).

Hooked Anchor Eurocode prohibits its use (hence the ones with washer and nut must be utilized) for anchor bolts yielding over 300 N mm^{-2} (about 44 ksi).

Anchor with Washer or Equivalent As mentioned earlier, please check [6, 9] for an exact evaluation of the actions in the reinforced concrete (pay attention to providing good concrete confinement) since this kind of analysis is not in the scope of this text. About the design resistance of the steel, similarly to the bolts in tension (see Section 3.4), the value can be calculated with the expression

$$\frac{k_2 f_{ub} A_s}{\gamma_{M2}}$$

where A_s is the bolt net area and $k_2 = 0.9$.

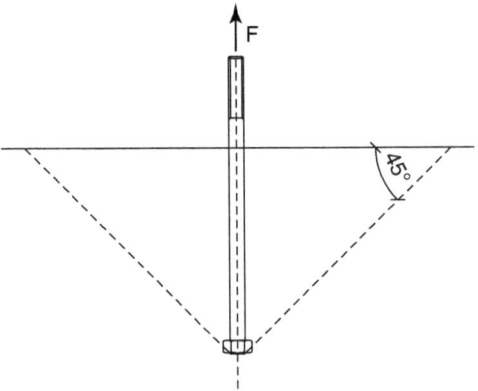

Figure 4.23 Cone of concrete radiating outward from the anchor.

Figure 4.24 Overlapping cones. Source: From Ref. [8].

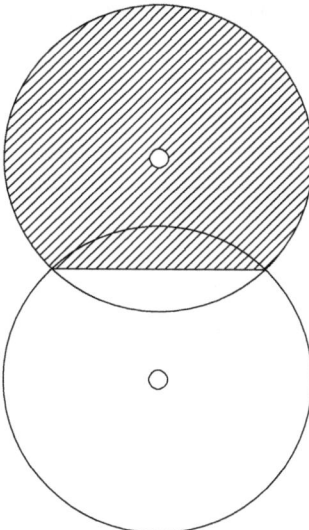

Figure 4.25 Eurocode anchor with washer.

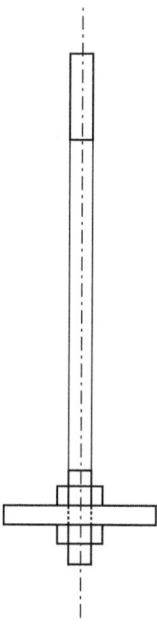

4.4.3.3 Other Notes

Table 4.2 seems interesting because it provides some indications about "standard" lengths of bolts in tension as used in the United Kingdom and the dimensions of plates (not washer–nut systems) to be applied to resist tension (based on the 4.6 or 8.8 class of the anchor bolt material). Table 4.2 is just a general reference and hence should not be applied without further evaluating the design conditions.

Table 4.2 Possible standard dimensions of plates inside concrete according to [10].

Diameter		M20	M24	M30
Length (mm) inside the concrete		300	375	
		375	**450**	450
		450	600	**600**
Plate inside the concrete (mm)	Dimension	100 × 100	120 × 120	150 × 150
	Thickness	12 (4.6)	15 (4.6)	20 (4.6)
		15 (8.8)	20 (8.8)	25 (8.8)

Note: Numbers in bold are the most commonly adopted values.
Source: Taken from Ref. [10].

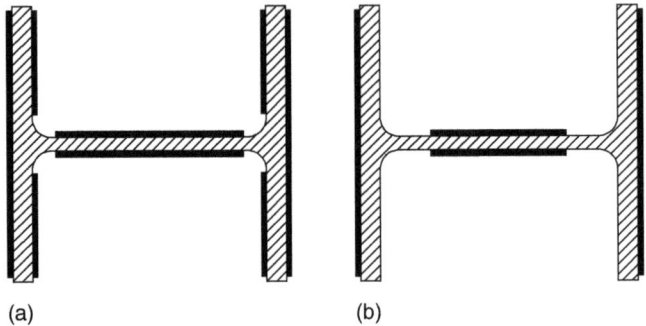

(a)   (b)

Figure 4.26 Possible welding between columns to base plates.

4.4.4 Welding

Welding must be checked according to usual rules. Figure 4.26 shows a "standard" welding detail on panel (a) and a possible cheap variant for low-stress applications on panel (b).

4.4.5 Shear Resistance

The base plate can resist shear by means of:

- Friction
- Anchor bolts
- Special shear lugs (shear keys), as plates or steel profiles welded below the base plate.

Both EC and AISC agree with the above.

4.4.5.1 Friction

If the column conveys a downward (gravity) load (it is in compression as in a classical scheme), the friction can resist all or part of the shear. Taking the AISC provisions [4] as reference, a possible value for the friction coefficient is 0.4. This is lower than what was considered in the previous edition of the AISC design guide 1 [8], that is, 0.55 for a (classical) situation where the contact between grout and base plate is above the concrete, 0.7 when the steel–grout (or concrete) contact is level with the foundation, and even 0.9 assuming the contact plane with the grout is below concrete of at least the plate thickness. However, AISC suggests careful assessment of the load combinations and, if necessary, lowering the available load for the friction. In contrast, the EC (Ref. [5] in particular) defines 0.2 as the reference value for the friction coefficient between the grout and the base plate.

Notice that the friction resistance can act in combination with another system, which can be designed to take only the "remaining" forces not borne by the friction. However, some standards forbid using the friction when calculating the resistance to seismic shear.

It is not trivial to remember that the load given by pretensioning the bolts (to be specified in detail in the design documents) could contribute to the frictional resistance.

4.4.5.2 Anchor Bolts in Shear

First, note that, historically, French standards have always opposed the assumption that anchor bolts can work in shear. Indeed, to make the anchor bolts work in shear some steps must be followed. For example, if the base plate has oversized holes (largely common), the anchor bolt–plate contact for the transmission of forces is not likely to happen simultaneously on all the bolts. Then, an assumption may be to consider only a certain number of anchors to really work (e.g. two according to [11]) or that anchor bolts can only resist a small force (about 10 kN each according to [12]). Alternatively, washers can be welded on-site to the base plate (which is sometimes done in large structures but it is not recommended for small and medium jobs because of site welding). If this latter approach is followed, some sources (e.g. Refs. [4, 6]) suggest applying some bending to the anchor bolts in the part inside the grout (assuming a double-curvature system since a cantilever would be too penalizing).

Eurocode recommends using the lower of two values as the shear resistance for anchor bolts: on one hand, the resistance in analogy to bolts, that is,

$$\frac{\alpha_v f_{ub} A}{\gamma_{M2}}$$

which can be seen in detail in Section 3.3; on the other hand, it must be evaluated also as

$$\frac{\alpha_{bc} f_{ub} A_s}{\gamma_{M2}}$$

where, similar to what has already been seen, A_s is the tension area of the anchor and α_{bc} is a coefficient defined as

$$\alpha_{bc} = 0.44 - 0.0003 f_{yb}$$

with f_{yb} the anchor bolt yield value (between 235 and 640 N mm^{-2}).

If shear and tension are simultaneously present, the check must follow the methods specified for bolts.

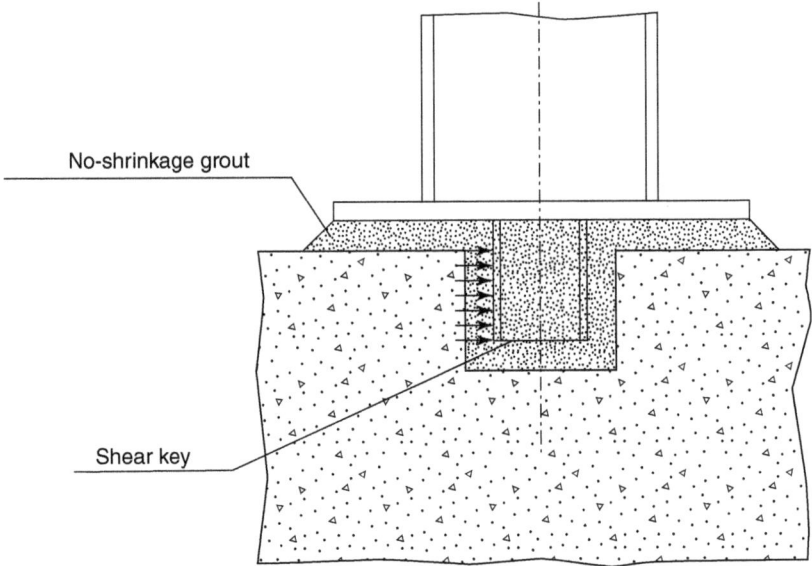

Figure 4.27 Contact pressure generated by the shear key.

4.4.5.3 Shear Lugs

Using a shear lug, consisting of plates or a real profile welded under the plate (the same size as the column or more frequently smaller), is an effective method to resist major design shear. The disadvantage is that it requires the preparation of a "cockpit" and a few additional difficulties during erection, which does not make it a preferred choice for small fabrication jobs.

The system (called by many different names, e.g. *shear key, shear stub,* and *shear nib*), pushed against the concrete around the pit, will react, and hence the contact pressure must be verified in addition to the profile itself working in shear and bending as a cantilever.

There are various approaches to checking the contact pressure: Ref. [8] gives $0.7 f'_c$ as the reference value in a confinement situation but then it advises to conservatively use $0.35 f'_c$ (unconfined value). Instead, [4] says to take either $0.85 f'_c$ or $0.55 f'_c$, then modifying it depending on the actions (see that design guide for details). For a solution in accordance with EC the engineer can instead follow the method already seen to calculate the contact pressure transmitted by the base plate (see Figure 4.27).

As the figure suggests, it is advised not to consider the part of the lug inside the top thickness of the grout for resistance.

4.4.6 Rotational Stiffness

The rotational stiffness can be assessed by means of Table 4.1 once it is established that

$$k_{C,l \text{ or } r} = k_{13}$$

$$k_{T,l \text{ or } r} = \frac{1}{\frac{1}{k_{15}} + \frac{1}{k_{16}}}$$

$$k_{13} = \frac{E_c \sqrt{b_{\text{eff}} l_{\text{eff}}}}{1.275E}$$

$$k_{15} = \frac{0.85 \ell_{\text{eff}} \, t_p^3}{m^3}$$

$$k_{16} = 1.6 \frac{A_s}{L_b}$$

If there is no prying action, k_{15} should be halved and k_{16} has 2 instead of 1.6 as the initial coefficient. The symbols were already seen when dealing with EC in the previous sections; in particular, t_p is the base plate thickness, E_c is the concrete elastic modulus, l_{eff} and b_{eff} are related to the concrete in compression and ℓ_{eff} to the end plate in tension, m is as in Figure 3.26, A_s is the tensile anchor bolt area, and L_b is the anchor elongation length. Furthermore, with regard to μ in Table 4.1, refer to Section 3.1.

While the right part of the plate is considered, the k values of the right-hand side will be taken (subscript r) and similarly in the analysis of the left part (subscript l).

When there is more than one row of bolts in tension, the calculation should be performed for each row of bolts.

4.4.7 Measures to Improve Ductility

The joint may be considered with a good degree of ductility if (from [4]) the limit state that governs the design is not some kind of failure in the concrete (in particular the failure of the cones because of the anchors in tension) or the failure of the welds. The other limit states (forming of a plastic hinge, plate mechanisms, anchor bolt yielding but also excessive contact pressure) allow a redistribution of the actions.

4.4.8 Practical Details and Other Notes

Figure 4.28 shows a typical sequence for the installation of the anchor bolts and the base plate. Figure 4.29 displays some alternative solutions, all of which allow some dimensional adjustments on-site.

It is highly advisable to group the bolts and place them with templates (Figure 4.30) to facilitate the work of masons and greatly improve the quality of the result.

In the unfortunately common event that anchor bolts are not properly installed, an immediate fix is necessary because the correct placement is critical to ensure that the structure is plumb and can be erected without issues. The following situations may arise:

- Placement is not correct but can be solved slotting the holes of the base plate; it could be a problem especially if the design assumes that the anchors resist base shear;

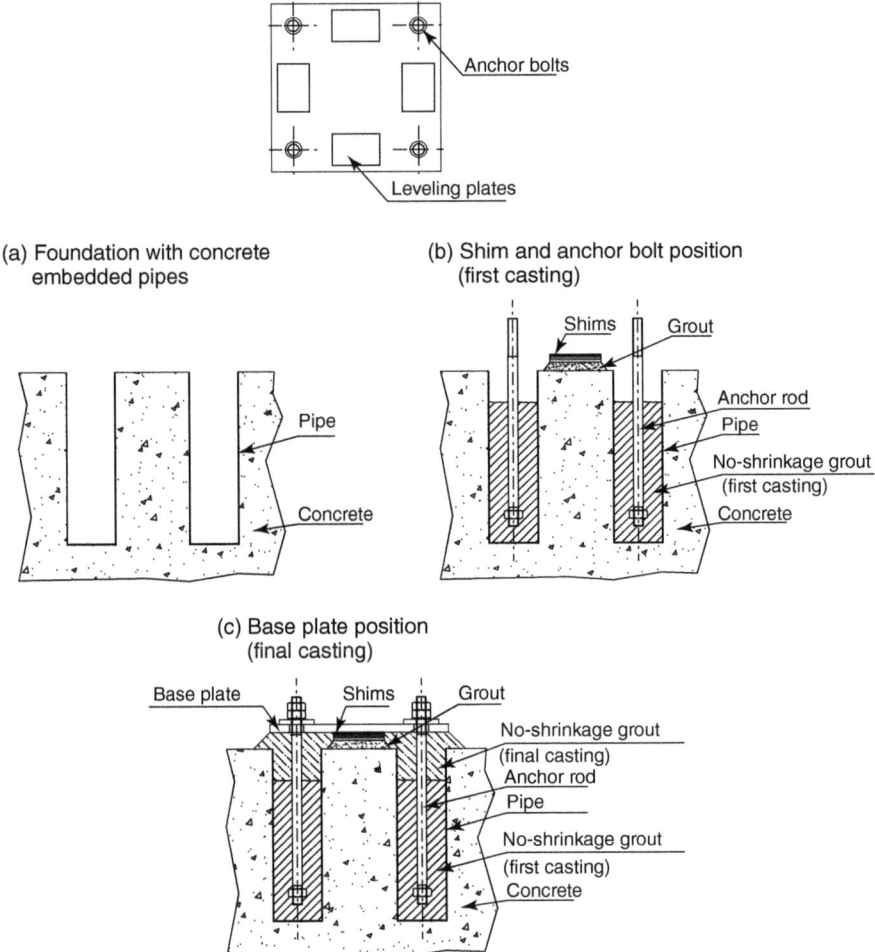

Figure 4.28 Procedure for positioning anchors and base plate.

- Positioning is substantially incorrect. If caught in time, it is convenient to change the geometry of the base plate or weld the plate offset whenever possible (the modified design resistance needs to be rechecked); it can also be a solution to cut the wrong bolts and install chemical or mechanical anchors (design to be rechecked); in an extreme situation the whole anchor group might need to be laid again and the relevant part of the foundation made again.
- Bolts are bent because some equipment (e.g. cranes or, this was seen too, snow ploughs!) passed over them. If the damage is contained and anchor bolts do not work near the limit, an attempt to straighten them could be done; if the damage is too heavy, one of the above solutions should be applied.

Finally, as discussed in Section 6.17, it is suggested to provide ad hoc access holes for casting and compacting the grout or the concrete below the base plate especially when the plate is wide or when shear lugs are provided.

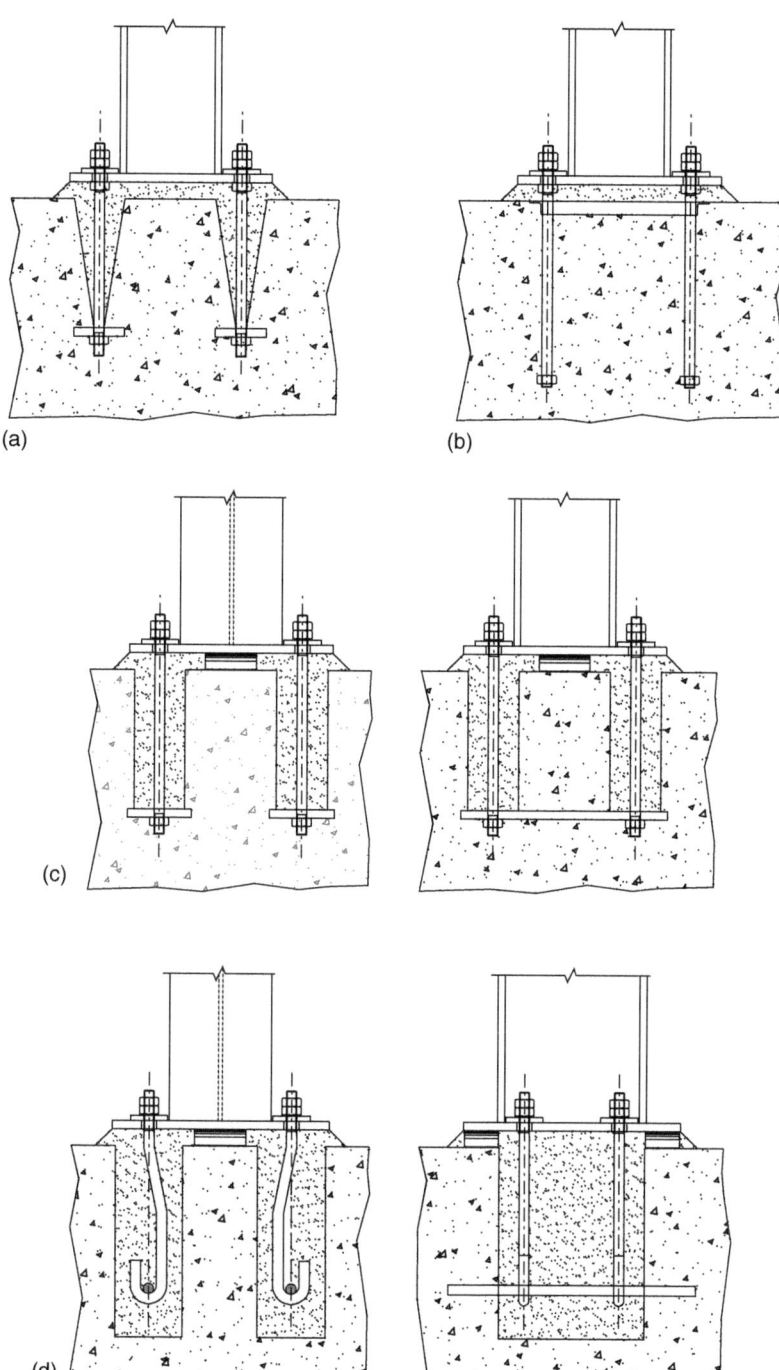

Figure 4.29 Other possible solutions for the base detail.

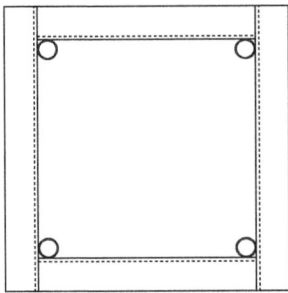

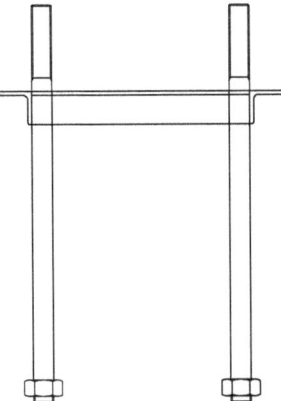

Figure 4.30 Possible anchor bolt template made of angles.

For the grout, Ref. [4] suggests a compression design resistance that is twice the concrete compression resistance.

If the foundation does not have grout (i.e. the base plate does not lean on concrete) and the anchor bolts work in compression, the engineer should check the compression resistance of the concrete below the anchor bolts, where the anchors concentrate the actions.

4.4.9 Fully Restrained Schematization of Column Base Detail

For the base of the column to be fully restrained, it is required that the reaction offered by the ground (as well as the type of concrete foundation) be compatible with the loads: A fully restrained schematization when the soil is not able to react in the requested range means having a different restraint (likely a pin connection) with obvious consequences that are very serious if the base rigid restraint is the only lateral resisting system in a certain direction. When possible, it is recommended not to consider the base detail as rigid: it improves the deflection and allows savings with the quantity of material but there are remarkable problems with the base plate and the foundation (both for the concrete and the soil), which makes the choice globally uneconomic.

It is to be avoided, except in cases of real necessity, to have columns work as cantilevers (inverted pendulum) on their weak axis.

4.4.10 Example of Base Plate Design According to Eurocode

This example designs, according to the EC (taking $\gamma_{M0} = 1.05$, $\gamma_{M1} = 1.05$, $\gamma_{M2} = 1.25$), the column base plate of an industrial building. The column is an HEA 240 and the governing load cases are the following, which are typical in similar kinds of buildings: The first one (SLU1) occurs when snow and wind (on the column strong axis of the column) act together and SLU2 originates from the wind load in the orthogonal direction that gives maximum uplift when the dead loads are factored as 1 to maximize upward forces. The lateral resisting systems are portals with rigid bases on one side and braces on the weak side (responsible for the remarkable uplift).

We consider S275 as the column material, S235 for the plates, and class 5.6 (yielding at 300 MPa, rupture at 500 MPa) for the anchor bolts.

SLU1:

$N_{Ed} = -250$ kN (compression)

$V_{\text{major Ed}} = 50$ kN

$V_{\text{minor Ed}} = 5$ kN

$M_{\text{major Ed}} = 55$ kN m

$M_{\text{minor Ed}} = 0$ kN m

SLU2:

$N_{Ed} = 110$ kN (uplift)

$V_{\text{major Ed}} = -5$ kN

$V_{\text{minor Ed}} = 120$ kN

$M_{\text{major Ed}} = -5$ kN m

$M_{\text{minor Ed}} = 0$ kN m

4.4.10.1 Uplift and Moment

If the concrete has $R_{ck} = 25$ N mm^{-2}, that is, $f_{ck} = 0.83 \times 25 = 20.75$, we get $f_{cd} = 20.75 \times 0.85/1.5 = 11.8$ N mm^{-2}. Assuming a ratio of 4 between the area of the plate and the foundation (it will very likely be more), the result is $f_{jd} = 0.67\sqrt{4} \times 11.8 = 15.8$ N mm^{-2}.

We consider a base plate thickness equal to 20 mm. This means that $c = 20\sqrt{[225/(3 \times 1.05 \times 15.8)]} = 42.5$ mm, and hence $l_{\text{eff}} = 240 + 2 \times 42.5 = 325$ mm, which is reduced to 300 mm (plate width), and $b_{\text{eff}} = 12 + 2 \times 42.5 = 97$ mm, so an effective area that is about 29 100 mm². SCS provides a higher value because it also considers the part below the web in the computation (conservatively neglected here).

The distance of the center of compression from the center of the plate on both sides is equal to the distance between the centerline of the flanges and the axis of the column, namely $z_{C,l} = z_{C,r} = 230/2 - 12/2 = 109$ mm, smaller in absolute value than the eccentricity, equal to $-55\,000/250 = -220$ mm, and thus we will have one side (the left-hand side) in tension and one (the right) in compression. The only effective area in resisting compression in this manual computation, is below the right flange and it bears $15.8 \times 29\,100 = 460$ kN. It must, however, be verified that this number is not higher than the value that the flange and the web of the column can take in compression: $F_{c,fb,Rd} = M_{c,Rd}/(h - t_f)$ and, conservatively assuming a class 3 section so that the elastic resistance is considered, we have $F_{c,fb,Rd} = 675 \times 10^3 \times 275/1.05/(230 - 12) = 811$ kN. SCS correctly takes the plastic resistance and gives us 895 kN.

In the tension zone, on the left, the plate must instead be checked for tension bending (T-stub, Figure 4.32) and the column web in tension near the flange.

Now, assuming the geometry in Figure 4.31, we run some checks noting that, in order to strictly apply the formulas of EC, the anchor bolts should not have a pitch that is greater than the width of the profile.

Alternatively, stiffeners as in Figure 4.33 could be welded to comply with hypotheses for T-stub equations in EC and allow moving the anchors externally. Another alternative (the method used by SCS) would be to make the calculation following the references cited in Section 3.10.7.

Assume a throat weld size of 6 mm for flanges, $m_x = 70 - 0.8 \times 6\sqrt{2} = 63$ mm. Then we have

$$\ell_{eff,cp} = \min(2\pi 63, \pi 63 + 150, \pi 63 + 2 \times 75) = 288 \text{ mm}$$

$$\ell_{eff,nc} = \min(4 \times 63 + 1.25 \times 65, 75 + 2 \times 63 + 0.625 \times 65, 0.5 \times 300,$$
$$0.5 \times 150 + 2 \times 63 + 0.625 \times 65) = 150 \text{ mm}$$

$$\ell_{eff,1} = \ell_{eff,2} \left(= \sum \ell_{eff,1} = \sum \ell_{eff,2} \right) = 150 \text{ mm}$$

Let us check if there is prying action: The elongation length of the anchor is eight times the diameter plus the thickness of the grout and base plate plus the washer and half nut (estimated about 15 mm); hence $8 \times 24 + 20 + 50 + 15 = 277$; this is to be compared with L_b (see Section 3.10.4), which according to EC becomes (A_s of M24 = 353 mm^2) $8.8 \times 63^3 \times 353 \times 2/(300 \times 20^3) = 647$ mm, and

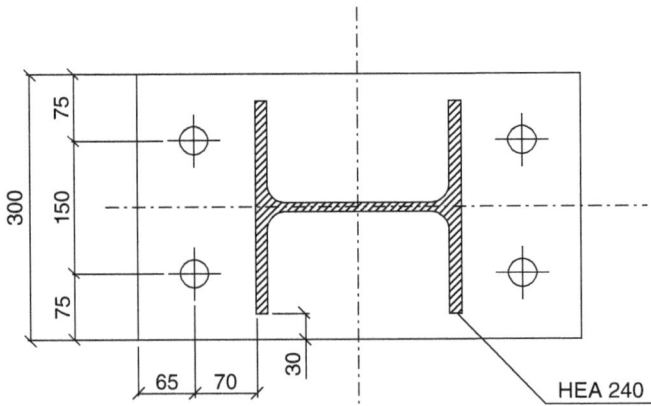

Figure 4.31 Geometry.

Figure 4.32 T-stub parameters.

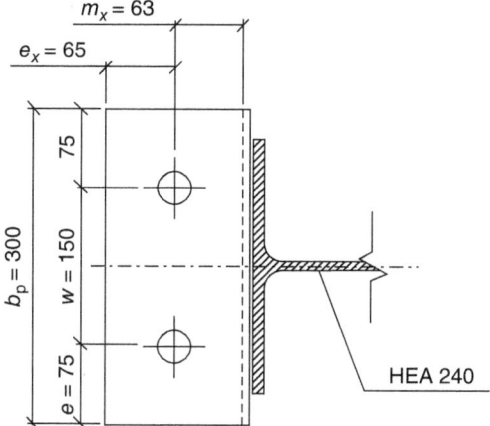

Figure 4.33 Possible solution (not used in the example) to comply with EC assumptions when anchors are outside the column width.

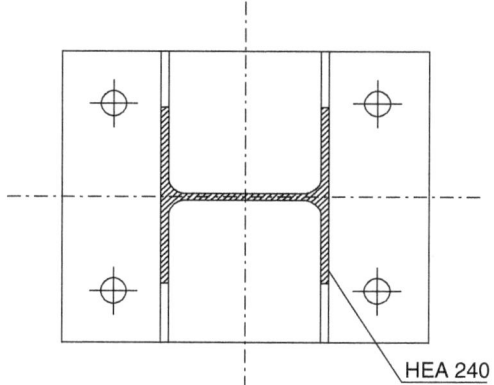

therefore prying action is possible. In fact, some sources (e.g. Refs. [4, 10]) deny the possibility that prying occurs in a base plate, so the engineer could directly apply this hypothesis. However, taking the worst possible situation as per recent instructions (see Section 3.10.4), we have: $F_{T,1-2,Rd} = 2(0.25 \times 150 \times 20^2 \times 225/1.05)/63 = 2(3.21 \times 10^6)/63 = 102$ kN.

We use here 225 MPa instead of 235 MPa because the thickness is >16 mm.

About the resistance of the anchor bolts (M24) on one side, it is $F_{T,3,Rd} = 2(0.9 \times 353 \times 500/1.25) = 2 \times 127.1 = 254$ kN. Then $F_{T,2,Rd} = (2 \times 3.21 \times 10^6 + 65 \times 2 \times 127.1 \times 10^3)/(63+65) = 179$ kN, and thus $F_{T,Rd} = \min(102, 179, 254) = 102$ kN. It is underlined that the engineer also has to check (not in the scope of this text) that the foundation and the soil can resist the design loads. The T-stub resistance in tension is therefore 102 kN and this is the actual reference value to take since the web column in tension does not govern the design.

The lever arm z is equal to $z_{C,r} + z_{T,l} = 109 + 185 = 294$, and hence the resisting moment is $M_{j,Rd} = \min(102 \times 294/((109/-220) + 1), -460 \times 294/((185/-220) - 1) = 59.4$ kN m (the tension side governs the design), bigger than the design action (93% design ratio).

Regarding the second combination (SLU2), both sides work in tension (eccentricity is −45 mm) and the resistant force is again 102 kN on each side. For the arm $z = 185 + 185 = 370$ mm, so $M_{j,Rd} = \max[102 \times 370/(185/-45 + 1), 102 \times 370/(185/-45 - 1)] = -7.4$ kN m, and hence the utilization ratio is 68%.

4.4.10.2 Shear

Since the uplift is remarkable and it will highly stress the anchors, we choose to let a shear key take care of the shear. The first combination has a shear value that is slightly over 20% of the axial force so the friction could absorb it. However, SLU2 cannot rely on friction because of the uplift and, therefore, we weld a profile on the bottom part of the base plate that will likely be placed only in base plates where braces land.

Let us assume a piece of HEA 120 that is 150 mm long. The part inside the grout thickness must not be considered in design so the effective depth to take is $150 - 50 = 100$ mm.

If we design a pit as in Figure 4.34 (that will have grout inside after installing the column) and we consider a ratio between the concrete area (facing the lug in the pit) and the steel area that is about 2.5 in both directions (the depth and the width of HEA 120 are quite similar), the resulting contact pressure is $f_{jd} = 0.67\sqrt{2.5} \times 11.8 = 12.5$ N mm^{-2}, which means a resisting force of about $12.5 \times 113 \times 100 = 141$ kN in the weak-axis direction and a little more (the flange is 120 mm wide) in the strong axis, both above the design loads (the exact calculation in SCS gives us a maximum design ratio of 83%).

Shear and bending in HEA 120 must also be checked. The design bending moment can be evaluated as (in the SLU2 case, which is the one supposedly governing) $120(50 + 100/2) = 12$ kN m, to be compared with a weak moment resistance of $49\,000 \times 275/1.05 = 12.8$ kN m. The eccentricity (which could be considered in different ways, even zero) has to be set manually in SCS.

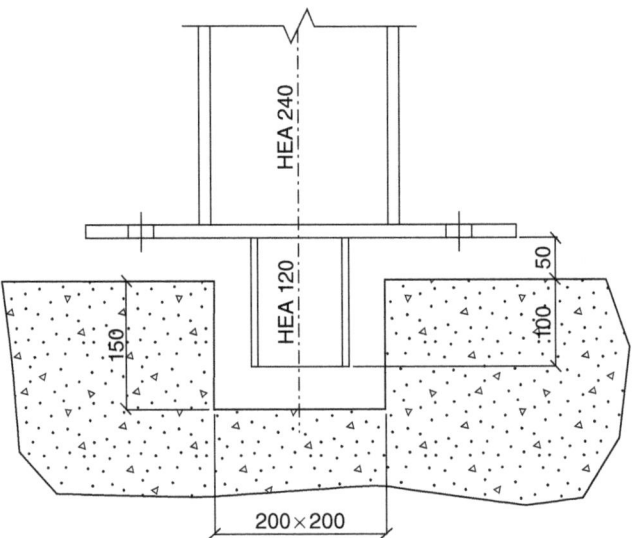

Figure 4.34 Shear lug pit detail.

It is left to the reader to complete the remaining checks.

From SCS calculations we note that the anchors could also work in shear (at 36%) because the interaction with tension is not a problem (61% exploitation, bolt eccentricity set equal to 0).

4.4.10.3 Welding

For the column, the engineer may recommend double fillets with a 6 mm throat for the flanges and 4 mm for the web. This is almost a full-strength weld, with resulting benefit in ductility.

Similarly, in a simplified manner, a 4-mm throat is designed in the weld all around the shear lug (which we just saw working with a good design ratio).

4.4.10.4 Joint Stiffness

The stiffness in the part in compression is $k_{C,r} = k_{13} = 30\,200\sqrt{(300 \times 97)}/(1.275 \times 210\,000) = 19.3$ mm. For the part in tension $k_{T,l} = 1/(1/k_{15} + 1/k_{16}) = 1/(1/(0.425 \times 150 \times 20^3/63^3) + 1/(2 \times 353/277)) = 1/(1/2.0 + 1/2.6) = 1.1$ mm. For SLU1, then, $e_k = (109 \times 19.3 \times 185 \times 1.1)/(1.1 + 19.3) = 93$ mm and $\mu = (1.5 \times 55/59.4)^{2.7} = 2.43$, and hence $S_j = 210\,000 \times 294^2/(2.43\,(1/19.3 + 1/1.1)) \times (-220)/(93 - 220) = 1.4 \times 10^4$ kN m.

For SLU2, some coefficients change: $\mu = 1$, eccentricity $e = -45.5$, $z = 370$, and $e_k = (185 \times 1.1 - 185 \times 1.1)/(1.1 + 1.1) = 0$ mm, and thus $S_j = 210\,000 \times 370^2/(1(1/1.1 + 1/1.1)) \times (-45.5)/(0 - 45.5) = 1.6 \times 10^4$ kN m.

4.4.10.5 Comparison with AISC Method for SLU1

Carrying out the design for the same base plate (500×300) according to AISC for the combination SLU1, we get $m = 141$ mm, $n = 54$ mm, $n' = 59$ mm, and hence, with $\lambda = 1$ (conservative), $\lambda n' = 59$ mm. The governing length is therefore 141. Assuming, as we just did, foundations with dimensions at least two times the base plate, $f_{p,max} = 0.6 \times 0.85 \times 0.83 \times 25\sqrt{4} = 21$ N mm^{-2}, and thus the critical eccentricity is $500/2 - 250\,000/(2 \times 21 \times 300) = 230$ mm. Since for SLU1 the eccentricity is $55\,000/250 = 220$ mm, the small eccentricity equations apply, so $h' = 500 - 2 \times 220 = 60$ mm and $f_p = 250\,000/(300 \times 60) = 14$ N mm^{-2}, and therefore the contact pressure is verified with a 66% approximate ratio.

We must now evaluate the bending moment on the plate, loaded by the contact pressure, as $(14 \times 60)(141 - 60/2) = 93\,200$ N mm/mm. When the material is S235, the required thickness becomes, $t = \sqrt{(4 \times 93\,200/(235 \times 0.9))} = 42$ mm, or, considering S275 as the material, 39 mm. As is clear by the numbers, the approach (the AISC method only considers the part in compression for small eccentricities) and the results are quite different. To get a better performance by an AISC-based design, the engineer could shorten the long side (500) of the plate, consequently lowering m and possibly widening the plate to make room for the anchors if geometrically necessary.

4.5 Chemical or Mechanical Anchor Bolts

Mechanical and chemical anchors ("Hilti" probably being the most famous producer) are usually adopted when connecting to a structure that already exists so that precasting bolts is not possible. "Small" parts such as stairs or door frames might also be fixed with this kind of anchor since they do not provide heavy loads. This has the additional advantage of a much wider tolerance of a precast anchor bolt group because the anchors are installed only at the end, with the steel part already in its position.

When connecting to vertical walls, the designer should keep in mind that threaded bars going through the walls can also be utilized.

From a design point of view, in addition to the limit states of the steel parts, that is, the anchors themselves and the plate (refer to the previous section about base plates), all the checks typical of the concrete must be performed since they are usually the ones governing the design.

In most cases it is advisable to follow the charts provided by anchor producers that sometimes even distribute free software to support the design of their products.

It is suggested to carefully evaluate the different types of anchors as anchor bolts with the same diameter and from the same vendor may have different bearing capacities. This clarification of the type, as well as the diameter and the length, must be clearly indicated in the design documents to prevent fabricators from buying just the cheapest available.

4.6 Fin Plate/Shear Tab

The *fin plate* (or *web side plate* or *single shear plate* or *shear tab*) is a connection made with a vertical plate usually welded to the main member (a column or a beam) and bolted to the secondary member (a beam). There are rare cases where the connection is made with two parallel plates with the web of the beam inserted between them (or there are two plates welded to the beam web that are bolted to the fin plate) but the erection issues are apparent though there is a design advantage (bolts work with two shear planes), so this combination is commonly avoided.

The secondary element is a secondary beam if the main member is a primary beam while it is a primary (or secondary) beam if the main member is a column.

The fin plate is recognized as a hinge even if it is able to develop small bending moments. The rotation capacity of the fin plate is not commonly checked (according to "traditional" methods) if the secondary member has deflections in the acceptable range (which is supposed to be the rule). According to the EC, the calculation of the rotation stiffness should, however, be done. Published by the influential and authoritative European Convention for Constructional Steelwork (ECCS) with design examples based on EC, Ref. [13] bypasses the verification ensuring the rotation capabilities as in Section 4.6.3 and coupling it with good ductility (see Section 4.6.4). This approach seems acceptable.

The shear tab in I- or H-shaped profiles (i.e. IPE, HE, W, UB, UC) connects the web of the secondary member and thus is highly effective in conveying shear while it is much less efficient with axial loads since the whole area is not effective (shear lag phenomenon, see Section 3.19.1), as shown in Figure 2.6, having the

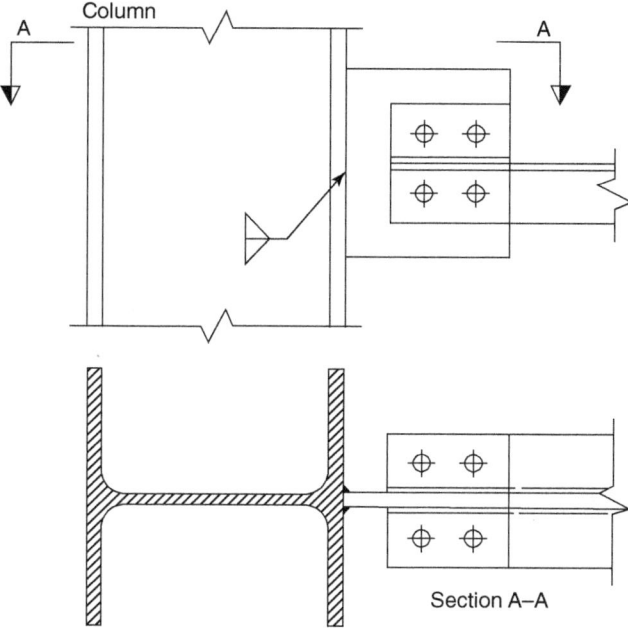

Figure 4.35 Angle brace or strut connected by using an additional angle (called "lug angle" in the EC) in order to transmit higher axial forces.

forces to converge into the web. However, having different shaped profiles (U and L type) as secondary members and possibly by adopting details such as bolting the second leg of the angles or the flanges of the channels, the fin plate becomes highly effective for large tensile and compression loads and it is widely used to connect bracings (Figure 4.35). It is actually effective for braces designed in compression even when only the flange or web is bolted as these braces do not usually transmit big actions, the instability being the element that governs the design of the brace itself.

4.6.1 Choices and Possible Variants

Here we discuss in detail the possible variants for this joint, such as the position of the pin (hinge), the location of the plate, and the notches (copes) in the secondary member. Most of the considerations can be applied later while discussing other connection types without mentioning the options in detail.

4.6.1.1 Pin Position

1. It is possible (actually the most frequently chosen option) to locate the theoretical pin at the axis of the main member. This scheme minimizes the stress in the column (only concentric axial load is transferred) or main beam (no torsion) but it creates a moment in the bolt group due to its eccentricity from the connection axis.
2. It is possible to locate the theoretical pin at the center of gravity of the bolt group. This scheme minimizes the stress in the bolts but worsens the one in

the main element for the occurrence of moments in the latter. If a beam is the primary member, possibly not balanced on the other side by another secondary, the induced torsion is likely a problem, and hence this choice is not recommended in such cases.
3. It is possible to locate the theoretical pin at any position within the range set by the options above. It might, for example, be convenient to take the axis at the contact point between the column flange and the fin plate when the column is connected with the strong axis: The eccentricity in the bolt group is smaller at the cost of a bending moment in the column, which, though, working with its strong axis, can oppose consistent strength (to check with numbers, obviously).

4.6.1.2 Location of Plate Welded to Primary Member
1. It is possible to weld the plate only to the web (if it is a beam or if it is a column oriented according to the weak axis) or to the flange (column strong axis); see Figure 4.36 for examples.

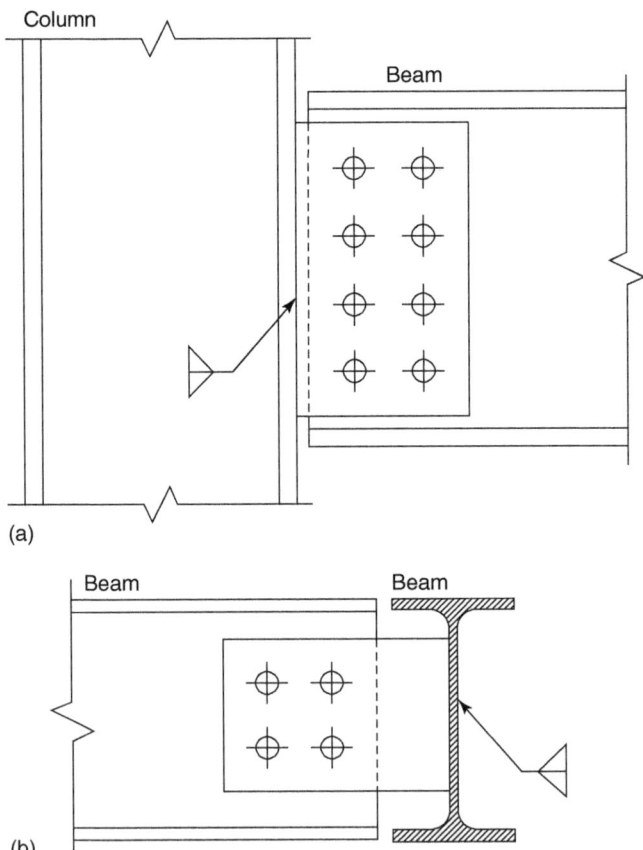

Figure 4.36 Classical solution with the shear tab welded to the primary member flange (a) or web (b).

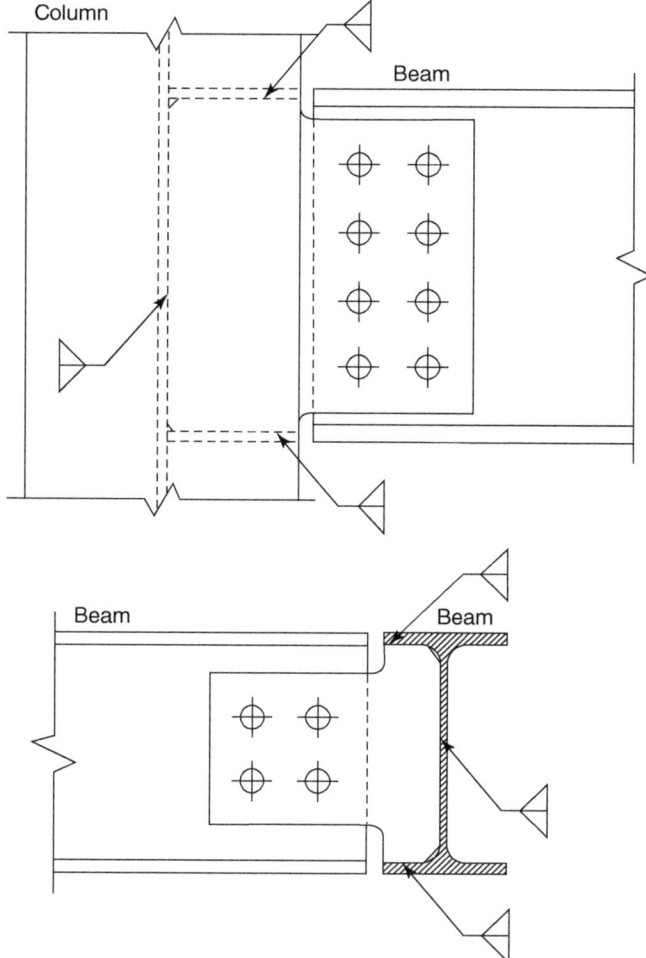

Figure 4.37 Shear tab welded also to the primary member flanges (or stiffeners).

2. It is possible to weld the plate to the web and both flanges (in columns and beams). To do this in columns, some additional stiffeners might be needed as in Figure 4.37.
3. It is possible to weld (Figure 4.38) the plate to the web and one flange (usually the top flange) of the main member (column or beam).

4.6.1.3 Notches (Copes) in Secondary Member

Sometimes, for the onset of problems related to the eccentricity, there is the need to bring the secondary element near the axis of the main one. Depending on the geometry (usually the top of the steel is the same but it is not the rule), it might be necessary to cope/notch the beam:

1. Only the top flange is notched (Figure 4.39, left side).

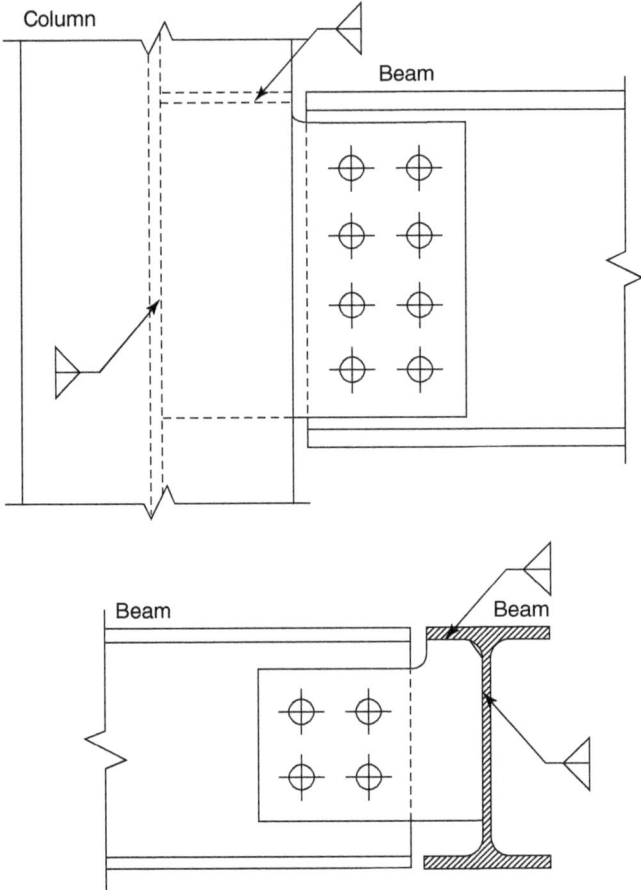

Figure 4.38 Shear tab welded to the primary member top flange (or stiffener).

2. Both the flanges are notched as in the right side of Figure 4.39, which represents an infrequent case of a secondary beam that is deeper than the primary beam, sometimes found when the primary beam is a wide-flange type or the primary is, say, part of a composite construction (so it does not need to be very deep).
3. Only half of the top and bottom flange is notched. This might help in inserting the beam during erection (Figure 4.40).
4. Both flanges have two half notches on each side, as in Figure 4.41. This will allow inserting the beam during erection.

Notching (to allow the secondary beam to be closer to the primary member so that eccentricity is lowered) brings additional costs and so should be avoided if the checks are satisfied without coping the beam.

4.6.1.4 Reinforcing Beam Web

It may happen that the verification of the secondary beam web is not satisfied because of bearing or other limit states, especially in the case of

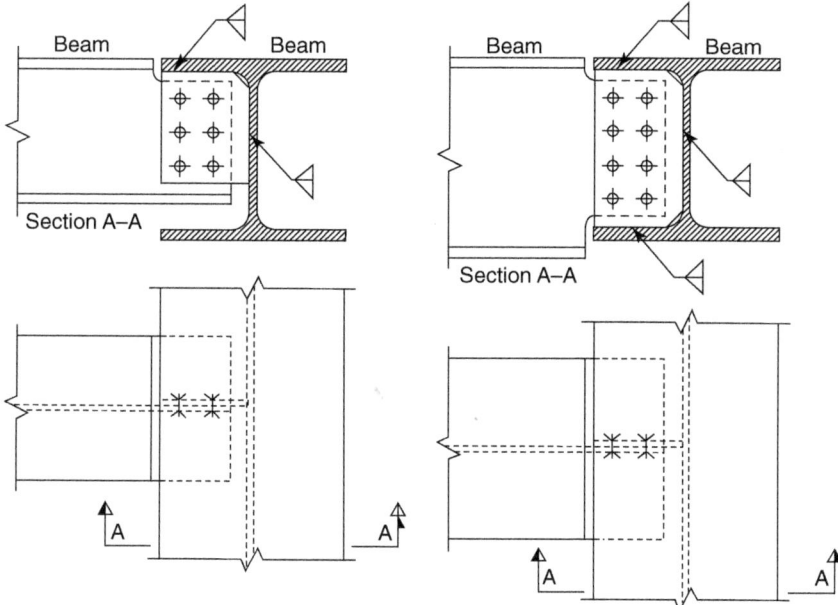

Figure 4.39 Notched configurations.

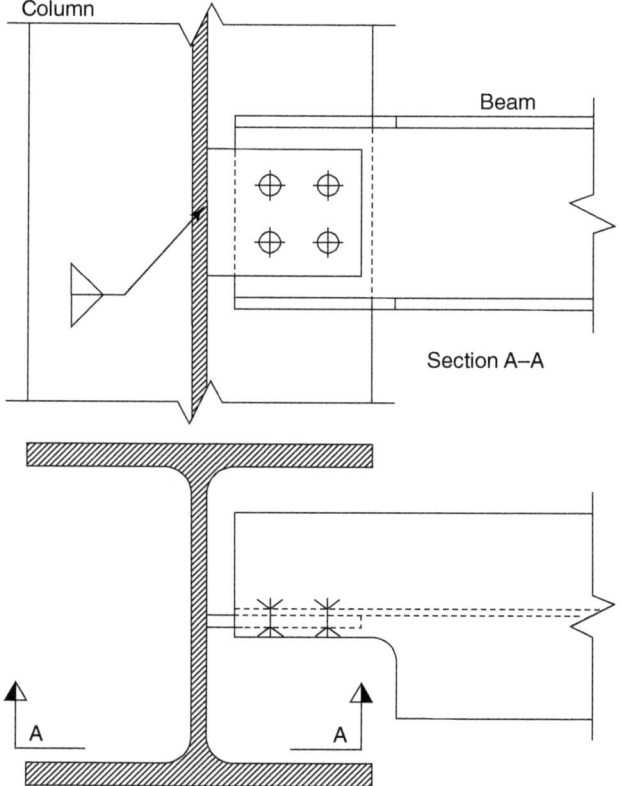

Figure 4.40 Notched flanges can help erection.

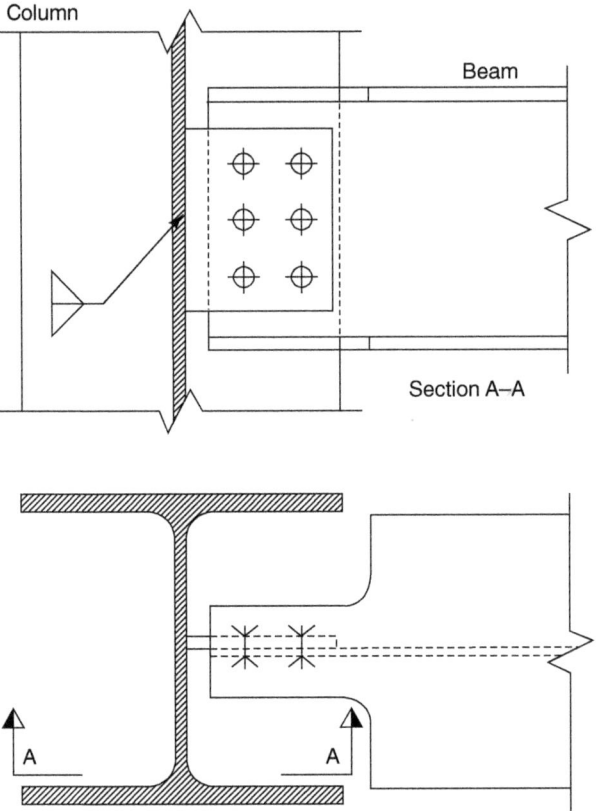

Figure 4.41 Both sides of the flanges are notched to allow positioning during erection.

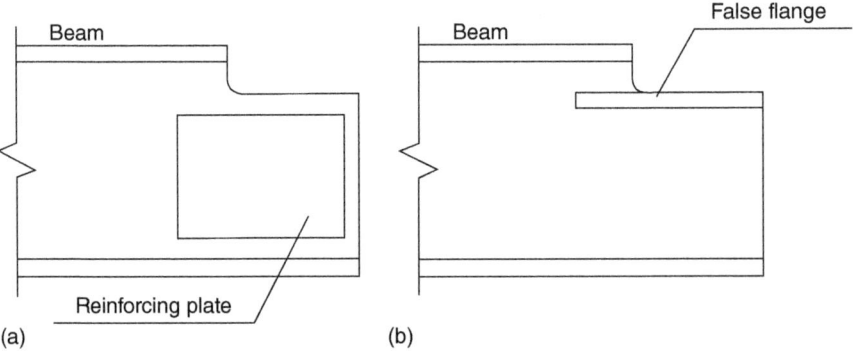

Figure 4.42 (a) Reinforcing plate and (b) false flange.

relevant horizontal forces or eccentricity. In these instances, a reinforcing plate (Figure 4.42a) can be welded to the web, taking into account that if the structures are later galvanized, suitable measures must be provided (Section 6.14).

If the beam is notched, Ref. [1] prescribes to horizontally extend the reinforcing plate beyond the limit of the notch for a length equal to the depth of the notch itself.

Another solution ("false flanges") is to weld plates perpendicular to the web in order to re-create flanges where the beam has been notched (Figure 4.42b).

4.6.2 Limit States to Be Considered

The design must deal with the following (taking into account eccentricities):

- Bolt shear
- Bearing for plate and beam web
- Block shear for plate and beam web
- Plate resistance
- Plate buckling (see Section 3.21, in particular Section 3.21.2)
- Secondary-member resistance taking into account bolt holes and possible notches/copes
- Local resistance of the main member
- Weld resistance.

4.6.3 Rotation Capacity

A method to check the rotation capacity in fin plates can be found in [13, 14]. The geometrical meaning is quite intuitive and consists of avoiding any contact between the parts.

The joint rotation capacity (Figure 4.43) can be considered satisfied if the following is verified:

$$z > \sqrt{(z-g)^2 + (l)^2}$$

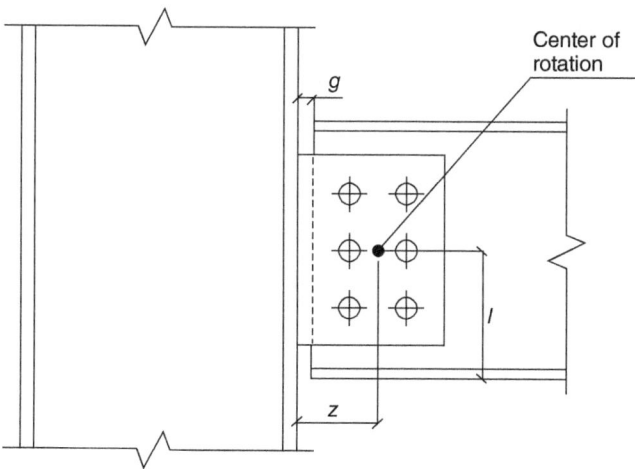

Figure 4.43 Symbols in the formulas to check the rotation capacity.

If the condition is not fulfilled, the rotation capacity (to then be compared with the design rotation requested by the calculation model) is

$$\Phi_{available} = \arcsen \frac{z}{\sqrt{(z-g)^2 + (l)^2}} - \arctg \frac{z-g}{l}$$

4.6.4 Measures to Improve Ductility

The limit states that may limit the redistribution of the forces are the failure of welds, the bolt shear, the rupture of the net section in any element (in shear or tension), the block shear, and the plate instability. If any of these limit states do not govern the design, the joint has good ductility.

4.6.5 Measures to Improve Structural Integrity

Fin plates/shear tabs can resist relevant horizontal forces in the same order of magnitude as the shear. The stress in the web of the primary member can become a limiting factor, though, and hence any stiffener connecting web and flanges in the primary element can be a solution. To numerically perform a check in the web of a column that has a welded shear tab transferring axial load, the web can be represented by a beam that is as deep as the web thickness and as long as the column depth minus the thickness of the flanges. Instead, Ref. [15] proposes (Figure 4.44) a less conservative and better performing plastic method that considers the subsequent formation of plastic hinges in the web of the column (whether this is an I- or H-shaped profile or a hollow steel section).

4.6.6 Design Example According to DIN

We try to design here a fin plate connecting two beams (HEB 300 and HEB 240) according to the old German standard (DIN 18800). A similar situation with heavy profiles might be found in an industrial mezzanine or in a commercial building where the beams should be as shallow as possible. Connections like fin plates apply more frequently to slender beams like IPE or UB, but this example shows that they can be acceptable also for larger beam types.

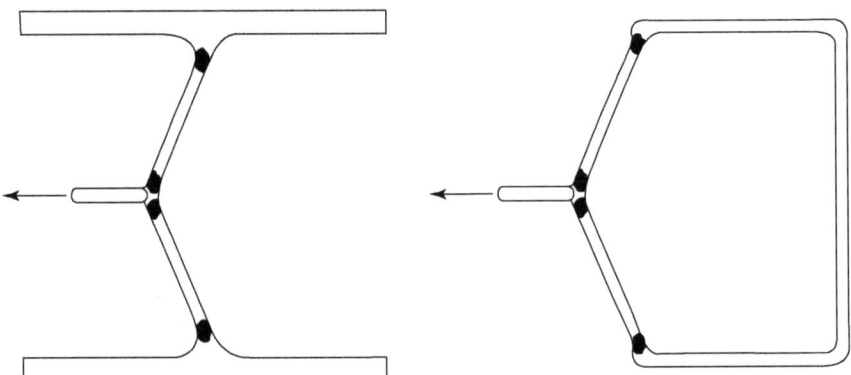

Figure 4.44 Plasticization of the column due to horizontal forces; see Ref. [15].

Following is the data in this problem:

- Main beam material: St 37-2 (equivalent S235, defined as per traditional German standards)
- Secondary beam material: St 37-2
- $N_{Ed} = 20\,\text{kN}$ (tension in the beam)
- $V_{\text{major Ed}} = 110\,\text{kN}$
- $V_{\text{minor Ed}} = 5\,\text{kN}$
- $M_{\text{major Ed}} = 0\,\text{kN m}$
- $M_{\text{minor Ed}} = 0\,\text{kN m}$

It is assumed that project specifications require class 10.9 bolts. By analogy with the beam material, we choose to take St 37-2 also for plates.

A tentative design configuration could be with 6M16 (three rows for each column), 55 mm pitch (about three times the hole diameter as typical "standard" in many joints). The geometry in Figure 4.45 involves a 3-mm overlap between the plate and the HEB 240 root radius, acceptable for the limits shown in Section 6.6.

It is considered, to avoid torsion and because this is likely the hypothesis in the calculation model, that the position of the theoretical axis of the connection is the same as the axis of the main beam.

With a geometry as in Figure 4.45 we have $e_{\text{bolt group}} = 55/2 + 30 + 35 = 92.5\,\text{mm}$.

4.6.6.1 Bolt Shear

The bolts must resist the following design bending moment coming from eccentricity and from the value given by the analysis model (0 in this case):

$$M_{\text{ecc,d}} = (e_{\text{bolt group}} V_{\text{major Ed}}) + M_{\text{major Ed}} = 110 \times 92.5 + 0 = 10\,175\,\text{kN mm}$$

A possible simplified approach (see Figure 4.46) for quick hand calculations might be to consider only the four external bolts as taking the bending moment:

- $n_{\text{bolts}} = 2$ (number of bolt pairs considered)
- $d = 123\,\text{mm}$ (distance between bolts in each pair)
- $V_{\text{ecc,d}} = \dfrac{M_{\text{ecc,d}}}{n_{\text{bolts}} d} = 10\,175/(2 \times 123) = 41\,\text{kN}$
- $\theta = 27°$ (angle that the shear, due to eccentricity, forms with the horizontal)

The maximum shear per bolt is obtained by combining the following actions:

$$V_{\text{s+ecc,d max bolt}} = \sqrt{\left(\frac{V_{\text{major Ed}}}{n_{\text{bolt}}} + V_{\text{ecc,d}} \sin\theta\right)^2 + \left(\frac{N_{\text{Ed}}}{n_{\text{bolt}}} + V_{\text{ecc,d}} \cos\theta\right)^2}$$

$$= \sqrt{\left(\frac{110}{6} + 41 \sin 27\right)^2 + \left(\frac{20}{6} + 41 \cos 27\right)^2}$$

$$= \sqrt{37^2 + 40^2} = 54\,\text{kN}$$

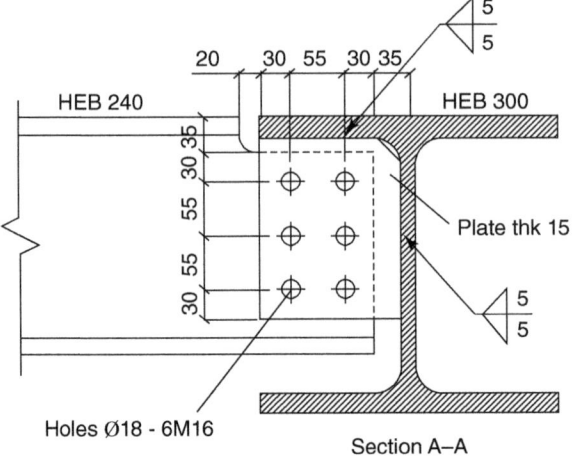

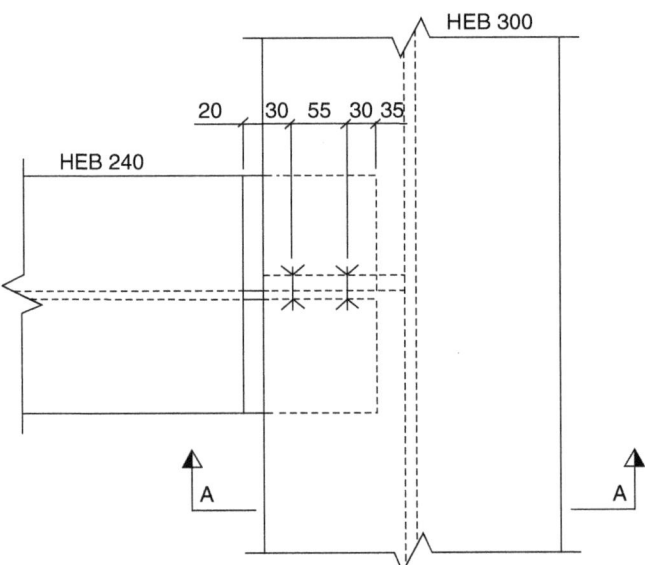

Figure 4.45 Possible configuration, to be checked.

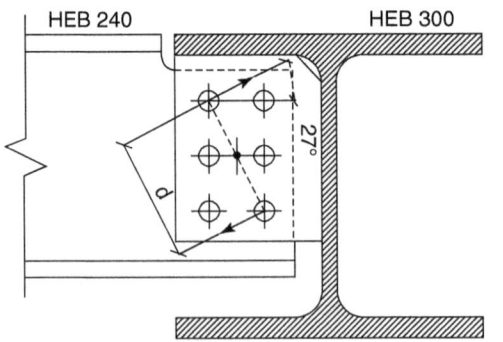

Figure 4.46 Simplified model that considers only the two external bolt pairs.

This will be compared with the bending shear resistance of each bolt, which can be calculated as

$$\tau_{a,Rd} = \frac{\alpha_a f_{u,b,k}}{\gamma_M} = 500 \text{ N mm}^{-2}$$

$$V_{Rd\ 1\ bolt} = \tau_{a,Rd} A n_{shear\ planes} = 79 \text{ kN}$$

$$\frac{V_{s+ecc,\ d\ max\ bolt}}{V_{Rd\ 1\ bolt}} = 0.68 \leq 1$$

Taking into account all six bolts, we would get (results from the software SCS) a 65% ratio with the elastic method and 55% with the instantaneous center of rotation. It should be noted that the width of HEB 300 is over the width limits (254 mm) of the SCS demo version, so if the user does not have a license, simulating the example results in a final sketch that is slightly different, with a reduced width for the primary member. This though has no effect over results since there is no check in the primary beam depending on its width except the top weld (and, very marginally, the bearing since the geometry is a little different).

4.6.6.2 Bearing

For the bearing check, interpolating from [16] (formulas in Chapter 3) we get $\alpha_1 = 1.53$. Then, the bearing design resistance stress according to DIN becomes

$$\sigma_{l,Rd} = \frac{\alpha_1 f_{y,k,pl}}{\gamma_M} = 335 \text{ N mm}^{-2}$$

and the resistant forces for the plate and the beam web can be obtained.

For the vertical forces, the total thickness of the plate(s) in the connection is

$$t_{pl\ tot} = t_{pl} n_{fin\ plates} = 15 \text{ mm}$$

The design shear limit for bearing is then, for each bolt,

$$V_{l,Rd} = \sigma_{l,Rd} t_{pl\ tot} d = 335 \times 15 \times 16 = 80 \text{ kN}$$

Comparing the value with the design action yields

$$V_{l,Ed} = \frac{V_{major\ Ed}}{n_{bolts}} + V_{ecc,d} \sin\theta = 37 \text{ kN}$$

$$\frac{V_{l,Ed}}{V_{l,Rd}} = 0.46 \leq 1$$

For the horizontal forces

$$V_{l,Ed} = \frac{V_{major\ Ed}}{n_{bolts}} + V_{ecc,d} \cos\theta = 40 \text{ kN}$$

$$\frac{V_{l,Ed}}{V_{l,Rd}} = 0.50 \leq 1$$

With SCS the results are 45% and 47%, respectively, with the elastic method and 52% and 52% with the instantaneous rotation method, which has bigger components along the axes.

For the bearing in the beam web (usually quite penalizing since the web is thin and it cannot be easily changed as is possible with the plate), the design bearing resistance per bolt is

$$V_{l,Rd} = \sigma_{l,Rd} t_{pl\ tot} d = 335 \times 10 \times 16 = 54\,\text{kN}$$

Action values are the same as above:

- Vertically $37/54 = 0.69 \leq 1$
- Beam web stressed by horizontal forces $40/54 = 0.74 \leq 1$

With SCS we have about 77% (both) as exploitation ratio with the instantaneous center method and 67% and 71% with the elastic method.

4.6.6.3 Block Shear

The block shear pattern that governs the design is the one shown in Figure 4.47 when there is a notch. The plate would have a similar failure line but it is thicker than the web (the material is the same), and hence it is useless to also check the plate.

It must be noted that DIN rules do not provide guidance to calculate block shear, so we choose to follow the EC method.

Let us calculate the reference area, starting from the vertical area:

$$A_{vert} = (p_{vert}(n_{rows} - 1) + a_{vert\ beam\ top} - d_0(n_{rows} - 0.5))t_{web\ tot.\ sec.}$$
$$= (55 \times 2 + 30 - 18 \times 2.5) \times 10 = 950\,\text{mm}^2$$

$$A_{horiz} = (p_{horiz}(n_{cols} - 1) + a_{horiz\ beam} - d_0(n_{cols} - 0.5))\,t_{web\ tot.\ sec.}$$
$$= (55 \times 1 + 30 - 18 \times 1.5) \times 10 = 580\,\text{mm}^2$$

The shear is clearly prevalent over the axial force and the block shear resistance can be evaluated as (we take $\gamma_M = 1.1$ as coherent with DIN)

$$V_{eff,Rd} = 235 \times 950/(\sqrt{3} \times 1.1) + 0.5 \times 360 \times 580/1.1 = 212\,\text{kN}$$

which, compared with the design shear action of 110 kN, gives a 52% design ratio.

If the geometry of the plate and the beam web are different but with similar thicknesses, the same formula can be applied for the plate. In addition, since the

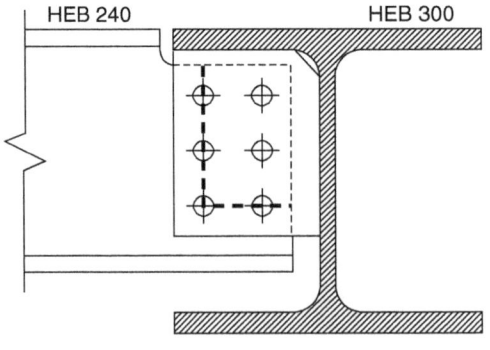

Figure 4.47 Beam web block shear pattern governing the design.

standards do not provide guidance for combined actions, it is safe to invert the shear and tension area and check the resulting value against N if the axial action is similar in value to the shear (not needed here). In fact, SCS runs several checks of different possible block shear patterns in the plate and in the beam web. For our case, where we perform only manual checks, some engineering judgment is suggested to narrow down the possibilities.

4.6.6.4 Plate Resistance

If we had a plate welded to only the web of the primary beam as in Figure 4.48, it would be necessary to evaluate the maximum eccentricity of the plate since this could be, depending on the connection axis location, at the primary connection or at the bolt group. If the maximum eccentricity is at the bolt group, usually the center of the bolt group is considered; however, taking the far bolt vertical row is more conservative (probably too much).

In our example, having welded the plate to the top flange too, any moment is negligible, even shear and tension.

From SCS we notice that if the plate was welded only to the web, its stress would have been significant: 53% for strong-axis moment, 27% for weak-axis moment, 51% for strong-axis shear, 2% for weak-axis shear, and 4% for axial actions.

4.6.6.5 Beam Resistance

Since the beam is notched, its local resistance must be assessed. Here, as in SCS, only a resistance check is performed; otherwise the discussion would extend to the entire structural analysis, not only the connections. Thus it is left to the engineer to evaluate for a possible interaction (luckily rare) with global instability issues in the beam (lateral and torsional buckling).

The notch in the section involves a reduction of the beam bending elastic modulus from 938 to 117 cm³. Taking as maximum eccentricity for the notched beam the distance between the connection axis and the edge of the notch (we prudently

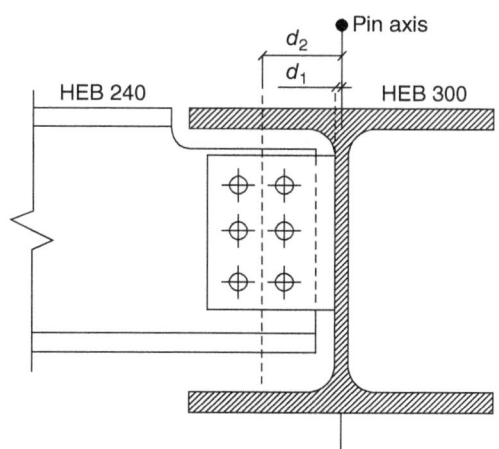

Figure 4.48 Possible different configuration.

neglect the notch radius), we get the moment

$$M_{Rd,str} \text{ of } 110 \text{ kN} \times 170 \text{ mm} = 19 \text{ kN m}$$

The elastic resistance is instead $117\,000 \text{ mm}^3 \times 235 \text{ N mm}^{-2}/1.1 = 25 \text{ kN m}$.

The DIN rules allow amplifying the elastic resistance of a factor that is the minimum between 1.25 and the ratio between plastic and elastic moduli. The net section is a T in our case and the ratio is over 1.25, so we take 1.25 and we obtain $25 \text{ kN} \times 1.25 = 31 \text{ kN}$ and thus an exploitation ratio equal to 61% (rounding brings 1% difference from the exact value in SCS).

The strong-axis shear design ratio (from SCS) is 45% (the notch does not affect the shear area much) while the other values can be neglected as presumable.

4.6.6.6 Plate Buckling

The plate being welded also to the top flange with the first column of bolts near the web of the primary beam, any buckling issue can be excluded.

4.6.6.7 Local Check for Primary-Beam Web

The resistant area is $0.9 \times 11 \times 170 = 1683 \text{ mm}^2$, hence a resistance of $1683 \times 235/(\sqrt{3} \times 1.1) = 208 \text{ kN}$, that is, a $110/208 = 53\%$ design ratio. Actually, the shear could also be considered going into the top flange (at least a part of it) and, therefore, this check could be omitted.

4.6.6.8 Welding

We prudently assign strong-axis shear and axial force to the web weld, also stressed by torsion (the plate and the secondary-beam web are not aligned as the view from top shows). We do not consider any eccentricity in the weld (which would be very small since the connection axis is at the primary-beam axis) because the weld to the top flange can take all the effects (as $\sigma_{\parallel}$). The top-flange weld will then be designed with the same thickness as the web weld. The weld on the top flange can take the weak-axis shear (here negligible for "hand" calculations) but it could also help taking some actions, in particular the axial load.

The symbols in the following equations are the ones used in DIN but the actions refer to the overturned (on the web of the primary) throat area (where a_s, l is the length):

$$\sigma_\perp = \frac{N_{Ed\,web}}{l \times 2a_s} = \frac{20\,000}{170 \times 2 \times 5} = 11.8 \text{ N mm}^{-2}$$

$$\tau_\perp = 0 \text{ N mm}^{-2}$$

$$\tau_\parallel = \frac{V_{major,Ed}}{l \times 2a_s} + \frac{M_{s,w,T}}{W_{R,w,T}} = \frac{V_{major,Ed}}{l \times 2a_s} + \frac{V_{major,Ed}(\text{pl.thk} + \text{web thk})/2}{(l \times 2a_s)(\text{pl.thk} + a_s)}$$

$$= \frac{110\,000}{170 \times 2 \times 5} + \frac{110\,000(15 + 10)/2}{1700(15 + 5)}$$

$$= 64.7 + 40.4 = 105.1 \text{ N mm}^{-2}$$

$$\sigma_{w,R,d} = \frac{\alpha_w f_{y,k}}{\gamma_M} = \frac{0.95 \times 235}{1.1} = 203\,\text{N}\,\text{mm}^{-2}$$

$$\sigma_{w,v} = \sqrt{\sigma_\perp^2 + \tau_\perp^2 + \tau_\parallel^2} = 106\,\text{N}\,\text{mm}^{-2}$$

Thus the exploitation ratio is 52% and, importantly, it does not govern the design. Also, a throat of 6 or 7 mm could be appropriate.

4.6.6.9 Rotation Capacity
We get

$$\sqrt{(z-g)^2 + l^2} = \sqrt{(30 + 55/2)^2 + 120^2} = \sqrt{57.5^2 + 120^2} = 133\,\text{mm}$$

when

$$\Phi_{available} = \text{arcsen}\frac{z}{\sqrt{(z-g)^2 + (h)^2}} - \text{arctg}\frac{z-g}{h}$$

$$= \text{arcsen}\frac{57.5}{127} - \text{arctg}\frac{42.5}{120} = 27 - 19.5 = 7.5°$$

which is to be compared with the value in the analysis model. The value (equivalent to 0.13 rad) is really large and so surely checked.

4.6.6.10 Ductility
The governing limit state is the beam web bearing, which means the connection has a good ductility since the less ductile limit states have lower exploitation ratios (the highest being the bolt shear at 65% calculated by the elastic method).

4.6.6.11 Structural Integrity
The structural integrity is to a certain extent already provided and checked since the design load combination has a relevant axial force in the beam. If an additional resistance wants to be reached, an easy way to accomplish this is to add a welded plate that connects web and flanges on the opposite side of the primary beam (unless a fin plate is actually present already when the connection is on both sides).

4.7 Double-Bolted Simple Plate

A type of connection that is interesting for ease of erection and because the secondary beam is just cut and (the holes in it) drilled (no welded parts in other words) is the one made by a double plate similar to a double-shear tab that is bolted on both parts: a bolted connection on the secondary beam, similar to a fin plate, and another bolted connection on the primary member (beam or column), more frequently to a stiffener welded to it (see Figure 4.49).

As mentioned, the erection looks easy because the length of the secondary member is less than the available clearance between the primary members. The

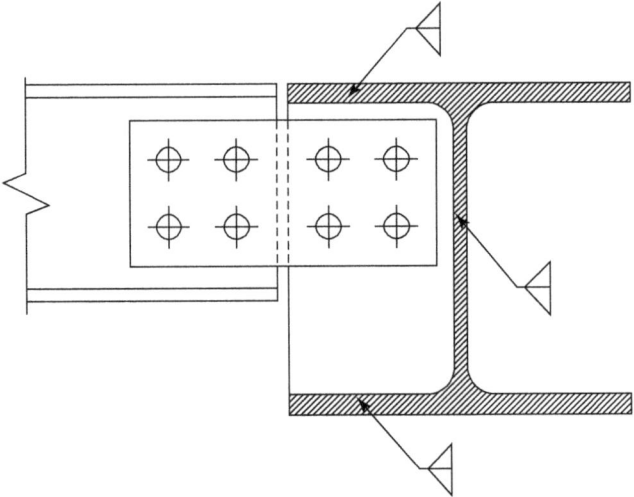

Figure 4.49 Double-bolted simple plate connection.

double-bolt group on each side makes the tolerance bigger because the available hole-to-diameter tolerance is now doubled too.

If the connection axis is, as it usually is, in the primary-member axis, the bolted group at the secondary is the more eccentric and, therefore, the more stressed. This translates into normally having the same or a bigger number of bolt columns than the group at the primary. Usually, the two bolt groups have the same bolt size, pitch, and gauge.

This kind of joint is often used for braces made of hollow steel sections, where a plate welded to the brace is connected to a plate welded in the beam/column corner with a pair of doubly bolted plates. Alternatively, though less efficient because only some parts of the braces are connected (shear lag phenomenon), the connection is also used for double channels or I/H-shaped sections (see Figure 4.50). The solution shown in Figure 4.50 is not very efficient (only the web is bolted) but it might be sufficient for a brace that is designed in compression, so the load transferred is small compared to the profile dimensions.

The connection has many aspects in common with the fin plate (refer to Section 4.6 for the details).

4.7.1 Rotation Capacity

The rotation capacity can be evaluated as for the fin plate (see Section 4.6.3) but is less of a problem unless the distance between parts is too small.

4.7.2 Ductility

The considerations given for the fin plate/shear tab also apply here.

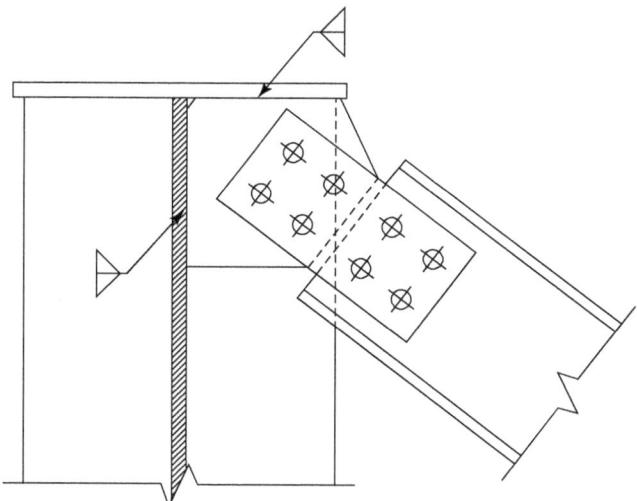

Figure 4.50 Double plate to connect an I- or H-shaped brace (HE, IPE, W, UB, UC sections).

4.7.3 Structural Integrity

The discussion for the shear tab is also valid here: The double simple plate can resist remarkable tension/compression actions, in the same order as the shear actions, and hence the guaranteed structural integrity is quite good.

4.7.4 Beam-to-Beam Example Designed According to Eurocode

Using EC, let us check (with $\gamma_{M0} = 1.1$, $\gamma_{M1} = 1.1$, $\gamma_{M2} = 1.25$) a connection between beams realized with a double-bolted simple plate. The main beam is IPE 360, the secondary is IPE 300, and they have the same "top of steel" (TOS). The beam material is S275. For the plates we will try to use S235, with class 8.8 bolts.

The loads are as follows (they might come from one single combination or they might be an envelope if this is not too penalizing):

$N_{Ed} = -30$ kN (compression in the beam)

$V_{\text{major Ed}} = 175$ kN

$V_{\text{minor Ed}} = 4$ kN

$M_{\text{major Ed}} = 0$ kN m

$M_{\text{minor Ed}} = 0$ kN m

The theoretical pin is assumed to coincide with the primary-beam axis because it is intuitive and quite likely it is also the same situation as the analysis model (though they might be different and still be acceptable).

Designing three rows of M20 as in Figure 4.51 could be a starting point to initiate the checks. Since the axis is on the main member, the external bolt group (here called group 1) will be more stressed than the one next to the primary (group 2), which is the reason for the initial choice of trying two columns of bolts in group 1.

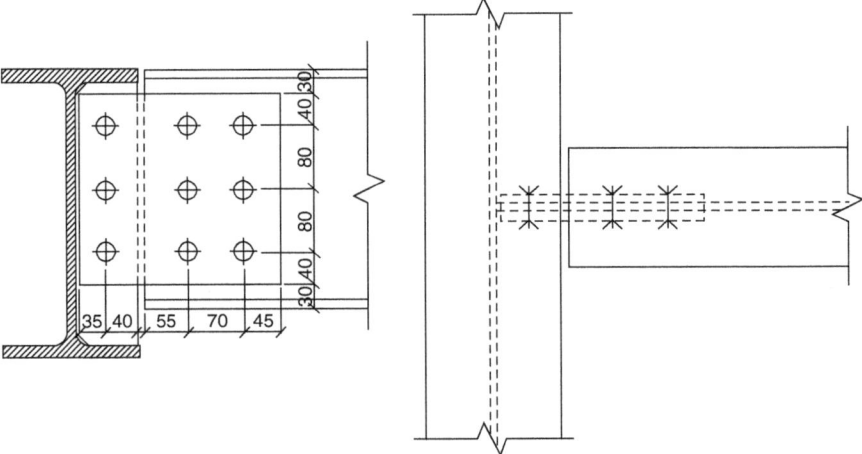

Figure 4.51 Connection geometry.

The group 1 eccentricity in this kind of joint is quite large and usually (as in this example) the most stressed limit state is the web bearing in the secondary, and thus it is recommended that the horizontal distance ("a horiz. beam") from the edge of external bolts be two to three times the hole (it is 55 mm as shown in Figure 4.51). The benefit in increasing the bearing resistance is more than the unfavorable effect of having increased the eccentricity (the distance from the axis is bigger and this causes a higher bending moment).

Note that an 8-mm-thick stiffener has been selected as the rib to be welded to the primary beam, necessary to then bolt the plates: this means a minimum difference with the thickness of the beam web (7.1 mm). If the difference were more (now 0.45 mm on each side is irrelevant), packing plates might be used.

4.7.4.1 Bolt Shear

Group 1 bolts have a moment given by the eccentricity that is equal to (the distance between the beams is 10 mm):

$$M_{ecc,d} = e_{bolt\ group} V_{major\ Ed} = \left(\frac{170}{2} + 10 + 55 + \frac{70}{2}\right) \times 175 = 32\ 400\ \text{kN} \times \text{mm}$$

The design resistance per bolt results, conservatively using the net area, that is, considering the threads inside the shear planes,

$$V_{Rd,\ 1\ bolt} = \frac{\alpha_a f_{ub}}{\gamma_{M2}} A_S n_{shear\ planes} = \frac{0.6 \times 800}{1.25} \times 245 \times 2 = 188\ \text{kN}$$

Using the symbols given in the example in Section 4.2.1:

$$T_{dx} = -30/6 = 5\ \text{kN}$$
$$T_{dy} = 175/6 = 29.2\ \text{kN}$$
$$I_p = \sum_{i=1}^{n=6}(c^2) = (4 \times 87.3^2 + 2 \times 35^2) = 32\ 900\ \text{mm}^4\ \text{mm}^{-2}$$

$$T_{mx} = 32\,400 \times 80/32\,900 = 78.8 \text{ kN}$$
$$T_{my} = 32\,400 \times 35/32\,900 = 34.5 \text{ kN}$$
$$T(\text{bolt 3}) = \sqrt{((5 + 78.8)^2 + (29.2 + 34.5)^2)} = 105.3 \text{ kN}$$

This means a 56% design ratio (the result in SCS). The instantaneous center-of-rotation method would deliver 48% (by SCS, which highlights convergence error, and so it is better to rely on the elastic method).

For group 2, we get (by SCS) 44% with the elastic method and 40% with the instantaneous center.

4.7.4.2 Bearing

Let us start from the secondary-beam web. Evaluating resistance in relation to horizontal forces, we have $p_1 = 70$ mm, $e_1 = 55$ mm, $p_2 = 80$ mm, and $e_2 = 70$ mm, and hence minimum values are

$$\alpha_b = \min\left(1, \alpha_d, \frac{f_{ub}}{f_u}\right) = \min\left(1, \min\left(\frac{e_1}{3d_0}, \frac{p_1}{3d_0} - 0.25\right), \frac{f_{ub}}{f_u}\right)$$
$$= \min\left(1, \min\left(\frac{55}{3 \times 22}, \frac{70}{3 \times 22} - 0.25\right), \frac{800}{275}\right) = 0.81$$
$$k_1 = \min\left(2.5, 2.8\frac{e_2}{d_0} - 1.7, 1.4\frac{p_2}{d_0} - 1.7\right) = 2.5$$

With a prudential approach we compare the maximum action $(5 + 78.8) = 83.8$ kN with the minimum resistance

$$F_{b,Rd} = \frac{k_1 \alpha_b f_u t d}{\gamma_{M2}} = 2.5 \times 0.81 \times 410 \times 7.1 \times 20/1.25 = 94.3 \text{ kN}$$

and hence a design ratio of 89%.

From SCS, we recognize that it was effectively necessary to position the holes at 55 mm horizontally as previously discussed because, if this distance was, say, 35 mm, the limit state would not be verified (1.21 ratio).

By proceeding analogously for vertical forces we find a 57% ratio.

The bearing check is widely acceptable for the double plate (the combined thickness is $12 \times 2 = 24$ mm, much above the 7.1 mm of the web, although the material is S235 instead of S275). SCS gives us a maximum utilization ratio equal to 35%.

Now we check the plate welded to the main beam. SCS provides a ratio of 86%: the horizontal 59.2 kN force is opposed by a 69.1-kN resistance, obtained positioning the hole at 40 mm from the edge. Leaving 35 mm on the other side (edge of the two plates), we cannot design a larger value because otherwise we would have a clash with the weld. If we want to increase the resistance, we could use an S275 plate, the same as the beam (this would also improve the weld quality but it would give a potentially relevant fabrication issue in having designed plates of different quality). Being a ductile limit state, this 86% exploitation ratio is acceptable.

The double plate works at 32% maximum when loaded in the bearing by bolt group 2.

4.7.4.3 Block Shear

As a rule, there are no notches in this kind of connection so there are no problems for the secondary. The block shear coming from tension (tearing out the web as in Figure 3.13, that is, Case 1 of Figure 3.12) could be a concern, except that our combination only has compression. Also the plates are adequate (see SCS).

4.7.4.4 Plate Resistance

First, we have to localize the most stressed part of the plate. If the tension is uniform, the (strong-axis) bending moment varies depending on the connection axis location. The most stressed sections might be, generally speaking, the external, but the load is transferred by bolts, and therefore we consider the section at the bolt group center. This bending moment is then the same as the one evaluated as 32 400 kN mm. The corresponding resistance is

$$M_{Rd,plates} = t_{pl} \frac{h_{pl}^2}{6} \frac{f_{y,pl}}{\gamma_{M0}} n_{fin\,plates} = 12 \times 240^2/6 \times 235/1.1 \times 2 = 49 \text{ kN m}$$

which means a 66% design ratio.

The other actions (axial, shear in both directions, weak-axis moment) give negligible effects, as is clear by SCS, but the combined action gives 99% (which is acceptable for ductility as per comments above).

4.7.4.5 Beam Resistance

There is no weakening of the beam except the reduced shear area because of the bolt holes. The check, even conservatively considering the reduced area, is widely within limits: from a 47% ratio (without holes) the check goes to 58% (deducting the holes).

4.7.4.6 Plate Buckling

For the plate welded to the primary beam, any instability can be excluded, being welded also to the flanges.

4.7.4.7 Primary-Beam Web Local Check

The plate is full depth so the check is useless.

4.7.4.8 Welding, Ductility, and Structural Integrity

The only weld is between the primary beam and the stiffener needed to bolt the plates.

This weld is minimally stressed (being near the axis of IPE 360, the eccentricity is practically zero), and therefore the check can be omitted. A double fillet with a 3-mm throat (4 mm maximum considering that an 8 mm thickness is welded) over the web and the flanges seems sufficient to guarantee not only the resistance but also good ductility. For additional comments about ductility and structural integrity, refer to the considerations of the fin plate example (Section 4.6.6).

4.8 Shear ("Flexible") End Plate

In this section, we consider an end-plate connection (see Figure 4.52) that is used when the engineer does not want to rigidly connect two members (refer to Section 4.12).

For this reason, this kind of end plate is called "flexible" and it is different from the other one that is "rigid." When some assumptions are verified (see Section 4.8.1), this end plate is considered as a simple hinge. However, EC classifies it as semirigid and the rotational stiffness should be evaluated and inserted in the analysis model although, as explained in the following section, there are loopholes.

4.8.1 Variants and Rotation Capacity

In the classical hinge scheme, there are different approaches to give the connection a certain rotation capacity.

The NTC classical approach, as taught by [3], is to use a plate that is approximately not thicker than 10 mm and to put bolts only inside the flanges.

The Anglo Saxon traditional joint, in addition to the thickness limitation (maximum 12 mm thick according to [15]) does not weld the flanges to the plate, getting a configuration also called "header plate" (Figure 4.53) or "shear end plate."

The British Constructional Steelwork Association [15] allows the plate to be welded to the flanges if the thickness is small. The approach of [13], which explains the EC recommendations, suggests instead using it inside the flanges so as not to be considered semirigid. Thus, even with a rigid application of the EC, a header plate that is designed internally to the flanges can be considered as a pin connection without further analysis with rotational springs.

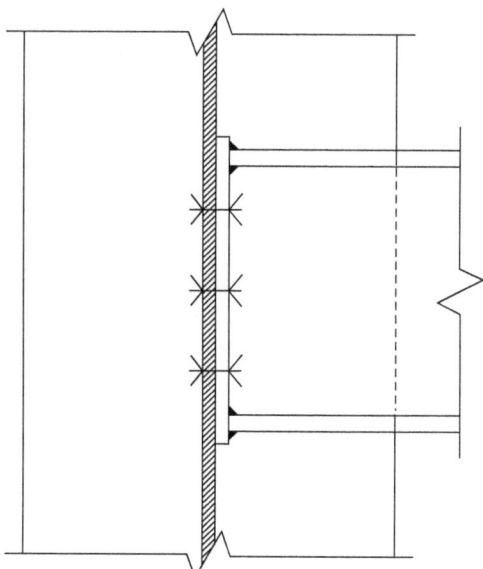

Figure 4.52 Possible connection to a column web.

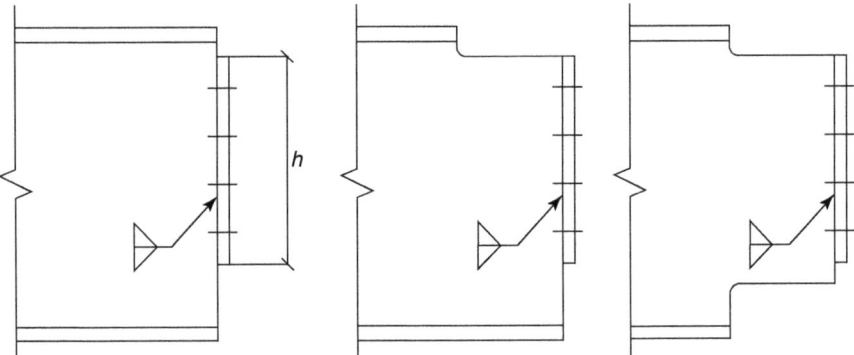

Figure 4.53 Header plate, possible configurations.

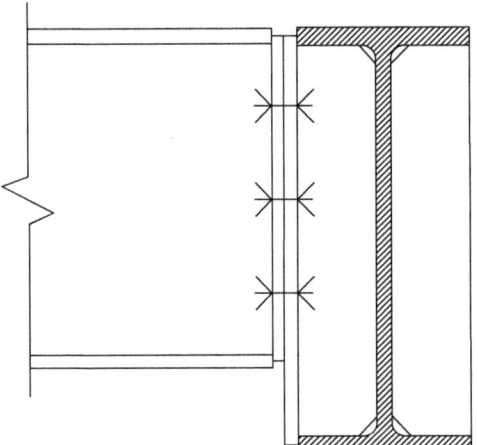

Figure 4.54 End plate connected to a plate welded externally to the primary beam.

A variant that is often utilized in practice but is quite difficult to find in manuals is that of Figure 4.54.

If the connection as in Figure 4.54 is on only one side and the pin axis is considered to be in the contact area of the plates, the primary beam is loaded by torsion (the engineer must verify its impact). If the pin connection is considered at the main beam, the plates and bolts especially will be loaded by an eccentricity moment that the designer must take into account. When the joints are on both sides, the loads might balance the torsional effects and this might help the calculations when applicable. Notches are possible, obviously, in the secondary beam, on both the top and bottom.

The shear end plate, generally speaking, might be applied in several situations (e.g. beam-to-beam or beam-to-column joints) on both the strong axis (connection to the flange) and the weak axis (connection to the web).

This joint would not allow any clearance for erection in the secondary-beam axis direction and hence the secondary is normally shortened by 1 mm or so on each side to allow its insertion on-site and to account for some possible additional overthickness given by, say, galvanization.

4.8.2 Limit States to be Considered

The analysis must consider the following:

1. Bolts in shear, possibly also in tension;
2. Bearing for the plate and its counterpart on the main beam;
3. Block shear for the plate and its counterpart on the main beam;
4. Resistance of the plate and its counterpart on the main beam, in particular, shear and bending moment for the plate (possibly also axial forces) and shear for the main beam (possibly also tension and bending moment);
5. Shear local resistance of the secondary near the header plate, possibly also resistance to tension;
6. Weld failure.

With reference to point 5, when using a header plate the part of the secondary beam web where the actions converge to transfer the forces to the plate is shorter than the web depth and equal to the plate depth, and hence the web resistance to shear has to be checked by means of the equation (EC symbols)

$$V_{Rd,sec} = t_{ws} h_{pl} \frac{f_{ys}}{\gamma_{M0}\sqrt{3}}$$

Concerning instead the stress in the end plate, it must be noted that there is a bending moment given by the fact that the force goes by the secondary-member web to the plate, and then it "moves" on each side (so halved) toward the bolts and lastly to the primary section. The shear will therefore generate on the end plate, where V_{Ed} is the design shear and p_{hor} the bolt gauge (assuming one column of bolts each side), a moment equal to

$$M_{pl} = \frac{V_{Ed}}{2} \frac{p_{hor}}{2}$$

If there are more columns of bolts, the arm will be increased considering the geometric center of the bolts.

If the plate is welded to the flanges, the stress is negligible (the beam supports the plate), but if there is a header plate, some include this bending moment in the design (it rarely governs though).

Similarly, the bending moment generated by an axial load can be evaluated (see example in Section 4.8.6).

Finally, we point out that welding the plate to the flanges (if the reference standards or design manuals allow it) will also improve both bearing (slightly) and block shear (remarkably).

4.8.3 Rotational Stiffness

According to the assumptions and limitations in Section 4.8.1, the connection is considered as a pin. If, for cases where the hypotheses are not verified and the rotational stiffness must be taken into account as per EC, see Section 4.12.

4.8.4 Ductility

As common to many joint types, the ductility can be improved by full-strength welds and not allowing the bolt shear to govern the design.

4.8.5 Structural Integrity

The capacity of a shear end plate in resisting horizontal forces is less than that of fin plates and clip angles, unless the thickness of the plate is increased, but this might compel to classify the end plate as rigid or semirigid as we discussed.

In order to satisfy structural integrity requirements [13] proposes for at least one side of the connection (i.e. the plate welded to the secondary on one side or the flange, web, or plate on the primary member on the other) that the following be verified:

$$\frac{d}{t_p} \geq 2.8 \sqrt{\frac{f_{yp}}{f_{ub}}}$$

where, as before, d is the bolt diameter and f_{ub} is the bolt ultimate tensile strength, while t_p and f_{yp} represent the thickness and the yield strength of the plate welded to the secondary or the web/flange/plate (depending on the type of connection) of the primary. Note that this criterion is also a measure of the ductility and rotation capacity of the joint. Since this check translates into keeping some additional capacity for opposing tension forces, Ref. [13] recommends designing the bolt group in shear by considering maximum 80% of its shear resistance (so that there is room for bolt tension).

The recent [17] proposes also some formulas (see Example 5.5 in [17]) that can be adopted for a check of the tie force according to AISC.

4.8.6 Column-to-Beam Example Designed According to IS 800

This example will design a column-to-beam connection according to the Bureau of Indian Standards LSD IS 800 [18] with Indian profiles on the column weak axis, that is, with a shear end plate bolted to the column web.

The column is ISHB250 and the beam is ISMB350. Usually, fin plate type connections are used for slender beams but shear end plates might be needed if the shear is high or simply because the fabricator recommends that kind of joint:

- Column and beam material: E300 IS 2062
- Plate material: E250 IS 2062
- $N = 15\,\text{kN}$ (tension for the beam)
- $V_{major} = 330\,\text{kN}$
- $V_{minor} = 7\,\text{kN}$
- $M_{major} = 0\,\text{kN m}$
- $M_{minor} = 0\,\text{kN m}$

After verifying that the erection space and sequence allow the beam to be inserted inside the column without notches or similar, a possible geometry is as given in Figure 4.55. We decide to weld the plate only to the web (header plate)

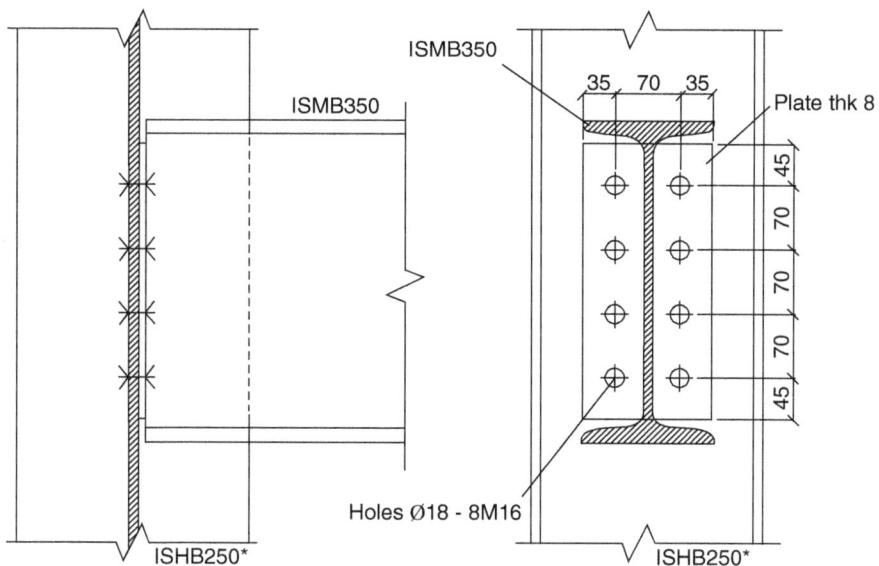

Figure 4.55 Header plate connection possible configuration.

to avoid semirigid joint considerations (although Appendix F in [18] is not clear about it, we have seen from many references that it is common to consider it as a simple connection). The plate chosen is 8 mm thick, similar to the beam and column webs and not too thick in order to help rotation if possible. Again, though, a rigid approach would mean also checking the connection rigidity.

4.8.6.1 Bolt Resistance

Even considering the threads inside the shear section, each bolt can resist (see Section 10.3.3 in [18]) $800 \times 157/(\sqrt{3} \times 1.25) = 58$ kN, and so $8 \times 58 = 464$ kN and a ratio of $\sqrt{330^2 + 15^2}/464 = 330/464 = 71\%$ (which looks acceptable since we know that bolt shear failure is not ductile so it is best if it is not too high and if it does not govern, as we will now check).

The tension request is negligible – from SCS, choosing a very simple method like the neutral axis, not through the center of gravity (see Section 4.3.1), it is only 6% of the capacity.

Let us then notice that the eccentricity is also next to zero (if we take the column axis as the connection axis, the eccentricity is half the web thickness, i.e. less than 5 mm).

4.8.6.2 Rotation Capacity and Structural Integrity

Let us decide to follow the instructions in [13] (note that IS 800 and EC are similar) and we deduce that the thicknesses are enough to guarantee rotation capacity and structural integrity:

$\frac{d}{t_p} \geq 2.8\sqrt{\frac{f_{yp}}{f_{ub}}}$, that is, $\frac{16}{8} \geq 2.8\sqrt{\frac{250}{800}}$ and so $2 \geq 1.56$, which is verified for the plate. It would be verified (though not necessary) for the column web too, where we have $\frac{16}{8.8} \geq 2.8\sqrt{\frac{300}{800}}$, and so $1.82 \geq 1.71$, which is acceptable.

4.8.6.3 Bearing

The plate bearing in the vertical direction (the horizontal shear is negligible) has $k_b = \min\left(1, \frac{e}{3d_0}, \left(\frac{pe}{3d_0} - 0.25\right), \frac{f_{ub}}{f_u}\right) = 0.83$.

Comparing the action $330/8 = 41$ kN with the resisting value $= \frac{2.5 k_b f_u t d}{\gamma_{mb}} = 2.5 \times 0.83 \times 410 \times 8 \times 16/1.25 = 87$ kN, we have a 47% exploitation.

The column bearing exploitation similarly results (from SCS) in 33%.

4.8.6.4 Block Shear

For block shear, setting $A_{tn} = 2(35 - 18 \times 0.5) \times 8 = 416$ mm², $A_{tg} = 2 \times 35 \times 8 = 560$ mm², $A_{vn} = 2(70 \times 3 - 18 \times 3.5 + 45) \times 8 = 3072$ mm², and $A_{vg} = 2(70 \times 3 + 45) \times 8 = 4080$ mm², the resistance is (from Section 6.4 in [18]) min($0.9 \times 410 \times 416/1.25 + 250 \times 4080/\sqrt{3}/1.1$, $0.9 \times 410 \times 3072/\sqrt{3}/1.25 + 560 \times 250/1.1$) = min(658, 651) = 651 kN, about twice the action of 330 kN.

4.8.6.5 Plate Check

We verify the shear in the plate, conservatively taking the net area (actually the code would also allow the gross area): $8(300 - 18 \times 4) \times 250/(1.1 \times \sqrt{3}) = 239$ kN. It has to be checked against half shear since it stresses the two sides of the plate and, therefore, the ratio is $165/239 = 69\%$.

On the weak side (from SCS) the design ratio is 5% only.

As we saw in Section 4.8.2, the moment given by shear forces for header plates is $330/2 \times 70/2 = 5.8$ kN m while the plate resists (elastically) in bending (strong axis): $8 \times 300^2/6 \times 250/1.1 = 27$ kN m, is obviously adequate (21%).

For the axial force, choosing a simplified method and multiplying half the axial force (the one acting on one side) by the horizontal distance between the bolts and the beam axis ($70/2 = 35$), we get a moment equal to $15/2 \times 35 = 0.26$ kN m. The plate can elastically (conservative) resist on the weak axis $300 \times 8^2/6 \times 250/1.1 = 0.73$ kN m (ratio 35%). Using SCS, the best way to calculate the stress is by the T-stub method (EC–BS option recommended) and the exploitation value becomes about 9% only.

4.8.6.6 Beam Shear Check

Since we are using a header plate, we check the shear only considering the minimum between the web depth and the plate depth: $8.1 \times \min(300, 350 - 2(14.2 + 14)) \times 300/(1.1\sqrt{3}) = 374$ kN for a ratio of $330/374 = 88\%$.

4.8.6.7 Column Resistance

The shear locally is resisted by the web (37% by SCS). For the axial force (actually also very little) transferred by the beam, it is taken by the central horizontal stiffener (Figure 4.56). Probably, the stiffener would have not been a good solution (it would have increased the stiffness too much) if, in the structural integrity check previously made, only the column web satisfied the inequality (and the plate did not).

Figure 4.56 Final design.

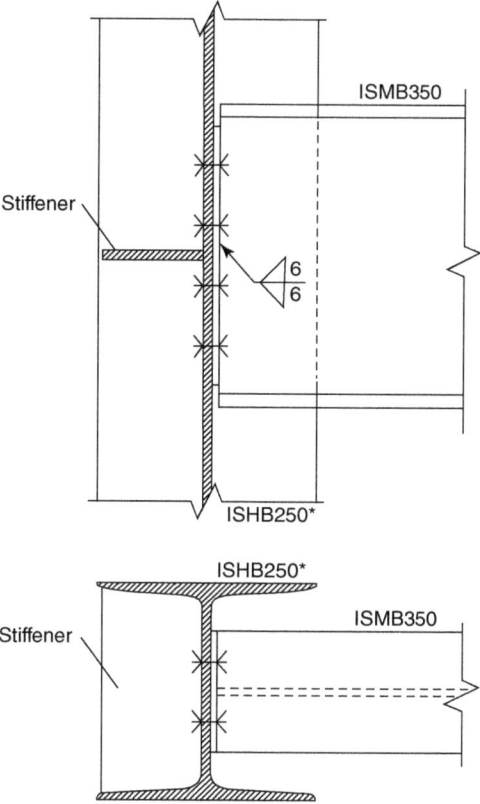

4.8.6.8 Welds
If the nominal size of the weld (and so the leg, which is the reference dimension in IS 800) is 6 mm, the total throat thickness is similar to the beam web thickness and we have (from SCS) the weld working a little below 80%.

4.8.6.9 Conclusion
The secondary web shear governs, which seems acceptable.

4.9 Double-Angle Connection

This type of connection is widespread because it is easy to standardize and because it does not require any welding.

It can be used in beam-to-beam connections as well as in column-to-beam joints (see Figure 4.57).

The joint is clearly a pin/hinge (only very limited moments can be taken), also known, for example, as *clip angles* or *web cleat connection.*

All the considerations for fin plates and shear end plates also apply here for a design that follows EC; that is, stiffness must be checked or details have to be adopted to make sure that the rotation capacity is acceptable: for example, limited

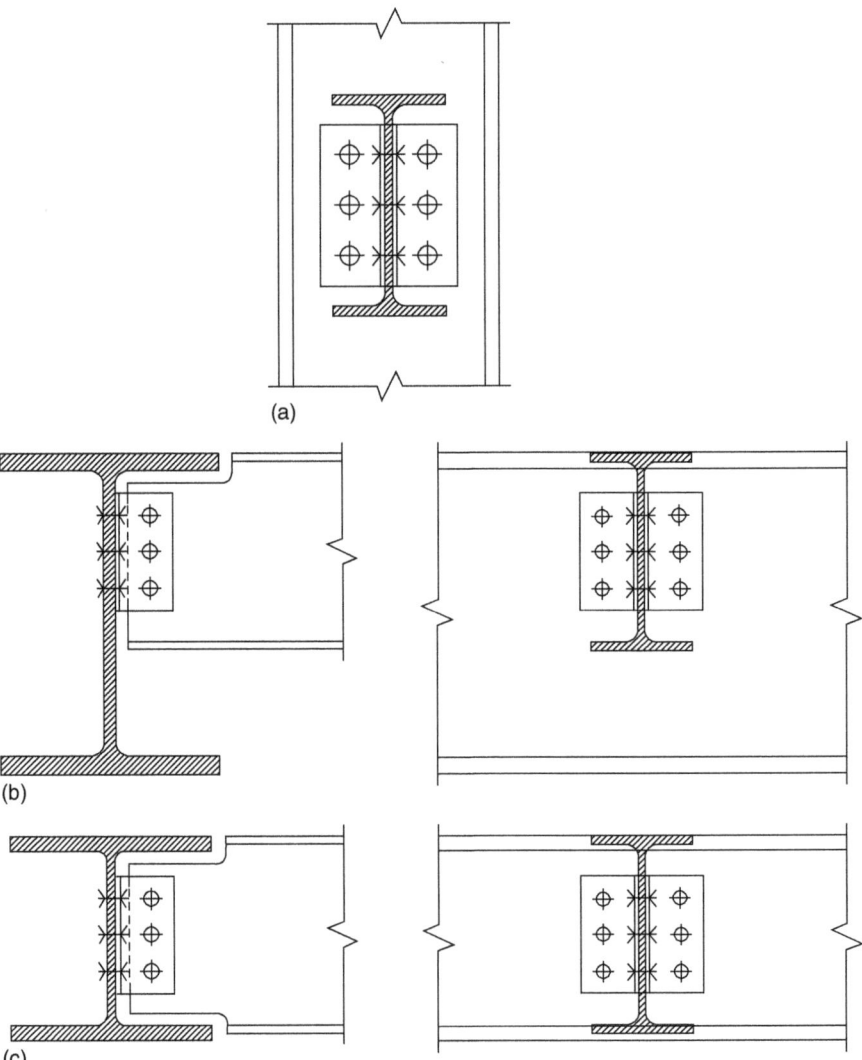

Figure 4.57 Examples (the first with a column-to-beam connection) of clip angles.

thickness for angles (possibly less than 10 mm, as in end plates) and a distance between members such that the rotation is allowed (as in fin plates; usually 10 mm is enough; AISC [1] set the limit to about 15 mm). Note that it is the angle profile that allows, deforming, the joint rotation. In order to help the rotation, it is also recommended that the distances between bolts not be short.

The angles can be fabricated by using hot-rolled angles (possibly with legs of different size) or by bending plates.

When the angles connect beams on both sides of the main beam, the erection can be dangerous because, to fix the second beam, the first one must be kept in position without being tightened. It is therefore sometimes convenient to add some seated connections as temporary support for erection operations.

4.9.1 Variants

It is also possible to weld the angles on one or both parts of the connection instead of using the bolts. This might make the shipping more troublesome (especially if the angles are welded to the secondary), and, although opinions differ on this, it could ease the erection. Generally speaking, though, the welded variant is not very popular.

Another (infrequent) alternative is to make the connection with only one angle, which is clearly less resistant (bolts work in only one shear plane) and causes more design issues (the bolt group on the primary member is also stressed by an eccentric moment in the bolt group plane) but there are evident advantages for erection. To eliminate the latter problem, T-shaped sections might be used instead of L-shaped sections.

4.9.2 Limit States to Be Considered

The design issues are similar to those discussed previously about the fin plate, and the reader is referred to Section 4.6, especially in relation to the secondary beam side, though here there are normally two resistant shear planes. The connection with the primary member shares many elements with the shear end plate (see Section 4.8) and the following must be evaluated:

- Bolt group, including eccentricities
- Actions in the angle
- Local actions on the main member, also weakened by bolt holes.

Compared to shear end plates, if the connection axis is taken at the bolt group in the secondary (not recommended but possible), the bolts also have an out-of-plane eccentricity (see Section 4.3). This assumption also brings a bending moment to the angles and the primary member (torsion, if it is a beam) and so does not appear to be the easiest way to check the joint.

4.9.3 Structural Integrity, Ductility, and Rotation Capacity

Here, the reader is referred to what has already been discussed for shear end plates and shear tabs. Generally speaking, double-angle joints have good ductility and can assure valid structural integrity due to the good capacity in horizontal deformation (the angles will stretch if overstressed), thereby allowing force redistribution. Since there are no welds, the only relevant fragile limit state to take care of is bolt shear.

4.9.4 Practical Advice

It is important not to position the bolts of the two groups too close to each other because the ones positioned first might hinder the insertion and tightening of the others. Section 6.4.1 provides some guidance on this.

The Steel Construction Institute [15] also recommends using angles that are at least 60% deeper than the secondary-beam depth.

4.9.5 Beam-to-Beam Example Designed According to AISC

The design check is similar to the one for the fin plate (and to the shear end plate for the parts connected to the primary member), and to differentiate the examples, the joint is dimensioned using design tables.

The connection to check is assumed to be part of a project for an American client (therefore AISC applies) but the steel is fabricated in Europe with European profiles. Our scope is to check that the proprietary tables used by the fabricator to dimension double-angle connections are also acceptable for the mentioned project. The fabricator, applying the "company tables," proposes to adopt 3 M20 for an IPE 400–IPE 300 (beam-to-beam) connection stressed mainly by a 120-kN design shear. Note a disadvantage when using design tables: there are usually significant assumptions, for example, taking for granted that axial load and/or weak-axis shear are negligible. Let us consider this simplification as acceptable here.

To check the proposal with the AISC tables (that are tailored for American sizes), we try to "translate" the available data: IPE 300 is 11.8 in. deep, so roughly 12 in. and 120 kN are 27 kips. Table 10-1 of [1] provides data for $3/4$-in. bolts (that have 19 mm as diameter) in groups of three bolts to connect W12 profiles (and others more or less deep so the approximation is fine). The tabulated values for the resistance of angles and bolts (the values already include resistance factors) are, prudentially considering A325 as the bolt class, 76.4 kips for a $1/4$ in. (about 6 mm) thickness and 95.5 kips for a $5/16$ in. (roughly 8 mm) thickness. For the angle material, the table assumes a 36-ksi (248-N mm^{-2}) yield value, quite similar to S235, and hence the values seem to be in the correct range.

The 6 mm thickness seems enough (even decreasing it by a 235/248 factor because of the yield strength) and the fact that increasing the thickness also increases the resistance suggests that the bolts in shear do not govern the design. Hence, 6 mm is chosen to guarantee good deformation capacity and ductility. It is also noted that summing the thickness of the two angles (the total is 12 mm) will go well over the web thickness of the IPE 300.

The values in the table have a vertical pitch of about 75 mm (3 in.) and minimum distances from the notch edge (vertically) of 32 mm. It is decided to adopt 70 mm as pitch (which slightly decreases the bolt group lever arm and thus its resistant capacity but the check margin is so ample that there is no concern) and 35 mm as the distance from the top edge for the angles (more than 32 mm and, therefore, safer). Consequently, an IPE 300 cope of 40 mm (more than 16 mm + flange thickness, which represents the minimum) seems a coherent choice (the bolt distance from the top edge becomes 40 mm). We also check that the angle will not overlap the root radius of the primary beam: the top of the steel distance will be $300/2 - (70 + 35) = 45$ mm, while for the IPE 400 it is 13.5 (flange thickness) + 21 (radius) = 34.5 mm.

For the secondary-beam web check, the tables provide reference for a 50-ksi yield material, which means approximately 345 N mm^{-2}. The tabulated result should therefore be reduced by multiplying it by 275/345, being the profile material S275 in this exercise. For standard holes in a beam with a top notch and a 45 mm horizontal distance of the holes from the free edge, the resistance is 200 kips/in., that is, 200 × 7.1/25.4 = 56 kips, widely above the required 27 kips.

Again from the AISC table, the bolt vertical axis is considered as 57 mm from the angle corner and we round this value to 60 mm: This worsens the eccentricity a little but it is feasible considering the ample margins (a 55-mm choice would be acceptable and more adherent to the table). The assumption for the bolt center distance from the free edge in the angles is 32 mm also horizontally, so 40 mm satisfies this. In the beam, the distance becomes 50 mm, higher than the 45 mm previously considered. Clearances seem good enough to guarantee the necessary access during erection (the bolts on the primary-beam web must not clash with the bolts on the secondary beam web).

The final discussed geometry is as in Figure 4.58. The angles have equal legs and each leg is 100 × 6 mm. Commercially, this does not seem a readily available profile and, therefore, it is decided to make it by bending plates. If desired, the above considerations might be changed to try to design a commercial L profile (maybe with unequal legs).

The AISC tables also give capacity values for the main beam. In this case, it is 526 kips/in., that is, 526 × 8.6/25.4 = 178 kips. Even when decreasing the result because of the different material (lower yield value as above), the safety margin is very wide.

If we reproduce the calculations by SCS, the governing limit state is the block shear (60%), in line with our considerations using the tables.

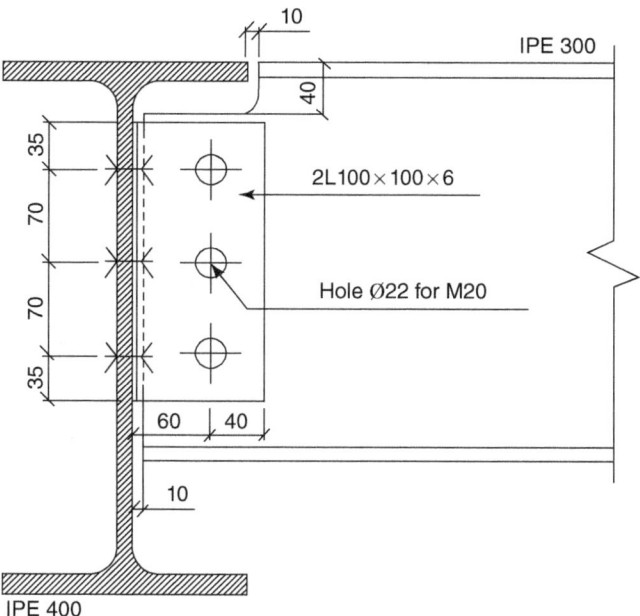

Figure 4.58 Final geometry.

In conclusion, the proposal is according to AISC rules and a design as in Figure 4.58 seems to optimize the resources with respect to ductility and cost-effectiveness.

4.10 Connections in Trusses

The connections in trusses (lattice girders) are usually either fully welded or bolted by using connections like fin plates (shear tabs) that are welded to the upper or lower chord and bolted to the diagonals/posts.

In order to make calculations, the instructions for welded connections in the first case or for fin plates in the second (similarly to braces) must be followed.

It is, however, necessary to make a few additional considerations related mainly to the connections involving angles, a widely used solution. If only one leg is connected, the bolt group axis is likely different from the angle axis of gravity, so a bending moment is supposed to be generated, given by the axial action times the distance between axes. This is the approach suggested also by [3]. Actually, only part of the angle transmits the force (*shear lag*; see Section 3.19.1) when only one leg is connected, and hence taking a bending moment as just described is probably too conservative.

4.10.1 Intermediate Connections for Compression Members

Members as double angles or channels in compression are commonly connected together to help stability (Figure 4.59). When this is needed (this is a design consideration related to members and, therefore, not in the scope of the book), there are two distinct situations as explained in detail by [3]. In the first case, the joint must not absorb shear forces but only perform a kinematic function that makes it impossible to buckle in the direction of the minimum radius of gyration. In fact, as the profiles are connected, they cannot buckle at the same time according to the minimum inertia axis (V–V, Figure 4.60) without distancing or interpenetrating. Buckling will therefore only happen according to X–X or Y–Y, which can be a substantial design advantage. This applies, for example, to double angles with

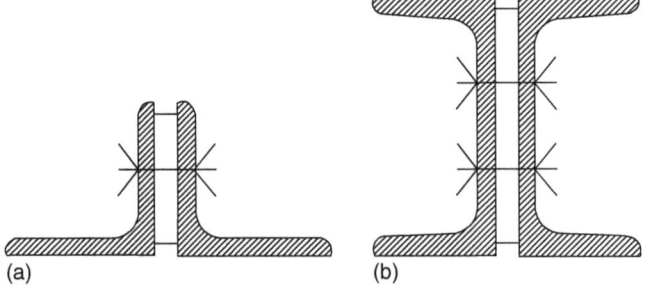

Figure 4.59 Examples of connections in angles (a) and channels (b).

Figure 4.60 Equal-leg angles, axes for slenderness calculations.

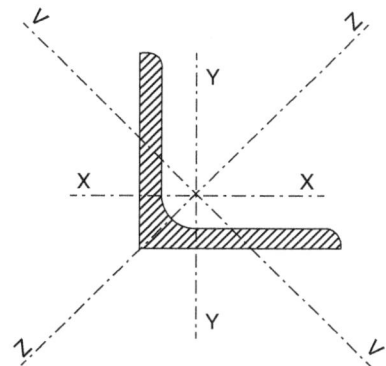

equal legs (Figure 4.59a) and the same buckling effective length in both directions. This is typical in trusses, so this situation applies in most cases. The connection is usually made by using a couple of bolts (likely the same size as the bolts in the end connections). The distance between consecutive intermediate connections is calculated so that its ratio with the minimum radius of gyration of the single angle (i.e. the gyration radius in relation to V–V as shown in Figure 4.60) is less than the slenderness of the single angle according to axis X–X (or Y–Y). AISC [1] prudentially recommends that the slenderness of the single angle not be more than 75% or 90% (depending on the shear deformation) of the two angles as a unit. Specifying the distance as 50 times the minimum radius of gyration is quite conservative in almost all cases (the global slenderness should be less than $50/0.75 = 67$ to make this slightly unconservative).

It is however different if the intermediate connections must guarantee that the two (angles might be even four) profiles have the same behavior as one single equivalent section. This is normally the case required when connecting two channels or two angles with unequal legs or in different schemes (two angles connected in a "butterfly" configuration, for example, mirrored with respect to the corner point so that they form an X-shaped section). In a similar situation, the connections (either intermediate or at the ends) must take shear forces without allowing movements inside the bolt hole (hence friction must resist or hole tolerances must be very limited). A comprehensive treatment of the forces to be considered in this kind of connection is given in [3] while not much exists in international standards. AISC [1] only provides some formulas of the equivalent slenderness (depending on friction or not) as well as tables with bearing capacity as a function of the intermediate links. Practically, also for this case, connectors at 50 times the minimum radius of gyration and bolts as discussed, similar to end-connection details, are normally seen. Again, the reader is referred to [3] for detailed calculation instructions. It is interesting to note what the old Italian standard UNI 10011 [19] used to prescribe in a simplified but effective way:

- Divide the compression member in at least three parts (therefore, at least two intermediate connectors).
- Connect profiles with at least two bolts along the member axis.
- Space the connectors a maximum of 50 times the minimum radius of gyration of the single member (actually 40 times for S355 equivalent material).

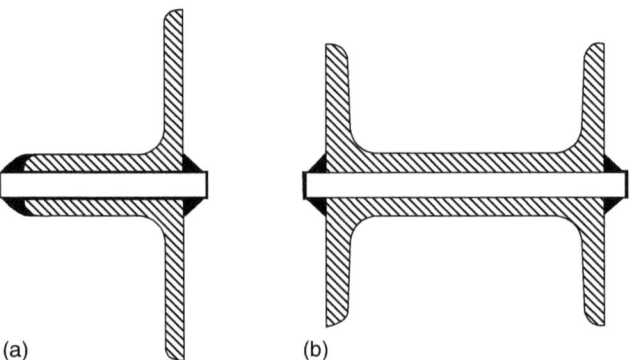

Figure 4.61 Welded intermediate connectors.

- Get the equivalent slenderness as the square root of the sum of the single-member slenderness and the composed member slenderness.

For the equivalent slenderness, recent instructions of [20] say, for example, that if the distance is less than 15 (!) times (70 for some butterfly configurations) the minimum radius of gyration of the single profile, the equivalent slenderness can be considered to be one of the composed member neglecting shear deformations while, with larger distances, these must be taken into account "using proven standards."

We finally note that connectors can possibly be realized by welding (Figure 4.61). In this case, the intermediate plate should be deeper than the members in order to ease operations (fillet welds are easily achievable).

4.11 Horizontal End Plate Leaning on a Column

Very few textbooks (possibly none) discuss what the neophyte sees as one of the top options in connecting a beam with a column, that is, welding a horizontal plate to the column and bolting it with the beam (directly if the bottom flange is wide enough or to a welded plate otherwise) laying on it. Relying on the direct support to transfer gravity forces and having bolts to resist other horizontal forces and possible bending moments provide an impression of simplicity and strength that draws interest. Even the erection, counting on the support, is facilitated.

The joint is efficient in transferring shear and tension/compression (similarly to a shear end plate) but care is needed (and this is the first reason why it is not classically much considered) when assessing the stability of the beam, restrained only on the bottom flange and hence not in good balance. Horizontal forces applied to the beam by secondary members or by plane braces might generate torsion. To help this aspect and improve the local strength, stiffeners can be welded (one central or a couple aligned with the beam flanges as outlined in Figure 4.62). The ribs however will increase the stiffness of the connection, which might partially invalidate (to check) a possible hypothesis of pin connection. At this point, the connection could be calculated as a rigid end plate (see the following section),

Figure 4.62 Beam leaning on column.

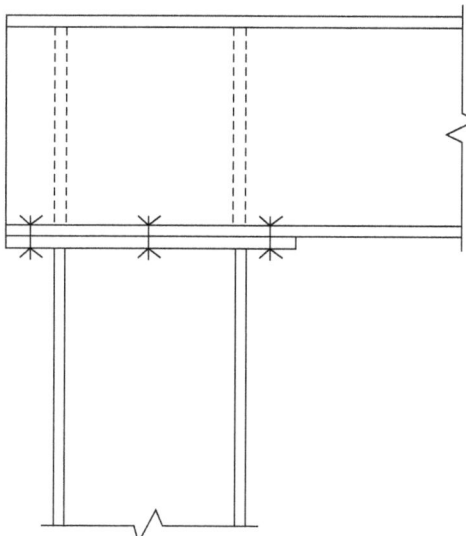

without overlooking the web panel shear check (see Sections 3.13 and 4.12.1), which here is actually in the beam web.

Generally speaking, we can then say that this kind of joint is not recommended for relatively important projects but it can be found in small jobs like mezzanines, agricultural structures, and small canopies.

It is different if the beam is on multiple spans and goes over the column but it is connected at the ends by, say, end plates (hence more stable): in this situation the beam is continuous and can transfer the bending moment, so the column might be used only as a simple support.

4.11.1 Limit States to be Considered

The joint check will take care of the following:

- Bolts in shear, possibly even in tension if stressed by a torsion or a bending moment transferred by the beam
- Bearing and block shear for plates
- Resistance of plates to shear and, if necessary, bending
- Local beam resistance
- Weld failure.

4.12 Rigid End Plate

The "proper" end plate, that is, the rigid one, is a critical joint, very useful since it can pass relevant bending moments, even up to the full strength of the beam. The most classic use is (Figure 4.63) in beam-to-column connections and end-plate splices that, when there is an angle between consecutive members, become the so-called apex connections. Another possible, less frequent, application is in

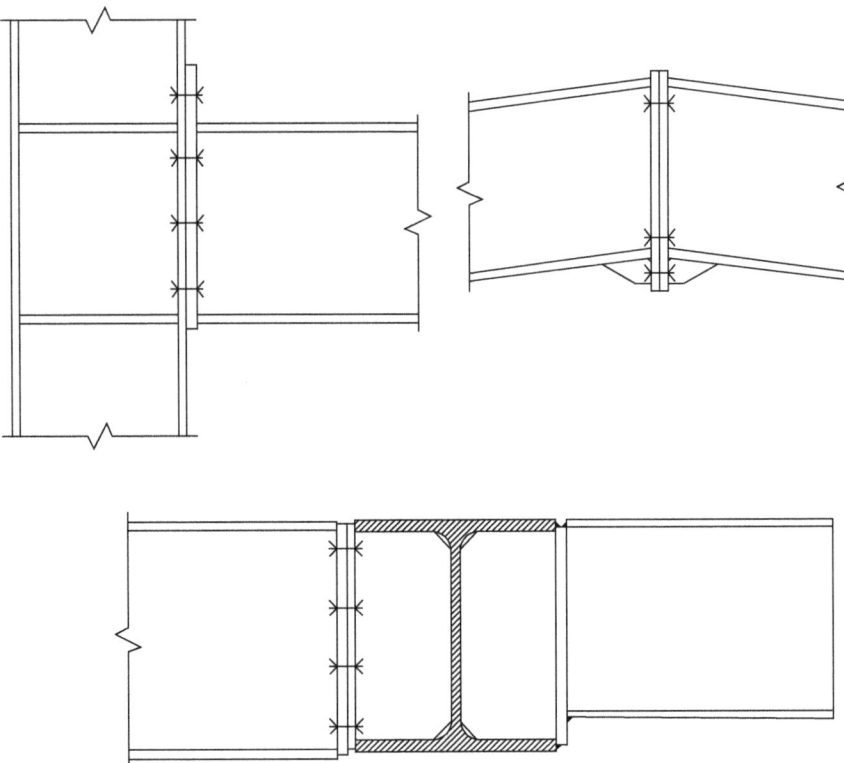

Figure 4.63 Different joints with rigid end plates.

beam-to-beam connections when the secondary beams come from both sides: in this case the rigid end-plate joint allows to take a bending moment, which might be necessary if there is a cantilever on one side or if it is needed to lower the beam actions (at the cost of a more expensive detail).

The end plate can be designed using a "flush" plate (Figure 4.64), in the sense that the plate does not extend over the beam flanges, when the bending moment in the joint is not remarkable. When the actions rise, it is then necessary to move to an "extended" type of end plate or to reinforce the beam end by increasing the beam depth, that is, adopting a "haunch" (see the considerations and checks in Sections 3.14 and 3.16). The basic concept of a haunch is that augmenting the beam depth we also increase the lever arm that is used to divide the bending moment in a tension on one flange and a compression on the other, so the forces decrease. The calculation scheme of an end plate is to transfer tension, near the flange in tension, through bolts and compression, near the compressed flange, by means of contact. The two forces, unless there is some additional axial action in the beam, are equal and opposite.

The calculation methods currently used by most international standards are based on plasticity because it describes the behavior of the joint better than that by classical elastic methods that rely on a triangular distribution of forces as in

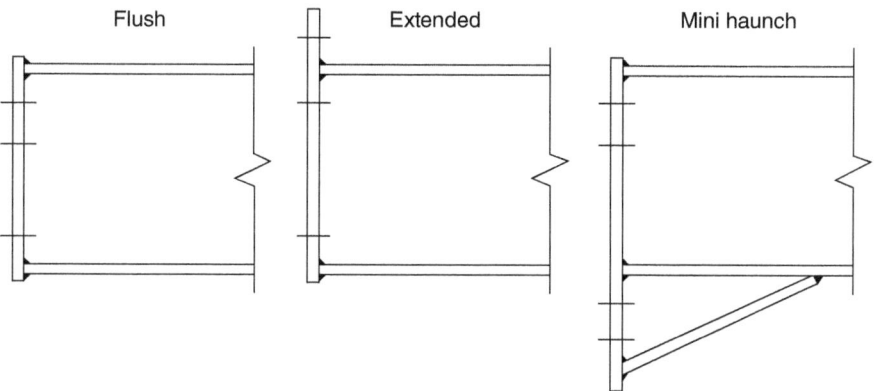

Figure 4.64 Some possible configurations.

Figure 4.65 Triangular (elastic) distribution of forces.

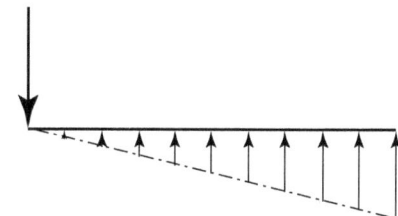

Figure 4.65. The reader is referred to the discussion in Section 4.3 about the simplified AISC methods (information about the AISC method in the design guides is in Section 4.12.12).

4.12.1 Column Web Panel Shear

If the portal frame is only on one side (top of Figure 4.66), the column web shear can be computed dividing the moment by the lever arm, which means it is numerically equal to the tension or compression. If the rigid end plate is on both sides and the bending moment is also the same with, say, both the tension flanges being the ones on top, the column web shear is zero. Instead, the shear is maximized when the bending moments on opposite sides have different signs, that is, tension on top on one side and tension on the bottom flange on the other. In this case, the shear is the sum in absolute value of the shears obtained on each side.

4.12.2 Lever Arm

Eurocode gives detailed instructions on the numeric value to consider in calculating the lever arm of the bending moment. In the compression zone the center of compression is usually taken in the middle of the compressed flange. It is instead taken as the center of the tension zone as follows:

- Middle of the flange in tension if the connection is fully welded
- Row of bolts in tension if there is only row of bolts in tension

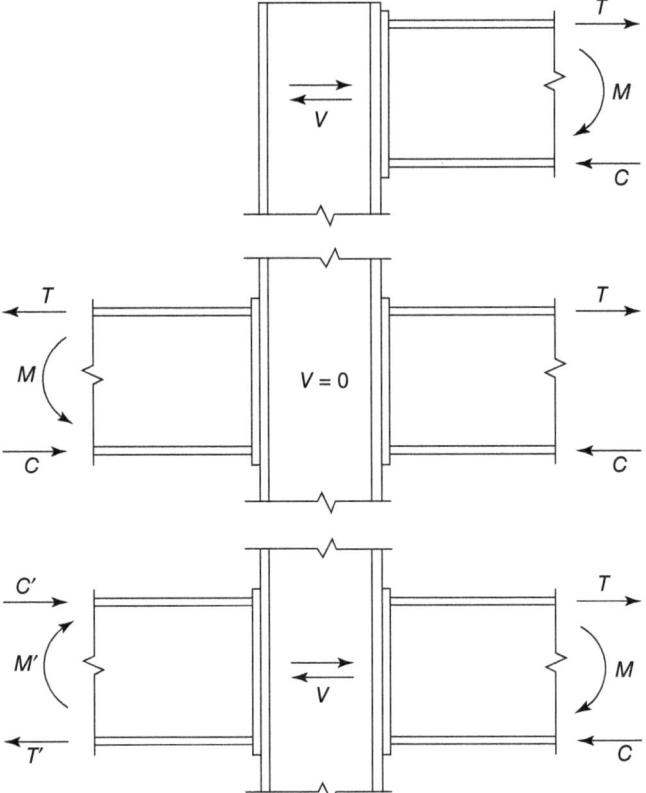

Figure 4.66 Column web shear, different cases.

- As an approximate value, a point midway between the farthest two bolt rows in tension (a more accurate value is determined by taking the lever arm as equal to z_{eq} obtained using the method given in Section 4.12.14, which usually requires a software like SCS as the procedure is quite long and tedious).

4.12.3 Stiffeners

It is likely that a rigid end plate, especially if on a column, will require some stiffeners. Taking a cue from the interesting discussion in [10], we show in Figure 4.67 some types of stiffeners and in Table 4.3 their usefulness.

It has to be noted that an alternative used in the United States for the supplementary web plates (called *web doubler plates* in the United States) is to weld a couple of plates at some distance from the web (as in Figure 4.68). This can help to avoid problems with galvanization (see discussion in Chapter 6) when this treatment is expected.

If necessary, the continuity plates (i.e. the horizontal stiffeners in the column that are the continuation of the beam flanges) can be realized, instead of full depth, only half depth, as ribs (with the result of not being effective against the web buckling according to [10, 21]).

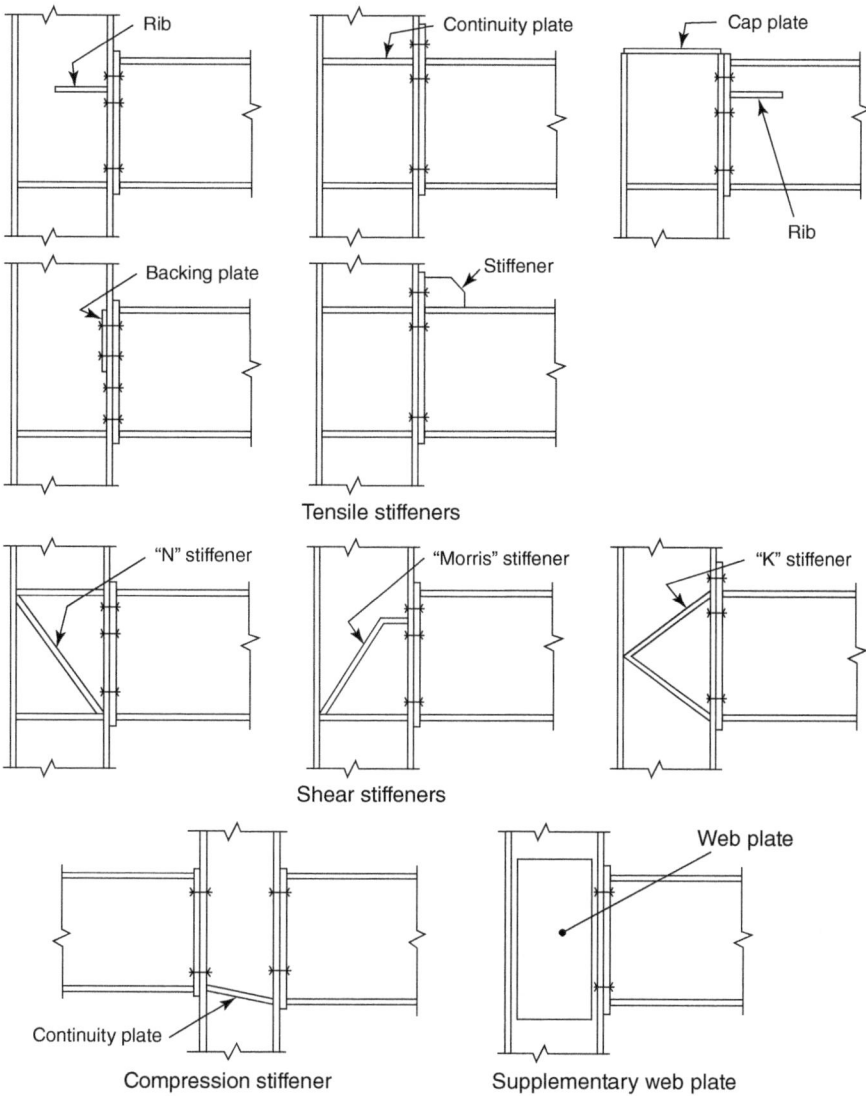

Figure 4.67 Reworking of a similar figure in [10] to show some types of stiffeners.

4.12.4 Supplementary Web Plate Check

The reader is referred to Section 3.13. For the welding detail, check Figure 4.69.

4.12.5 Check for Column Stiffeners in Compression Zone

The Steel Construction Institute [10] provides precise directions on how to size stiffeners (precisely, plates welded to column web and flanges), such as continuity plates and diagonal stiffeners in this (compression) case and in the next (tension and shear) two of the following Sections 4.12.6 and 4.12.7. Without going into

Table 4.3 Stiffener usefulness in helping limit states.

Stiffener type	Improved limit states						
	Column					Beam	
	Web in bending	Web in tension	Web in shear	Web in compression	Web buckling	Plate in bending	Web in tension
Flange backing plate	X						
Horizontal continuity plate	X	X		X	X		
Local stiffener	X	X		X		X	X
Plate welded to the web		X	X	X	X		
Diagonal (N, K) stiffener			X				
"Morris"-type stiffener	X	X	X				

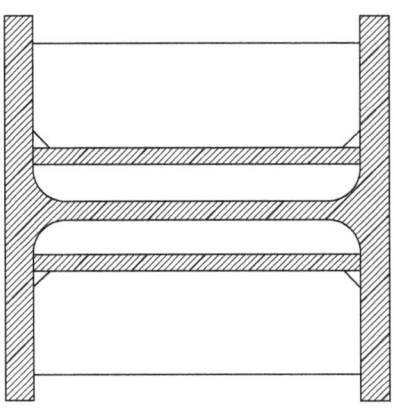

Figure 4.68 Possible alternative for web doubler plates on both sides. Source: From Ref. [1].

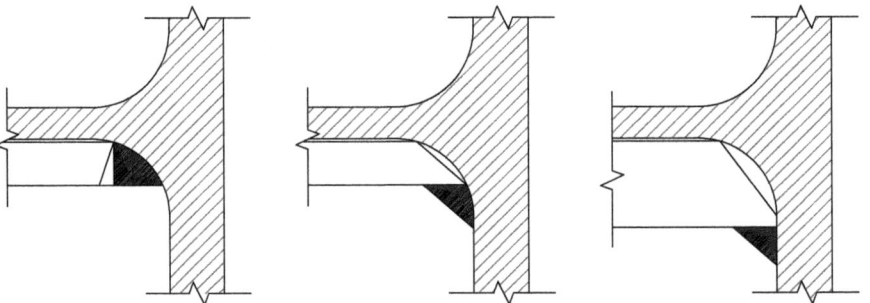

Figure 4.69 Welding details from [21].

too much detail because it would also be necessary to consult the tables of the BS, consider doing the following:

- Use a plate with a thickness not less than 1/19 of the width of each stiffener (the stiffener usually has a width that is slightly less than half the column flange width).
- As the stiffener effective width take a maximum 13 times the thickness.
- Design stiffeners to resist at least 80% of the compression force.
- Consider possible buckling issues in thin stiffeners of deep columns; using a thickness smaller than the column web is not recommended.

The initial tentative thickness (as in the tension zone discussed in the next section) is preferably similar or slightly bigger than that of the connected beam flanges.

4.12.6 Check for Column Stiffeners in Tension Zone

Following again [10], the net area of the stiffener (A_{sn}) must be

$$A_{sn} \geq \max \left(\frac{F_{ri} + F_{rj}}{f_y} - (L_t t_{wc}), \frac{m_1}{f_y} \left(\frac{F_{ri}}{m_1 + m_{2L}} + \frac{F_{rj}}{m_1 + m_{2U}} \right) \right)$$

where F_{ri} and F_{rj} are respectively the forces in the upper and lower bolt rows with reference to the stiffener (see Figure 4.70), f_y is the lowest yield value between the plate and the column web, t_{wc} is the web thickness of the column, L_t is the web width obtained spreading the force at 60°, and m_1, m_{2U}, m_{2L} are as in Figure 4.70. To adapt the formula to EC, we might substitute the term $L_t t_{wc}$ with $\omega b_{eff,t,wc} t_{wc}$ (refer to Chapter 3).

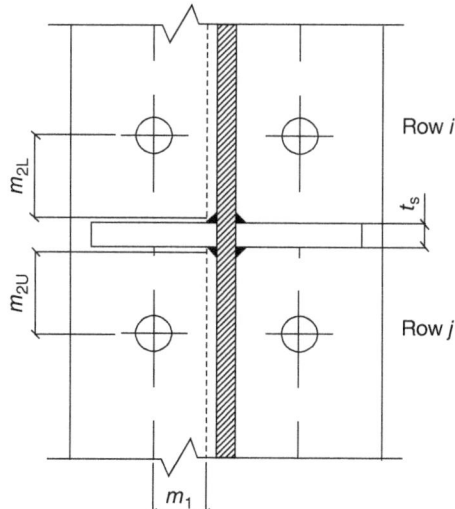

Figure 4.70 Symbols for formulas in [10].

4.12.7 Check of Column Diagonal Stiffener for Panel Shear

Once again from [10] we have that the stiffener area (A_{sg}) should be

$$A_{sg} = 2b_{sg}t_s \geq \frac{\Delta F}{f_{yw} \cos \theta}$$

where ΔF indicates the difference between the design action and what was taken by the web (i.e. the remaining part to resist), f_{yw} the column web yield, and θ the angle with the horizontal stiffener.

4.12.8 Shear Due to Vertical Forces

What is discussed now is the "normal" shear due to vertical forces, not to be confused with the shear on the column web panel just discussed given by the bending moment.

It is convenient to choose a specific and simplified model, which is different from the shear end-plate model. The design practice does not apply the shear to the bolts working in tension, especially those in the external areas of the connection, because they are supposed to work at their maximum tension capacity. Suffice to say that, by applying the EC formula for combined shear and tension, a bolt committed to 100% of its tension capacity has an "availability" of resources for shear that is only 28% of its maximum shear capacity. It is therefore convenient and easier for design to assign the shear to the bolts in the compression zone, which is otherwise lightly loaded.

4.12.9 Design with Haunches

As mentioned, haunches essentially allow increasing the lever arm that is used to distribute the bending moment, resulting in lower actions for bolts, end plates, column webs, and flanges. Haunches might also be utilized to enhance the local beam resistance.

The haunch angle is usually 60° (preferably not less than 45°) and the plates corresponding to the haunch web and flange are thicker or at least the same size as their respective beam parts.

It must be verified that the force concentration where the haunch ends at the beam does not require a vertical stiffener. The force (symbols as in Figure 4.71) is evaluated by

$$C' = \frac{C}{\tan \theta}$$

The compression force distributes through the flange and root radius with an angle in the range between 45° and 63° (depending on sources).

If the web does not withstand the force, a stiffener should be inserted.

4.12.10 Beam-to-Beam Connections

When connecting beams (in apex-like joints), the same considerations apply, with the advantage that all the limit states related to the column need not be

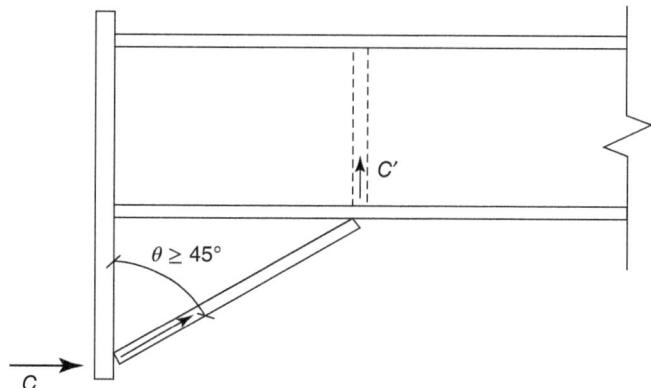

Figure 4.71 Haunch and its compression force.

evaluated. The engineer must then focus in designing the end plate, bolts, and welds. Even here the end plate can be flush or extended (the extension being in the positive moment part, usually the bottom flange in apex connections) and some local ribs can be used to lower the end-plate thickness (e.g. see the cases in Figure 4.72).

4.12.11 BS Provisions

The approach of the BS (as described by [10]) is similar to EN 1993-1-8, with a few relevant differences in addition to what has already been discussed:

- There is a limit thickness for the plate (or the column flange) beyond which the bolt plastic resistance cannot be used and the classic concept of the triangular elastic distribution has to be adopted; the basic concept is that the bolts, to have a plastic behavior, need some deformation of the plate or the flange, which does not happen if they are both too thick.
- If the column web compression is too high, it can be assumed that the compression is taken not only by the compressed flange of the beam but also partially by the web (the part necessary to satisfy the check), as long as the bolt tension is updated by considering the center of compression in the new position.

4.12.12 AISC Approach

The guides [21–23] discuss moment end-plate connections. In particular, Ref. [22] gives some formulas for end plates and bolts (as does [23], but paying particular attention to seismic applications), and Ref. [21] deals mostly with column limit states. For the part that is perhaps numerically more delicate in the design of the end plate, that is, the thickness of the plate itself and the size of the bolts, Ref. [22] has useful equations: Although the formulas are "long," once arranged on a spreadsheet, they can be very useful because they are dedicated to

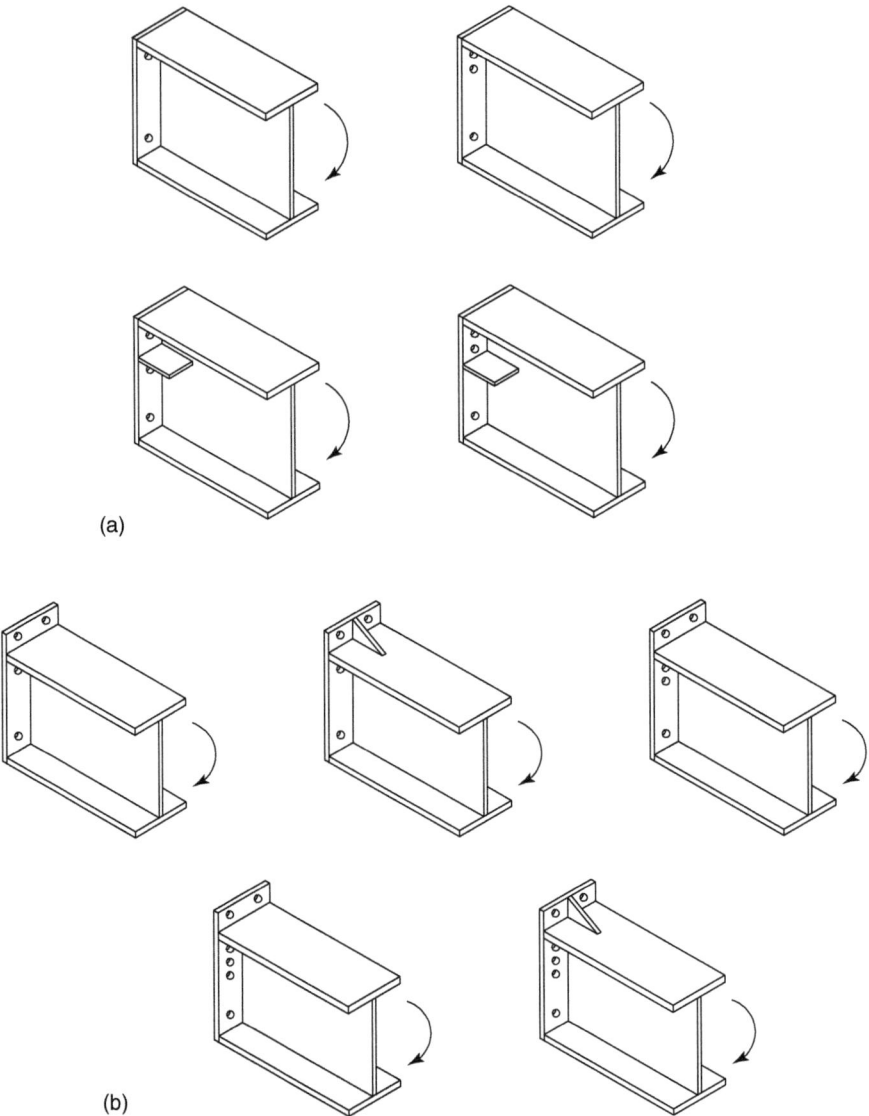

Figure 4.72 Cases with formulas in [22].

particular cases and the engineer does not have to concentrate on which steps to take as in the EC; he or she can simply apply the formula to the specific situation. Paraphrasing the concept, EC3 provides a precious general method that requires, for the T-stub design, many steps and comparisons with minimum values to apply. The AISC method, albeit made and illustrated for "few" cases (but actually the ones mainly used), provides clear equations that do not require any engineering judgment but simply require their execution.

The reader is referred to the sources [21–23] for detailed formulas.

4.12.13 Limit States to Be Considered

Figure 4.73 (from [10]) graphically summarizes the various limit states that should be verified as follows:

1. Bolt tension
2. End-plate bending
3. Column flange bending
4. Beam web tension
5. Column web tension
6. Beam flange to end-plate weld tension
7. Beam web to end-plate weld tension
8. Column web panel shear
9. Beam flange compression
10. Beam flange weld compression
11. Column web buckling
12. Beam web to end-plate weld shear
13. Bolt shear
14. Bolt bearing.

The designer will possibly have to evaluate additional actions, for example, on the beam weak axis, if they are not negligible.

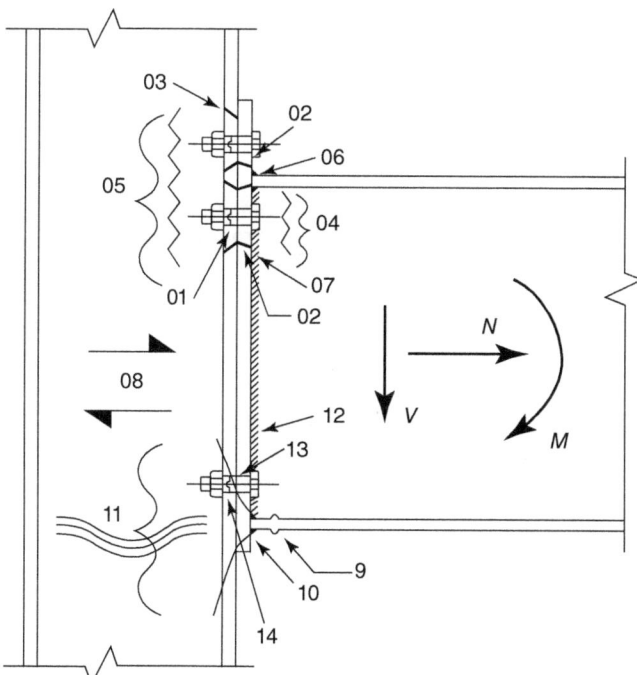

Figure 4.73 Figure proposed by [10] to illustrate the various limit states of an end plate.

4.12.14 Rotational Stiffness

The stiffness of a bolted end plate must be calculated according to [5] using Tables 4.4 and 4.5 (the various contributions will add up in the formula to find S_j seen in Section 3.1).

When there are two or more bolt rows in tension, an equivalent coefficient comes into play and it is calculated as

$$k_{eq} = \frac{\sum_r k_{eff,r} h_r}{z_{eq}}$$

where h_r indicates the distance of the bolt row r from the center of compression, while the other terms can be taken as

$$k_{eff,r} = \frac{1}{\sum_i \frac{1}{k_{i,r}}}$$

$$z_{eq} = \frac{\sum_r k_{eff,r} h_r^2}{\sum_r k_{eff,r} h_r}$$

where $k_{i,r}$ is the stiffness of the component i in the bolt row r.

When the connection is beam to column, k_{eq} is based on (and replaces) the following:

- k_3, which represents the column web in tension
- k_4, which represents the column flange in bending
- k_5, which represents the end plate in bending
- k_{10}, which represents the bolts in tension.

Table 4.4 Stiffness coefficients to evaluate when designing beam-to-column end plates according to [5].

Beam-to-column end plate	Bolt rows in tension	Stiffness coefficients K_i to consider to get S_j
Only on one side	1 row	$k_1\ k_2\ k_3\ k_4\ k_5\ k_{10}$
	2 or more	$k_1\ k_2\ k_{eq}$
On both sides with equal and opposite bending moments	1 row	$k_2\ k_3\ k_4\ k_5\ k_{10}$
	2 or more	$k_2\ k_{eq}$
On both sides with different bending moments	1 row	$k_1\ k_2\ k_3\ k_4\ k_5\ k_{10}$
	2 or more	$k_1\ k_2\ k_{eq}$

Table 4.5 Stiffness coefficients to evaluate when designing beam-to-beam end plates.

Beam-to-beam end plate	Bolt rows in tension	Stiffness coefficients K_i to consider to get S_j
On both sides with equal and opposite bending moments	1 row	k_5 (right) k_5 (left) k_{10}
	2 or more	k_{eq}

If the end plate connects two beams (apex), k_{eq} is obtained as follows (replacing them):

- k_5, which represents the end plate in bending
- k_{10}, which represents the bolts in tension.

It has to be highlighted that if the connection is not symmetric (which is quite common), the positive bending moment stiffness is different than the negative bending stiffness, which can be a significant practical complication if the stiffness coefficient has to be input into the software analysis model.

4.12.15 Simplifying the Design

The design of the bolt size and the end-plate thickness requires the biggest effort, but this process can only be partially simplified. According to [10], it might be convenient to start from the column side considering the two top rows in tension as reacting with the same value (as a group) in the case of an extended end plate. The contribution of the lower bolt rows is then assessed by taking, in a simplified manner, ℓ_{eff} as the vertical pitch of bolts (p in Section 3.10).

To shorten the design time, it is a good starting point to take, for example, an initial end-plate thickness to begin the checks that is about the same size as the column flange, likely larger if the plate is not extended and stiffened. Also, an end-plate width that is smaller than the column flange will likely require a thicker plate.

The continuity plates in the tension and compression zones of the column should also be initially taken with a thickness very similar to the beam flanges. If the beam is working near full capacity, prescribing a full-strength weld (also by fillet welds, not necessarily by complete penetration) might shorten the calculation procedure and be a good (ductile) solution.

4.12.16 Practical Advice

As already discussed in the context of the shear end plate, the end plate could be realized (see Figure 4.74) with a small tolerance (1 mm, sometimes also 1.5–2 mm) between the connecting plates/flanges by making the beam a little shorter. This eases the erection (note that superficial treatments such as hot-dip galvanization add some volume and might hinder the insertion of the pieces if the length is exact with no tolerance; even thermal dilatations might create the same issue) and does not represent a special problem from a design point of view. Finger shims to be inserted during erection can also be contemplated.

For the beam-to-end-plate welding, fillet welds should preferably be designed, but if the throat results in more than approximately 8–10 mm, it is perhaps more convenient to use partial (or complete) penetrations.

4.12.17 Structural Integrity, Ductility, and Rotation Capacity

An end plate designed to take the full strength of the connected beam (maybe even including an overstrength factor because of the material or the special situation) guarantees excellent ductility. According to [5], a joint that is designed

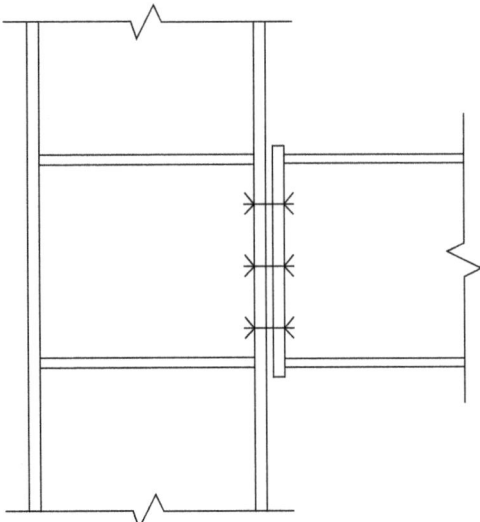

Figure 4.74 Tolerance for end plates.

for 1.2 times the beam capacity must not be checked for rotation capacity in a plastic analysis. If this hypothesis is not verified, refer to [5] for methods to check the rotation capacity. For the rest, the considerations generally valid for the other connections with moment resisting capacity (splices, angles bolted to the flange in tension) also apply here; that is, bolts in shear and welds should not be the critical limit state. The structural integrity of an end plate is outstanding since it can effectively transmit any type of action (even torsion and weak-axis moment).

4.12.18 Beam-to-Column End-Plate Design Example According to Eurocode

A small industrial building has HEA 160 columns rigidly connected (portal frames as lateral resisting system) to IPE 300 beams sustaining the roof.

The scope is to design the beam-to-column connection as an end plate that follows EC prescriptions, assuming that $\gamma_{M0} = 1.1$, $\gamma_{M1} = 1.1$, and $\gamma_{M2} = 1.25$. The profiles are in S275 while the plates should be S235, with class 8.8 bolts.

The beam has a 5.5° angle (about 10% slope) but this is not a problem and the calculation model can still be applied.

The most burdensome combination to check the connection is considered to be as follows:

$$N_{Ed} = -4 \text{ kN (beam compression)}$$
$$V_{\text{major Ed}} = 48 \text{ kN}$$
$$V_{\text{minor Ed}} = 1 \text{ kN}$$
$$M_{\text{major Ed}} = 54 \text{ kN m}$$
$$M_{\text{minor Ed}} = 1 \text{ kN m}$$

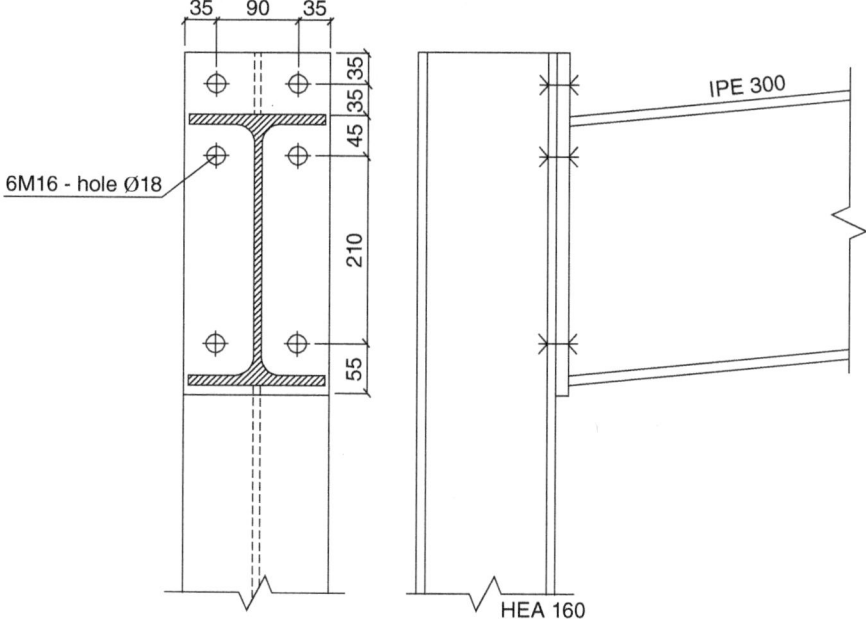

Figure 4.75 Initial geometry.

The weak-axis moment and shear can be considered as negligible. The compression is also negligible and we will not take it into account when checking tension limit states but will include it for the compression checks.

An acceptable geometry for the connection (and so a good starting point) could be the one shown in Figure 4.75. We note that it is very helpful for the calculations to have an extended end plate, but the chance of realizing it is actually to be verified because sometimes the top of the steel must be flush with the top of the beam flange. In this case, we assume that the purlins lean on the beam (but not in the zone where the beam connects to the column) and the roof panel is above them so the end-plate extension is applicable and is not an obstacle to other construction details. In other words, the purlins can be positioned right before the column connection (or right after, on the other side, as in Figure 7.58) so the proposed configuration is acceptable.

As per EC instructions, it can be assumed that in a beam-to-column bolted end-plate prying action develops.

Having designed a plate width that is the same as the column flange width but trying to avoid the top stiffener between the end plate and the beam flange to save labor (and allow more space for the purlins to be positioned), the end-plate thickness will likely be bigger than the column flange (if we had the stiffener, this would simulate the web action and the plate would be almost as thick as the column flange, probably 12 mm thick considering that the material is S235 and not S275), and hence we start with a 15 mm thickness to kick off our checks.

4.12.18.1 Column Flange Thickness Check for Bolt Row 1

As a first step, the tension resistance must be checked. We begin from the column, checking the tension of the bolts, chosen as M16 (it is not a rule but quite often the bolt diameter is larger than the plate thickness). The top bolts on the column side have the following dimensions:

$$m = 90/2 - 6/2 \times 0.8 \times 15 = 30\,\text{mm}$$

$$e = 35\,\text{mm}$$

$$e_{\min} = 35\,\text{mm}$$

$$e_1 = 35\,\text{mm}$$

$$n = \min(e_{\min}, 1.25m) = 35\,\text{mm}$$

$$\ell_{\text{eff,cp}} = \min(2\pi m,\ \pi m + 2e_1) = \min(188, 164) = 164$$

$$\ell_{\text{eff,nc}} = \min(4m + 1.25e,\ 2m + 0.625e + e_1) = 117\,\text{mm}$$

$$\ell_{\text{eff,1}} = \min(\ell_{\text{eff,nc}}, \ell_{\text{eff,cp}}) = \min(117, 164) = 117\,\text{mm}$$

$$\ell_{\text{eff,2}} = \ell_{\text{eff,nc}} = 117\,\text{mm}$$

$$F_{T,3,Rd} = 2 \times 90 = 180\,\text{kN}$$

$$M_{\text{pl,1,Rd}} = 0.25 \times 117 \times 9^2 \times 275/1.1 = 592\,\text{kN}\,\text{mm}$$

$$F_{T,1,Rd} = 4M_{\text{pl,1,Rd}}/m = 4 \times 592/30 = 79\,\text{kN}$$

$$M_{\text{pl,2,Rd}} = 0.25 \times 117 \times 9^2 \times 275/1.1 = 592\,\text{kN}\,\text{mm}$$

$$F_{T,2,Rd} = \frac{2M_{\text{pl,2,Rd}} + n\sum F_{t,Rd}}{m + n}$$

$$= (2 \times 592 + 35 \times 180)/(30 + 35) = 115\,\text{kN}$$

$$F_{T,Rd} = \min(79, 115, 180) = 79\,\text{kN}$$

Using SCS (with the "EC pure" option as the calculation method), we fully confirm the above values (press the button T-Stub Notes on the Bolts tab of the results to generate a dedicated report).

4.12.18.2 Column Web Tension Check for Bolt Row 1

Preliminarily we get

$$b_{\text{eff,t,wc}} = \ell_{\text{eff}} = 117\,\text{mm}$$

$$A_{vc} = 3880 \times 2 \times 160 \times 9 + 9(2 \times 15 + 6) = 1324\,\text{mm}^2$$

The connection is on only one side and, therefore, the transformation parameter β is 1; hence, $\omega = \omega_1 = 1/(\sqrt{1 + 1.3(117 \times 6/1324)^2}) = 0.86$ and then $F_{\text{t,wc,Rd}} = 0.86 \times 117 \times 6 \times 275/1.1 = 151\,\text{kN}$.

The value is higher than the one previously obtained and so it does not affect the design.

4.12.18.3 Beam End-Plate Thickness Check for Bolt Row 1

$$e = 35\,\text{mm}$$
$$b_p = 160\,\text{mm}$$
$$w = 90\,\text{mm}$$
$$e_{min} = 35\,\text{mm}$$
$$e_x = 35\,\text{mm}$$
$$m_x = 35 - 0.8 \times 5\sqrt{2}$$
$$\quad = 29\,\text{mm (a 5-mm throat weld has been assumed)}$$
$$n_x = \min(e_x, 1.25 m_x) = 35\,\text{mm}$$
$$\ell_{\text{eff,cp}} = \min(2\pi 29, \pi 29 + 90, \pi 29 + 2 \times 35) = 161\,\text{mm}$$
$$\ell_{\text{eff,nc}} = \min(4 \times 29 + 1.25 \times 35, 35 + 2 \times 29 + 0.625 \times 35, 0.5 \times 160,$$
$$\quad 0.5 \times 90 + 2 \times 29 + 0.625 \times 35)$$
$$\quad = 80\,\text{mm}$$
$$\ell_{\text{eff},1} = \min(161,\ 80) = 80\,\text{mm}$$
$$\ell_{\text{eff},2} = 80\,\text{mm}$$
$$F_{T,3,Rd} = 2 \times 90 = 180\,\text{kN}$$
$$M_{\text{pl},1,Rd} = 0.25 \times 80 \times 15^2 \times 235/1.1 = 961\,\text{kN mm}$$
$$F_{T,1,Rd} = 4 \times 961/29 = 133\,\text{kN}$$
$$M_{\text{pl},2,Rd} = 0.25 \times 80 \times 15^2 \times 235/1.1 = 961\,\text{kN mm}$$
$$F_{T,2,Rd} = (2 \times 961 + 35 \times 180)/(29 + 35) = 128\,\text{kN}$$
$$F_{T,Rd} = \min(133, 128, 180) = 128\,\text{kN}$$

Except for some rounding, the results are the same as in SCS.

4.12.18.4 Beam Web Tension Check for Bolt Row 1
Since bolt row 1 is an extension of the plate, there is no web tension here.

4.12.18.5 Final Resistant Value for Bolt Row 1

$$\min(79, 151, 128) = 79\,\text{kN}$$

4.12.18.6 Column Flange Thickness Check for Bolt Row 2 Individually

$$m = 30\,\text{mm}$$
$$e = 35\,\text{mm}$$
$$e_{min} = 35\,\text{mm}$$
$$n = 35\,\text{mm}$$
$$\ell_{\text{eff,cp}} = 2\pi m = 188\,\text{mm}$$
$$\ell_{\text{eff,nc}} = 4m + 1.25e = 164\,\text{mm}$$

$\ell_{\text{eff},1} = \min(188, 164) = 164$ mm

$\ell_{\text{eff},2} = 164$ mm

The verification will then match the procedure seen for row 1:

$F_{T,3,Rd} = 180$ kN

$M_{pl,1,Rd} = 830$ kN mm

$F_{T,1,Rd} = 111$ kN

$M_{pl,2,Rd} = 830$ kN mm

$F_{T,2,Rd} = 122$ kN

$F_{T,Rd} = 111$ kN

4.12.18.7 Column Web Tension Check for Bolt Row 2 Individually

Similarly to bolt row 1, this does not affect design (from SCS, 187 kN).

4.12.18.8 Beam End-Plate Thickness Check for Bolt Row 2 Individually

$m = 90/2 - 7.1/2 - 0.8 \times 3\sqrt{2} = 38$ mm

$m_2 \text{(to then calculate } \alpha\text{)} = 45 - 10.7 - 0.8 \times 5\sqrt{2} = 29$ mm

$e = 35$ mm

$b_p = 160$ mm

$w = 90$ mm

$n = 35$ mm

$\ell_{\text{eff,cp}} = 2\pi 38 = 237$ mm

$\lambda_1 = 38/(38 + 35) = 0.52$

$\lambda_2 = 29/(38 + 35) = 0.40$

$\ell_{\text{eff,nc}} = \alpha m = 6 \times 38 = 228$ mm

$\ell_{\text{eff},1} = \min(237, 228) = 228$ mm

$\ell_{\text{eff},2} = 228$ mm

$F_{T,3,Rd} = 2 \times 90 = 180$ kN

$M_{pl,1,Rd} = 0.25 \times 228 \times 15^2 \times 235/1.1 = 2740$ kN mm

$F_{T,1,Rd} = 4 \times 2740/38 = 288$ kN

$M_{pl,2,Rd} = 2740$ kN mm

$F_{T,2,Rd} = (2 \times 2740 + 35 \times 180)/(38 + 35) = 161$ kN

$F_{T,Rd} = \min(288, 161, 180) = 161$ kN

4.12.18.9 Beam Web Tension Check for Bolt Row 2 Individually

The beam flange is inside the diffusion zone of the force; therefore, checking the web is not necessary.

4.12.18.10 Column Flange Thickness Check for Bolt Row 2 in Group with Bolt Row 1

$m = 30$ mm
$e = 35$ mm
$e_{min} = 35$ mm
$e_1 = 35$ mm
$n = 35$ mm
$p = 80$ mm

Following EC rules carefully, we obtain

$\ell_{eff,cp} = \min(\pi m + p, \ 2e_1 + p) + 160 = \min(174, 150) + 160 = 310$
$\ell_{eff,nc} = \min(2m + 0.625e + 0.5p, \ e_1 + 0.5p) + p = \min(122, 75) + 80$
$\quad = 155$ mm
$\ell_{eff,1} = \min(\ell_{eff,nc}, \ \ell_{eff,cp}) = \min(310, \ 155) = 155$ mm
$\ell_{eff,2} = \ell_{eff,nc} = 155$ mm
$F_{T,3,Rd} = 4 \times 90 = 360$ kN
$M_{pl,1,Rd} = 0.25 \times 155 \times 9^2 \times 275/1.1 = 785$ kN mm
$F_{T,1,Rd} = 4 M_{pl,1,Rd}/m = 4 \times 785/30 = 105$ kN
$M_{pl,2,Rd} = 0.25 \times 155 \times 9^2 \times 275/1.1 = 785$ kN mm
$F_{T,2,Rd} = (2 M_{pl,2,Rd} + n \Sigma F_{t,Rd})/(m + n)$
$\quad = (2 \times 785 + 35 \times 360)/(30 + 35)$
$\quad = 218$ kN
$F_{T,Rd} = \min(105, 218, 360) = 105$ kN

Let us notice that extending the column above the plate, that is, increasing e_1 in EC symbols (e_x in [10]), would provide a benefit.

Subtracting the bolt row 1 resistant value to the group result, we get $105 - 79 = 26$ kN.

4.12.18.11 Column Web Tension Check for Bolt Row 2 in Group with Bolt Row 1

Being that the influence zone is bigger than the row 1 individually, the result will be more than 187 kN and, therefore, irrelevant (224 kN from SCS, $\gg 26$ kN).

4.12.18.12 Beam End-Plate Thickness Check for Bolt Row 2 in Group with Bolt Row 1

This check is not applicable because on the beam side there is a flange in the middle and so the bolt rows do not create a group.

4.12.18.13 Beam Web Tension Check for Bolt Row 2 in Group with Bolt Row 1

As above, this is not applicable.

4.12.18.14 Final Resistant Value for Bolt Row 2

This is the minimum value found for the row 2 individually and as a group with row 1, and thus min(111, 187, 161, 26) = 26 kN.

We now check if the connection hereby designed could provide the necessary moment resistance:

$M_{Rd} = 79(300 + 35 - 10.7/2) + 26(300 - 45 - 10.7/2) = 26 + 6.5 = 32.5$ kN m,

which is well below the required 54 kN m. If we decide not to extend the column further as previously hinted, or add some backing plates (since mode 1 governs T-stub design, backing plates might bring some benefit), we could insert a horizontal continuity plate aligned with the beam top flange as in Figure 4.76. In physical terms, the stiffener insertion keeps the combined action of rows 1 and 2 from stressing the column flange beyond its limit.

By inserting it (say 10 mm thick) and performing calculations by means of SCS, we get 88 kN for the first bolt row and 120 kN for the second and thus a resisting bending moment $M_{Rd\,bolts} = 88(300 + 35 - 10.7/2) + 120(300 - 45 - 10.7/2) = 29 + 30 = 59$ kN m.

So far, we have not considered if the compression part of the joint can develop the forces necessary to have the above moment, so we will check this now.

The solution so far verifies the forces in tension and seems appropriate and easy to fabricate, so we can move on to the next design steps.

4.12.18.15 Vertical Shear

We assume that only the two bolts in the compression zone will take the shear.

From Table 3.7, one M16 can resist 60 kN in shear so the check is largely satisfied.

Since the bearing checks are negligible we omit them (the plate is thick and welded to both flanges): For similar numerical examples see the step-by-step exercise in Section 4.7.4 or otherwise SCS.

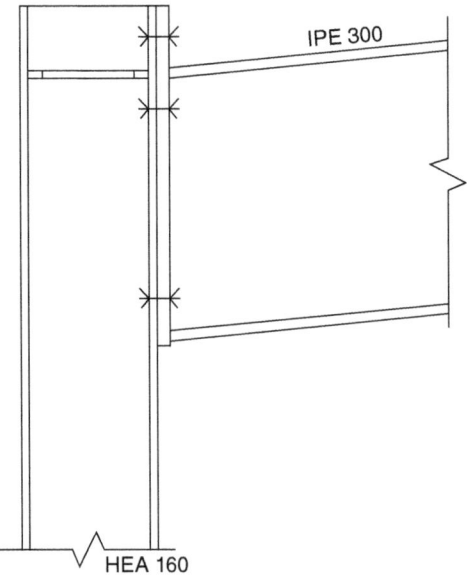

Figure 4.76 Horizontal stiffener added.

4.12.18.16 Web Panel Shear

The resistant capacity is $0.9 \times 275 \times 1324/(\sqrt{3} \times 1.1) = 172$ kN, less than the required 186 kN. We might then weld a plate to the column web as reinforcement (ensuring that with some precautions hot-dip galvanization will not cause problems) or we might otherwise design a diagonal stiffener. The diagonal stiffener could, however, interfere with the lower bolts. A remedy would consist in moving these bolts below the bottom flange, as in Figure 4.77. A different solution could be using a Morris stiffener. Its fabrication being a little more complex, we decide to follow Figure 4.77.

4.12.18.17 Column Web Resistance to Transverse Compression

By applying EC equations (see Chapter 3) we get

$A_{vc} = 1324$ mm² (previously calculated)

$b_{eff,c,wc} = 10.7 + 2\sqrt{2} \times 5 + 5(9 + 15) + 15 + (10 - \sqrt{2} \times 5) = 163$ mm

Let us notice that the last term in $b_{eff,c,wc}$ above takes into account that the plate only goes 10 mm beyond the bottom flange (conservative, since we just said that we will extend the plates below in order to then move the bolts downward):

$\beta = 1$, hence $\omega = \omega_1 = 1/(\sqrt{(1 + 1.3(163 \times 6/1324)^2)} = 0.76$

$\lambda_p = 0.932\sqrt{(163 \times (152 - 2(15 + 9)) \times 275/(210\,000 \times 6^2))} = 0.73$

$\rho = (0.73 - 0.2)/0.73^2 = 0.99$

$k_{wc} = 1$

$F_{c,wc,Rd} = \min(1/1.1, 0.99/1.1) \times 1 \times 0.76 \times 163 \times 6 \times 275 = 186$ kN

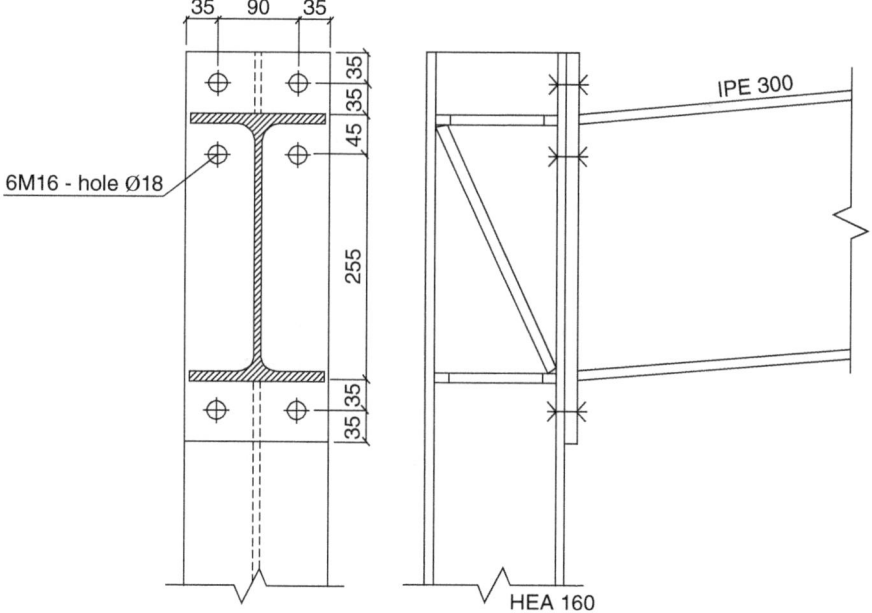

Figure 4.77 Final configuration.

The compression force is given by the bending moment (the additional design compression would be −4 kN, but when we compose it with the shear by considering the actual beam inclination, it becomes +0.6 kN, neglectable), which is $54\,000/z_{eq}(=300 − 45 + (35 + 45)/2 − 10.7/2) = 186$ kN, so the check is near the limit. The value we just found is based on an approximate z_{eq} (see later in this example for a more precise evaluation).

Since there is a continuity plate on top, we then decide to put another one on the bottom, which makes the design widely within limits: From SCS and exploiting the guide of [10], the additional compression given by the stiffener is $2 \times 10 \times 235(77 − 15)/1.1 = 265$ kN, the buckling not being an issue, for a total of 331 kN.

4.12.18.18 Stiffener Design

The tentative thickness for the check is similar to the beam flange, that is, 10 mm. We start from the horizontal stiffener and calculate its minimum required area following the discussed equations. We conservatively take F_{ri} and F_{rj} equal to the maximum value just found:

$$\max[(111 + 111) \times 1000/235 − 0.76 \times 164 \times 6,$$
$$30/235(111/(30 + 35) + 111(30 + 35) \times 1000)] = 436 \text{ mm}^2$$

With a 10 mm thickness and a b_{sg} width for each stiffener of 75 mm (which is less than 13 times the thickness) to be reduced, considering a 20×20 corner clip, to $b_{sn} = 75 − 20 = 55$ mm, we get a 1100 mm² area, well above the minimum previously found.

For the diagonal stiffener, a 75 mm width coupled with a 10 mm thickness satisfies the requested equations: $2 \times 75 \times 10 = 1500 \text{ mm}^2 > (188 − 172) \times 1000/(235 \cos(67°)) = 174 \text{ mm}^2$.

4.12.18.19 Welds

The IPE 300 does not work near the connection at its maximum strength and, therefore, welds for the flanges and the web near the beam full strength (double fillets with throat thickness of 5 and 3 mm, respectively) are presumably sufficient and abundant (good for ductility). A quick check confirms our assumption: The double fillet on the top flange can sustain $5 \times 360/(0.8\sqrt{3} \times 1.25) \times 150 \times 2 = 312$ kN. Multiplying it by the lever arm $(300 − 10.7)$ we have a 90-kN m moment, which is quite generous. The same applies to the web weld (the reader can check it by SCS). Actually, the welds also work transmitting T-stub forces near the top flange and 3.5 mm would be the optimum (full strength) for the web but checking the maximum force that the bolts transmit through the flange and the web (SCS does this automatically and eventually warns the user if the check is not satisfied) 3 mm is acceptable.

4.12.18.20 Rotational Stiffness

We have two bolt rows in tension and, therefore, we must also evaluate k_{eq}. The values are the following (considering only one washer per bolt, to set up in SCS in the general options):

$$k_1 = k_2 = \text{infinite (being all stiffened)}$$

Row 1:

$k_3 = 0.7 \times 130.3 \text{ (from SCS)} \times 6/152 = 3.6 \text{ mm}$
$k_4 = 0.9 \times 130.3 \times 9^3/30^3 = 3.16 \text{ mm}$
$k_5 = 0.9 \times 80(= \min(0.5 \times 160, 2 \times 29.3 + 0.625 \times 35 + 0.5 \times 90, 2 \times 29.3$
$\qquad + 0.625 \times 35 + 35.4 \times 29.3 + 1.25 \times 35), \text{from SCS})$
$\qquad \times 15^3/29.3^3 = 9.66 \text{ mm}$
$k_{10} = 1.6 \times 157/(10/2 + 9 + 15 + 4 + 13/2) = 6.36 \text{ mm}$
$k_{\text{eff,r}} = 1/(1/\infty + 1/3.6 + 1/3.16 + 1/9.66 + 1/6.36) = 1.17 \text{ mm}$

Row 2:

$k_3 = 0.7 \times 177.7 \text{(from SCS)} \times 6/152 = 4.91 \text{ mm}$
$k_4 = 0.9 \times 177.7 \times 9^3/30^3 = 4.32 \text{ mm}$
$k_5 = 0.9 \times 215(= 6 \times 35.8, \text{from SCS}) \times 15^3/35.8^3 = 14.3 \text{ mm}$
$k_{10} = 6.36 \text{ mm}$
$k_{\text{eff,r}} = 1/(1/\infty + 1/4.91 + 1/4.32 + 1/14.3 + 1/6.36) = 1.51 \text{ mm}$

Hence

$z_{\text{eq}} = (1.17 \times 330^2 + 1.51 \times 250^2)/(1.17 \times 330 + 1.51 \times 250) = 290 \text{ mm}$
$k_{\text{eq}} = (1.17 \times 330 + 1.51 \times 250)/290 = 2.63 \text{ mm}$
$S_{j,\text{ini}} = (210\,000 \times 290^2)/(1(1/\infty + 1/\infty + 1/2.63)) = 4.65 \times 10^4 \text{ kN m}$

The SCS value is a little higher (4.68×10^4) because the software also includes the (negligible) contribution of the third row, which is also the reason for z_{eq} and k_{eq} being slightly different.

Assuming a 10-m IPE 300 length, no braces, and a 6-m interstory, we have

$I_b/L_b = 8360 \times 10^4/(10 \times 10^3) = 8360 \text{ mm}^3$
$I_c/L_c = 1670 \times 10^4/(6 \times 10^3) = 2783 \text{ mm}^3$
$(I_b/L_b)/(I_c/L_c) = 3$

Thus the limit for complete rigidity is $25 \times E \times I_b/L_b = 4.39 \times 10^4$ kN m, E being the steel elastic modulus.

The limit for a pin connection is instead 8.78×10^2 kN m, so the connection can be considered as rigid. If we had a semirigid result, we should change the analysis model including the stiffness value as per EC instructions, which might change the results (deflection and action distribution). Note however that if we take [10] as reference, the joint can be taken as rigid since it is a single-story portal and it is "well proportioned for strength."

4.13 Splice

This type of joint can resist not only axial and shear forces but also the full bending strength of the connected members. Note that sometimes end plates that can restore the full bending between consecutive sections are called splices.

The splice is commonly used in the following two situations:

- When the designer wants to restore the strength of a column which, for transport problems (usually the limit is around 40 ft/12.5 m), cannot be fabricated as a single piece; although not the rule, in those cases the splice is normally realized so that the top part leans directly on the bottom part in contact (e.g. see Figure 4.78).
- When the designer wants to restore the strength of a beam which, as above, must be fabricated as several pieces and has remarkable bending moment at the connection point; although again not the rule, in those cases the solution is more frequently a splice with no contact between parts.

There are several possible variants, some of which are shown in Figure 4.78.

The solution on the bottom of Figure 4.79 might be useful when it is required to have a uniform TOS for some specific reasons, for example, if it is a runaway girder for cranes. The bending moment capacity of a similar connection is very good for positive bending moments, taken by the lower cover plate, but quite limited for negative moments (i.e. with tension in the top flange) that are resisted by the flush end plate. The splice position of a similar joint must therefore be chosen with care depending on the envelope of the design moments.

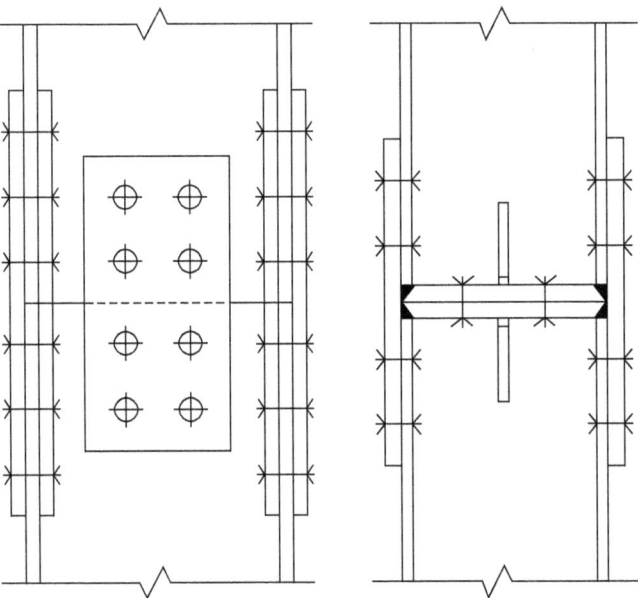

Figure 4.78 Column splices: classical configuration (with the part leaning on the bottom) and another with flush end plate and cover plates only externally on flanges.

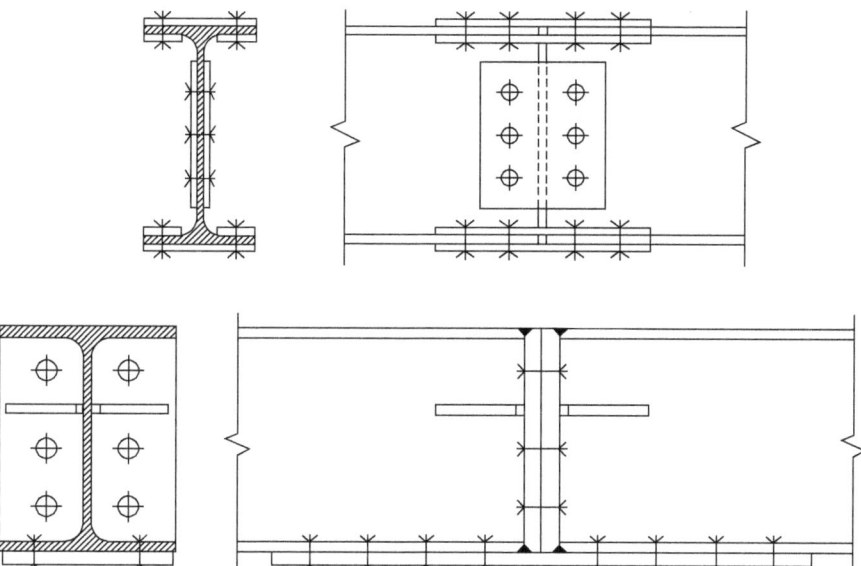

Figure 4.79 Classical beam splice (top) and with uniform top of steel (bottom).

Other variations might consist of some welded parts, to be realized either in the shop or on-site. For example, some plates as in Figure 4.79 might be welded on one side in the shop; though this solution might mean a difficult positioning of the beams on-site, mostly there is a likely chance of damage occurring to the plates during shipping. Another option when a full-strength connection is requested might consist in using a couple of plates to temporarily preassemble the parts, but then the members are fully welded (e.g. by complete penetration). The disadvantages of such a design are the usual ones involving field welds.

Other situations and adaptations come into play if columns of different sizes must be spliced. Some ideas and examples to solve the problem are illustrated in Figure 4.80.

4.13.1 Calculation Model and Limit States

This is generally the design model:

- If there is contact (direct or through plates) between the parts, the compression is transmitted by contact; the tension instead will stress the plates based on the areas of the various parts (in the sense that the action in the flange plates will be different than the web plates); other approaches (as in [10]) assign the axial force only to the flanges.
- The strong-axis shear is given to the web connection by either end plates or classic cover plates.
- The weak-axis shear is given to the end plate if there is one; otherwise it is given to the plates connecting the flanges.
- The strong-axis bending moment is broken up into two axial actions, a compression and a tension, by dividing it by the flange center distances, that is,

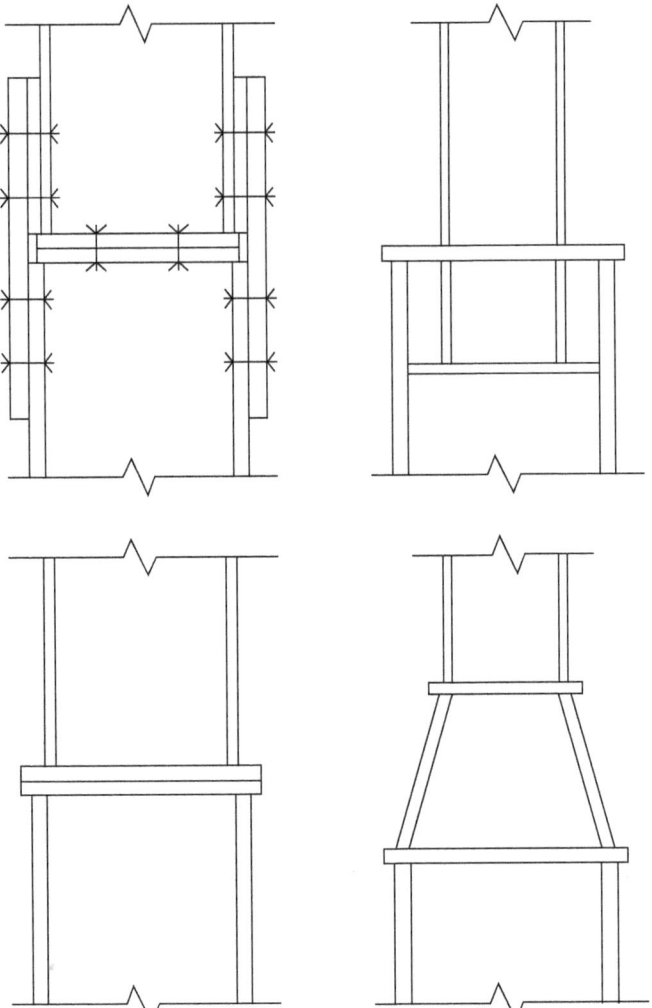

Figure 4.80 Alternative systems for splicing columns of different sizes.

the member depth minus the flange thickness; this action will be algebraically added to the concomitant axial force in the flanges; if, as shown in the bottom of Figure 4.79, there is no cover plate on one flange (the top in the figure), the end plate takes the actions.
- The weak-axis bending moment is given to the flange plates (the external in particular) or, in the second instance, to the end plate if present.

Once the calculation model is defined, the following limit states will be checked:

- Bolt shear (attention to the number of resistant sections); in cases such as the bottom of Figure 4.79, also bolt tension
- Bearing and block shear (the latter usually not governing)
- Plate resistance to shear, bending, axial forces

- Buckling of plates in compression
- Member resistance because of holes
- If present, weld failure.

The Steel Construction Institute [10] recommends designing the connection with the bolts in friction (at least to service loads) in order to avoid rotations due to the bolts slipping inside the holes.

4.13.2 Structural Integrity, Ductility, and Rotation Capacity

The splice is classically considered capable of restoring the capacity of the joined member and hence its continuity, so the rotation capacity is usually not checked. The splice also guarantees, generally speaking, a fine structural integrity and the ductility can as well be taken as good as long as the bolts in shear are not the most critical limit states.

4.13.3 Column Splice Design Example According to AS 4100

We need to design the splice of a column of a multistory structure that is 20 m long and hence it is necessary to divide it for transportation. The structure could be part of an industrial or residential project. Even an external safety stair could deal with a problem like this.

The column is best interrupted, for erection reasons, about 1–1.5 m above a floor according to [1]. If we decide to follow the advice (after checking that the zone does not have high bending moments), the actions in the separation point (modeled as rigid) result, with an envelope of the most unfavorable combinations to simplify our checks:

$N_c = -870$ kN (compression)
$N_t = 365$ kN (tension)
$V_x = 77$ kN
$V_y = 6$ kN
$M_x = 92$ kN m
$M_y = 4$ kN m

The materials are as follows:

- Columns 250UC72.9: 300 3678 AS/NZS
- Plates: 250 3678 AS/NZS
- Bolt class: 8.8 AS/NZS 1252 ($f_u = 830$ MPa).

The decision is to make contact between the parts: a simple indication of "machining" (to make the contact surfaces smooth as necessary) is enough (some sources as [15] say that a "normal" surface after the cut is effective to transfer compression forces). It is also preferred to use double plates on both the flanges and the web (to guarantee extra strength to the columns). Inside the flanges the plates are divided to avoid any clash with the web (and the root radius). A 100-mm internal plate avoids any interference (see Table 6.6 for some allowable interference).

As allowed by AS 4100 [24], the nominal diameter of the hole will be 2 mm larger than the nominal bolt diameter (valid for sizes smaller than M24).

4.13.3.1 Flanges

The strong-axis bending moment is divided by the distance between the centerlines of the flanges to translate it into the compression and tension forces acting on each single flange connection, so $N^*_{Mom} = 92\,000/(254 - 14.2/2 \times 2) = 384$ kN.

Each flange takes a portion of the design axial forces of $254 \times 14.2/9322 = 38.7\%$, that is, $870 \times 0.387 = 337$ kN in compression and $365 \times 0.387 = 141$ kN in tension, which summed with the value given by bending brings the total to 721 and 525 kN, respectively. As mentioned, though, the compression is by contact and there are no special checks for it (note that the contact pressure between flanges is 200 N mm^{-2}).

The double plate on each flange must therefore withstand 525 kN, as well as the moment and shear on the weak axis, which we assign to the external plates (more rigid). The moment on the external plate will then be half the design weak-axis bending moment plus the bending given by the eccentricity times the weak-axis shear (the joint axis is considered at the contact of the column parts). The additional moment will also stress the bolts. SCS shows that the moment is negligible, adding only a few percentage points of additional stress in both the bolts and plates. The flange bolts have a shear equal to (we have two resistant sections) $525/6/2 = 43.8$ kN, that is, 47% of the capacity ($=0.8 \times 0.62 \times 830 \times 225 = 92.6$ kN) of an M20 with threads in the shear plane. The most conservative and precise approach (which SCS follows) would be to consider that the two resistant sections do not split the shear in equal parts but split it proportionally to the areas of the plates (the external has an area bigger than the sum of the internal plates) but also in this case the check would be largely acceptable, the exploitation ratio (from SCS) being 54%. Adding the weak-axis actions, SCS gives a 56% ratio.

As a rough predimensioning of plates, it is recommended, generally speaking, that the sum of the plate areas be larger than the flange area, independent of the actions. This comes from the following evaluation: the plate material is usually worse or the same quality as the steel members; the axial forces act differently, the flanges being connected to the web and plates being much more susceptible to buckling (although the latter does not apply to this example, the compression being by contact).

AS 4100 does not require us to check block shear (although it would not govern here, the weak-axis shear being negligible, so $k_{unif} = 1$, and being the area of the weakest element, the internal plate, the same as the area in tension).

Let us then check the tension for the internal plate to which we assign, consistently with the resistant sections in the bolt group check, one quarter of the tension force, that is, 131 kN. As already mentioned, another approach is to assign the action depending on the net areas, which means getting $525/(250 - 22 \times 2 + (100 - 22) \times 2)(100 \times 22) = 113$ kN. The tension resistance is (Chapter 7 of AS 4100) $\Phi N_t = 0.9 \times \min(100 \times 10 \times 260,\ 0.85 \times (100 - 22) \times 10 \times 410) = 234$ kN, vastly acceptable (the design ratio is 48% or 56% depending on the assumption). SCS distributes the actions according to the net areas (verifying the external plate instead of the internal plate) and the result is similar (51%), not perfectly identical because the gross area resistance governs here.

Checking the bearing in the column flange we can omit verifying the plate bearing since the total thickness of the plates (thought the steel quality is slightly worse) is quite large. We have (Section 9.3.2.4 in [24]) $0.9 \times \min(3.2 \times 20, 50 - 22/2 + 20/2) \times 14.2 \times 430 = 269$ kN, well above the action per bolt, $525/6 = 88$ kN, and hence a 33% ratio. SCS includes weak-axis shear and bending and, therefore, gives 34%.

4.13.3.2 Web
The check for the web follows the check for the flange (for a double-bolted simple plate, see the example in Section 4.7.4). The web is stressed by the strong-axis shear only (and by the moment the eccentricity generates) and with four M20 bolts (the same bolt type used for flanges) and a double 8-mm-thick plate we have a maximum exploitation ratio (by SCS) near 50%.

4.13.3.3 Conclusions and Final Considerations
The design (see Figure 4.81) is quite conservative. The choice is made to let this key element of a multistory building have an ample safety margin, avoiding solutions that might look too thin. The engineer should remember that if the parts were not in contact, the plates should also be checked for compression, including buckling coefficients. The reference length for instability can be taken as the maximum distance between consecutive rows of bolts (the governing situation is usually the distance between the bolt rows across the joint axis, that is, the end surfaces).

4.13.3.4 Possible Alternative
The connection might also be realized (Figure 4.82), the forces not being excessive, with a shear end plate welded to the web and designing only an external cover plate for the flanges (maybe a little thicker and with an extra row of bolts to keep a good additional safety margin).

4.14 Brace Connections

As discussed, braces can be connected in many ways: the brace being classically pinned at its ends, the simple shear connections studied at the beginning of this chapter are commonly used.

The standard solution is probably the fin plate, which well suits both vertical braces and horizontal (floor) braces; in the latter case, the shear tab being normally horizontal but sometimes even vertical as in the former. The reader is referred to the considerations in Section 4.10.

Braces might also be connected by means of angles (called *lug angles* in EC) that are added to the fin plate in order to create a connection with good ductility (see Sections 4.24 and 3.19.1).

An end plate can be used as a brace connection (mainly if it is an H-shaped section like HE/W/UC or similar), but if it guarantees a fine response in compression (the force is transmitted by contact), it does not provide the

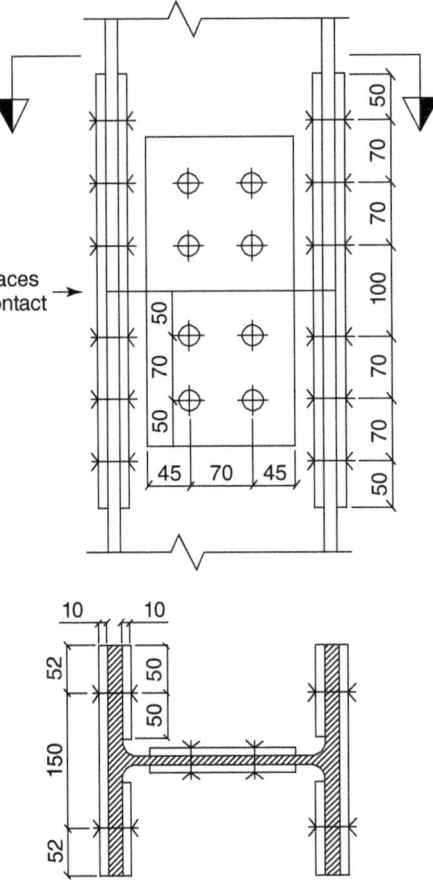

Figure 4.81 Designed splice.

same efficiency in tension where the plate and bolts work following the T-stub model and the plate might become quite thick. In addition, this solution does not provide much erection clearance and for bracings even small foundation differences or slightly out-of-plumb columns would make it troublesome to bolt the diagonals on-site. To give more ample tolerances during erection, fabricators sometimes ask the engineer to put double-bolted plates (see Section 4.7) or to design oversized holes (or slots). This calls for different computations like, for example, calculating the shear resistance by friction.

Another possibility, but one that is not used much, is the "kidney" slot (Figure 4.83). It is proposed by [15], with the following recommendations:

- Use it with only two bolts.
- Consider a shear capacity equal to 1.6 times the capacity of one single bolt and 1.5 times the bearing capacity of one bolt in the same plate.
- Adopt a slot width that is the same as the normal hole size and a length of $3d$ (where d is the bolt diameter).
- Keep a distance of at least $2d$ from the slot to the edges.
- Space the hole and slot at least $2.5d$.

4.14 Brace Connections | 219

Figure 4.82 Possible alternative design.

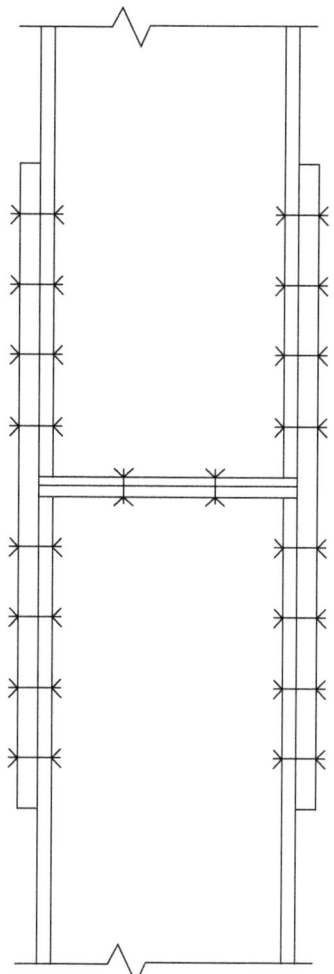

Figure 4.83 "Kidney" slot.

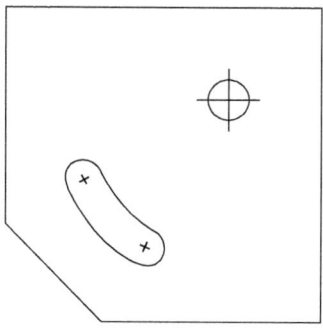

The advantages of a similar slot are a larger rotational tolerance during erection and the possibility that the same plate can be used for braces with a different angle.

Regarding connection of the brace gusset plate to beams and columns, this is usually "combined" with another connection (Figure 4.84) and hence it is likely that this latter joint must be dimensioned for the forces also coming from the brace, as, for example, in the case of Figure 4.85 (this concept was already discussed in Sections 2.5 and 2.16).

In order to design the gusset plate the engineer must also evaluate the Whitmore section (effective width) and the effective length for buckling (Section 3.21).

If the braces also work in compression and are made of double angles or channels, see the considerations in Section 4.10.1.

4.14.1 AISC Methods: UFM and KISS

This section will discuss AISC design methods for brace connections at the intersection of beams and columns, as shown in Figures 4.84 and 4.85.

The methods are for vertical bracing systems and may require adjustments in order to be applied to floor braces or, to a lesser extent, truss connections.

As mentioned in Section 2.2, the system shown in Figure 4.86 (and following figures) is highly undetermined and there are infinite equilibrium conditions. The scope is to determine a set of balanced actions for the vertical connection (which includes gusset-to-column and beam-to-column joints) and the horizontal connection (gusset-to-beam). The equilibrium of the forces must be granted in order to assign any bending moments to one or more of the connections in the system.

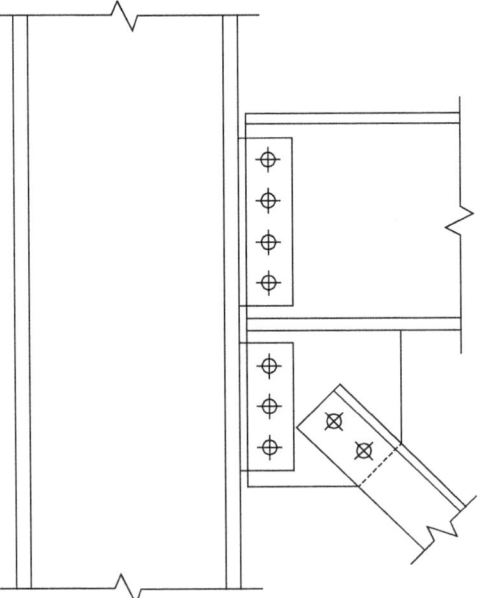

Figure 4.84 Brace example with a fin plate (shear tab) that connects both the beam and the brace gusset to the column.

Figure 4.85 Brace example with an end plate that connects both the beam and the brace gusset to the column.

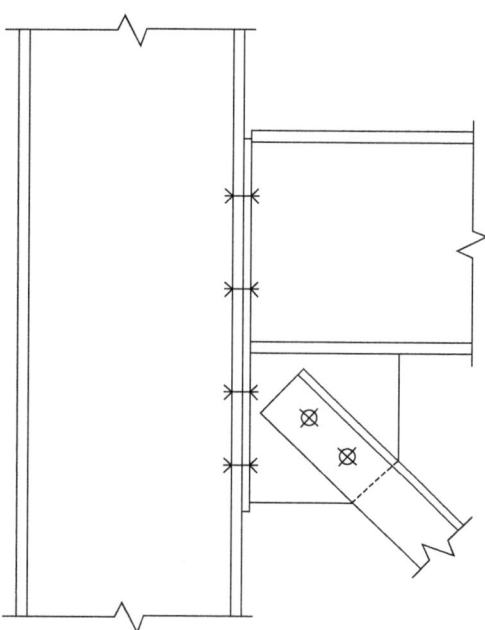

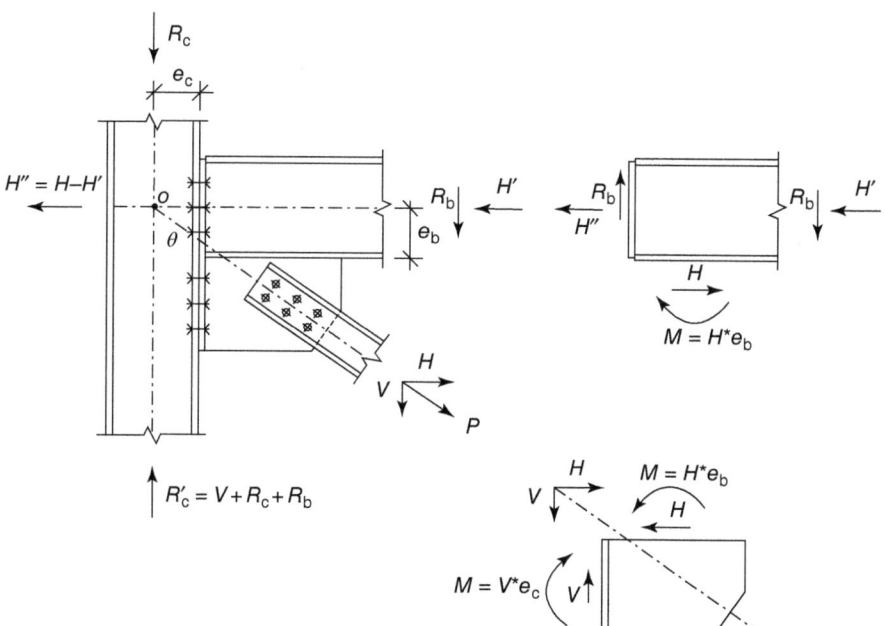

Figure 4.86 KISS method force distribution.

The two methods discussed here are as follows:

- The KISS (Keep It Simple Stupid) method brings a conservative design but is easy to apply.
- The UFM (Uniform Force Method) allows us to dimension the various connections without bending moments and hence to deliver good cost competitiveness.

4.14.1.1 KISS Method

The force distribution assumed by the KISS method is illustrated in Figure 4.86: The gusset-to-beam and gusset-to-column connections are designed not only for their components (simply, the vertical to the vertical joint, the horizontal to the beam) but also for the bending moments given by the respective eccentricities.

The beam-to-column connection must instead bear an R_b shear and an H'' axial force. For congruence, the vertical connection (bolted as shown in Figure 4.86) should also balance the moment $(V + R_b) e_c$ but often designers omit this check if the connection is quite deep (as it usually is) and thus with a good lever arm to oppose a bending moment. The same consideration is valid for the UFM.

4.14.1.2 Uniform Force Method

This method, the most used and the "official" one in [1], is effective for design (and hence economy) since, by applying some geometry rules, all the connections in the system can be dimensioned without bending moments.

There are indeed no bending moments in any of the joints (beam-to-column, gusset-to-beam, gusset-to-column) if their respective centers of mass satisfy the equation

$$a - b \tan \theta = e_b \tan \theta - e_c$$

where, given the symbols in Figure 4.87, a is the distance between the beam-to-gusset weld center and the gusset corner and b is the same for the gusset-to-column bolted connection. Being e_b, e_c, and θ given, there are infinite pairs of a and b values that verify the equation: graphically, it is sufficient that the intersection of the vertical line passing through the beam-to-gusset connection center and the horizontal line passing through the column-to-gusset connection center meet at the brace axis.

Once a and b are determined, the forces in the connection can be computed as

$$V_c = \frac{b}{r} P$$

$$H_c = \frac{e_c}{r} P$$

$$V_b = \frac{e_b}{r} P$$

$$H_b = \frac{a}{r} P$$

$$r = \sqrt{(a + e_c)^2 + (b + e_b)^2}$$

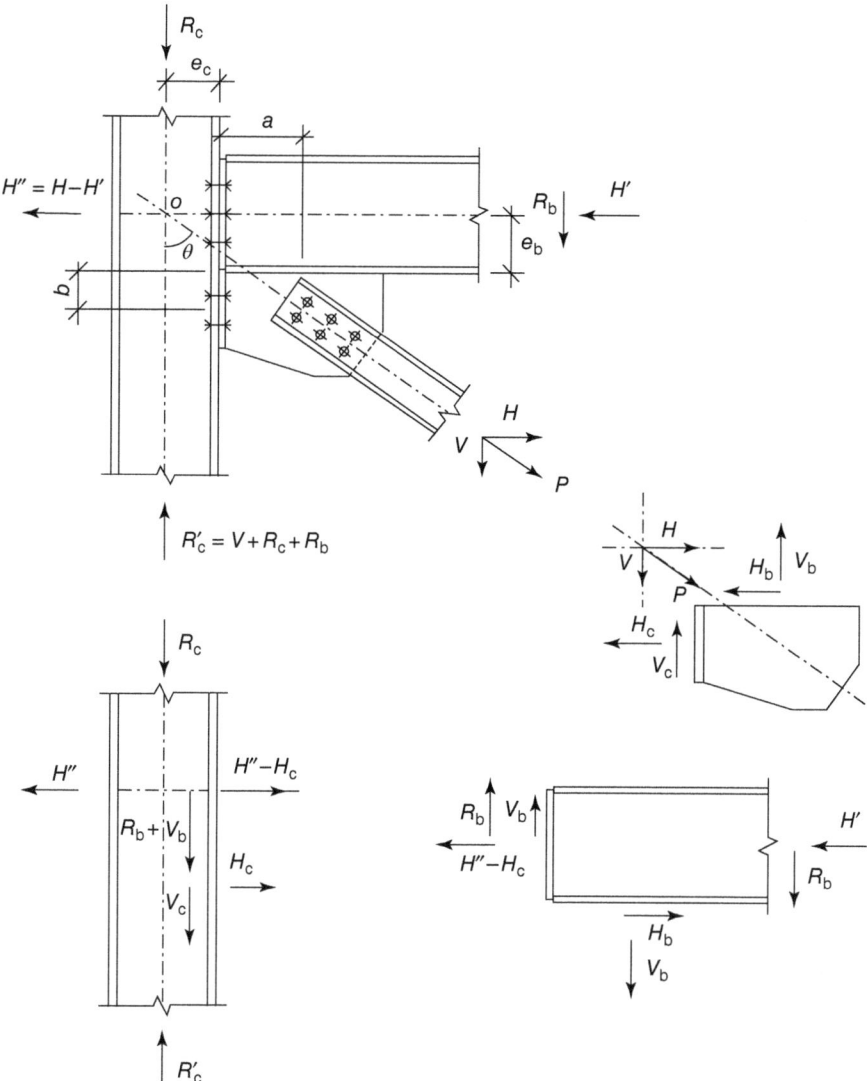

Figure 4.87 Uniform force method.

The beam-to-column connection is instead designed for the shear $R_b + V_b$ and the axial action $H'' - H_c$.

4.14.1.3 UFM Variant 1

If the geometry is chosen so that the brace axis converges at the intersection point O as shown in Figure 4.88, the gusset-to-beam and gusset-to-column design is simplified but there is a moment to be considered on the column and beam.

The bending moment acting on the beam can be taken as $M_b = H_b e_b$ while on the column it can be considered, if the joint is in an intermediate level of

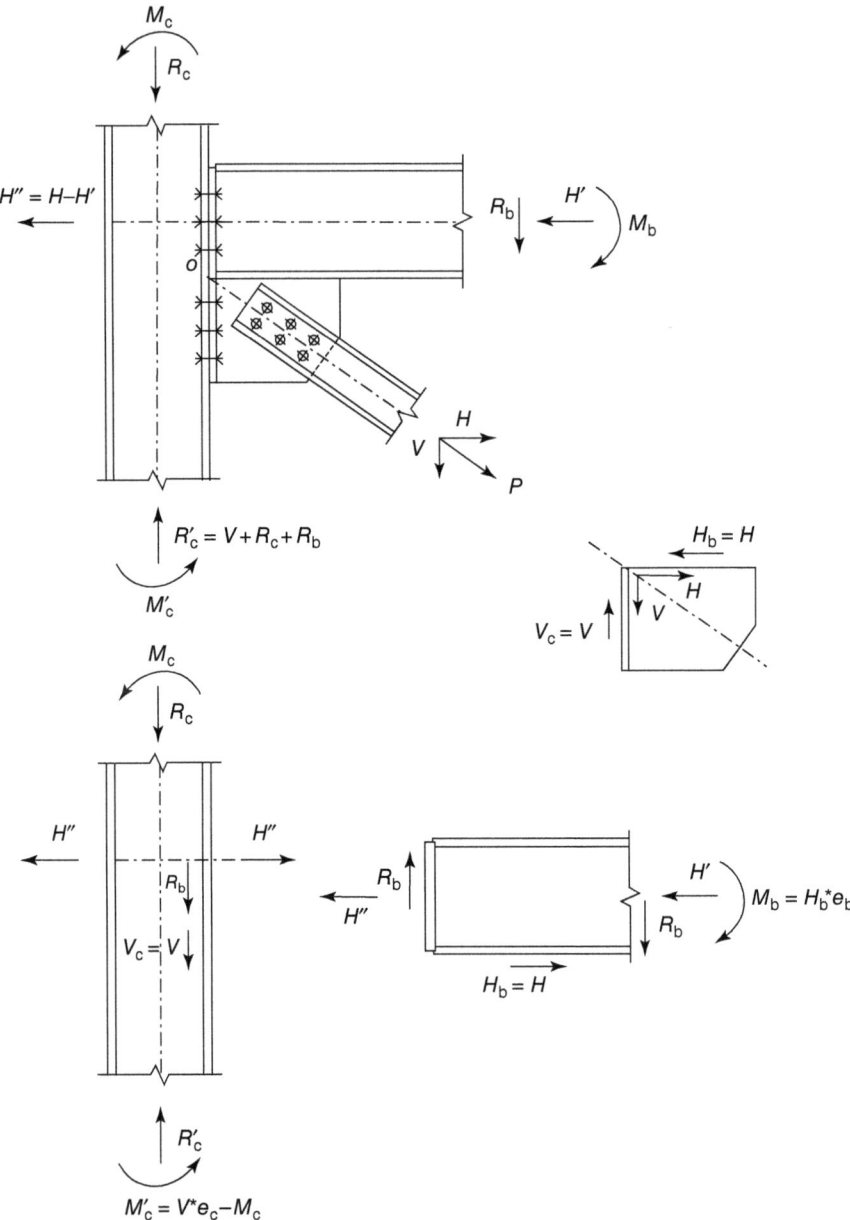

Figure 4.88 UFM, variant 1.

a continuous column, as $M_c = 0.5 V_c e_c$ with the same value for M'_c. If it is a single-story building, we have $M_c = 0$ and hence $M'_c = V_c e_c$.

4.14.1.4 UFM Variant 2

If the shear on the beam-to-column connection is too high (also for the beam web yielding), the calculation scheme (Figure 4.87) can be changed by reducing the action V_b of a chosen value ΔV_b (as high as needed) and by increasing V_c at the same time. This also means that in the gusset-to-beam design a bending moment $M_b = \Delta V_b a$ is introduced.

4.14.1.5 UFM Variant 3

In some geometric situations (very deep beam, steeply sloped brace, connection on column web, etc.) it may be economically convenient to realize the joint with the brace connected only to the beam and not directly to the column (Figure 4.89). This is equivalent to having $b = 0$ and, if the connection is on the column web, also $e_c = 0$. The a value that voids the bending moment on the gusset-to-beam

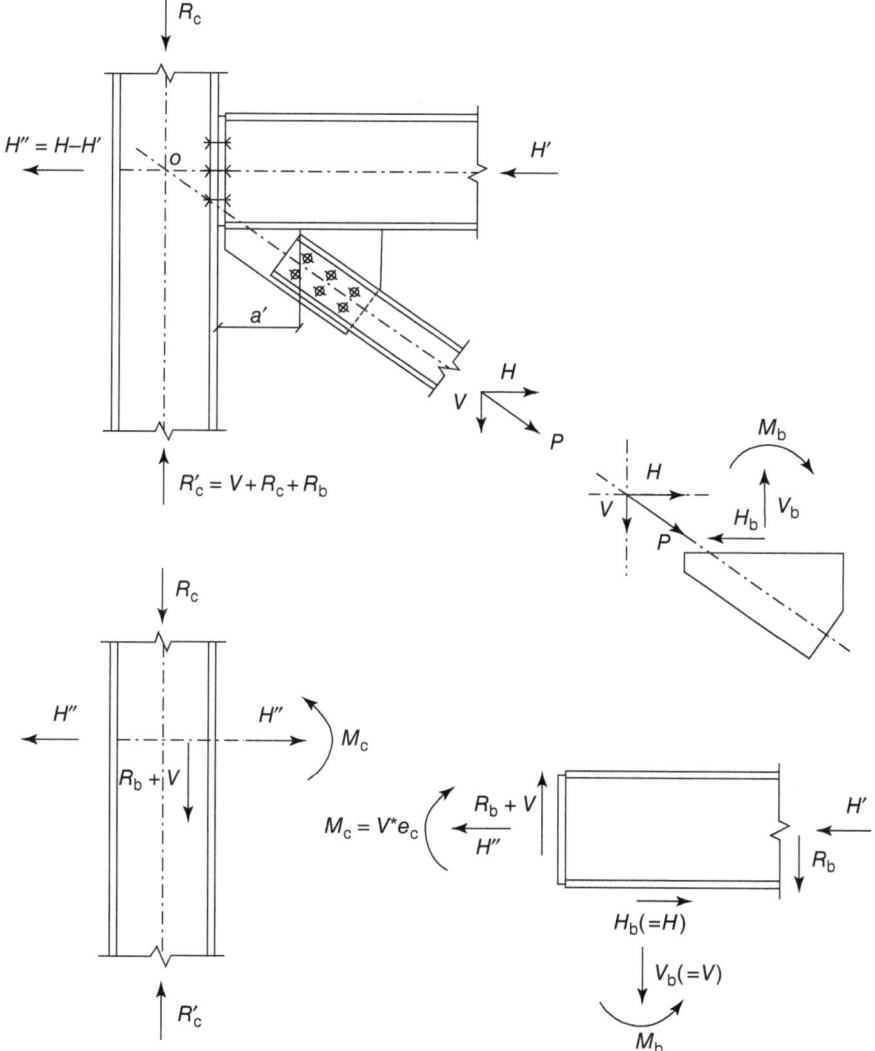

Figure 4.89 UFM, variant 3.

connection will then be

$$a = e_b \tan\theta - e_c$$

If the center of the gusset-to-beam connection is not a but a', the resultant bending moment becomes

$$M_b = V_b(a' - a)$$

It has to be noted that AISC recommends that the beam-to-column connection (since the balancing action H_c is not there anymore) also considers the additional moment

$$M_c = V e_c$$

This moment can be neglected (it is practically zero) if the connection is to the web of the column (i.e. to the column weak axis) because it is $e_c = {\sim}0$, and, therefore, the method can become very interesting in those cases.

4.14.1.6 UFM Adapted to Existing Connections

The relation between a and b that makes the moment zero is not always verified or applicable, as, for example, for existing connections.

In those applications there is a moment to be applied somewhere in the joint. According to [1], the moment is arbitrarily assigned to the stiffer connection. In the frequent case of a joint welded to the beam and bolted to the column, the moment is assigned to the gusset-to-beam connection. In a similar case, given that $b = b'$ (a' and b' denote the effective values of a and b that do not void the moment), we get the a value that would make the moment zero from the known equation:

$$a = e_b \tan\theta - e_c + b \tan\theta$$

The moment in the connection will thus be, similar to variant 3,

$$M_b = V_b(a' - a)$$

If the moment is given to the gusset-to-column connection, we would have

$$M_c = H_c(b' - b)$$

If the stiffness of the parts is similar and/or the designer wants to distribute the moment, Ref. [1] again recommends calculating the moments considering the following values for a and b:

$$a = \frac{K' \tan\theta + K\left(\frac{a'}{b'}\right)^2}{D}$$

$$b = \frac{K' - K \tan\theta}{D}$$

where

$$K' = a'\left(\tan\theta + \frac{a'}{b'}\right) \quad D = \tan^2\theta + \left(\frac{a'}{b'}\right)^2$$

4.14.2 Practical Recommendations

For slender braces designed as tension only, it is appropriate to prevent any vibration issue or large-deflection (gravity-induced, especially in horizontal braces) problem. In this regard, the engineer may specify to use suitably sized turnbuckles (normally commercially available) or the like.

It is instead an AISC practice (if the braces are "light") to reach some pretensioning by shortening the fabricated length of the diagonal brace from its theoretical length. This shorter length is $3/16$ in. (4.5 mm) if the diagonals are longer than 35 ft (about 10 m), $1/8$ in. (3 mm) between 20 ft (6 m) and 35 ft, and $1/16$ in. (1.5 mm) between 10 ft (3 m) and 20 ft (no reduction below 10 ft).

4.14.3 Complex Brace Connection Example According to CSA S16

We analyze here a brace connection of an industrial building. The brace (a single C130×13, the equivalent of a C5×9 American channel) frames into the W200×52 column weak axis and there is also a W250×25.3 beam at the intersection (see Figure 4.90). The column also has two (10-mm-thick) stiffeners because on the column strong side there is a moment connection that requires them.

The profile material is 350W, the plate material is 300W, and the bolts are A325M (M stands for metric). On the left of the column (not represented in the figures), there is another W250×25.3 beam.

The brace works in tension and compression with the following maximum values:

Case SLUlat

$$P_{f\,brace} = \mp 150 \text{ kN}$$
$$V_{f\,major\,beam} = 15 \text{ kN}$$
$$P_{f\,beam} = \mp 10 \text{ kN}$$

Maximum shear and tension for the beam are instead in another case:

Case SLUlive

$$P_{f\,brace} = -20 \text{ kN}$$
$$V_{f\,major\,beam} = 33 \text{ kN}$$
$$P_{f\,beam} = 15 \text{ kN}$$

4.14.3.1 Friction Connection for Brace

We start dimensioning the brace-to-gusset plate connection. The request by the fabricator (and erector) is to have oversized holes for bolts to ease the erection. The connection will then be designed as slip resistant at serviceability (adequate for this structure). The relevant load SLSlat is approximately

$$P_{f\,SLS\,brace} = \mp 100 \text{ kN}$$

Using M16 bolts, the slip resistance of one bolt (friction coefficient taken conservatively as 0.33 as per class A of [25]) is (Equation 13.12.2.2 in S16)

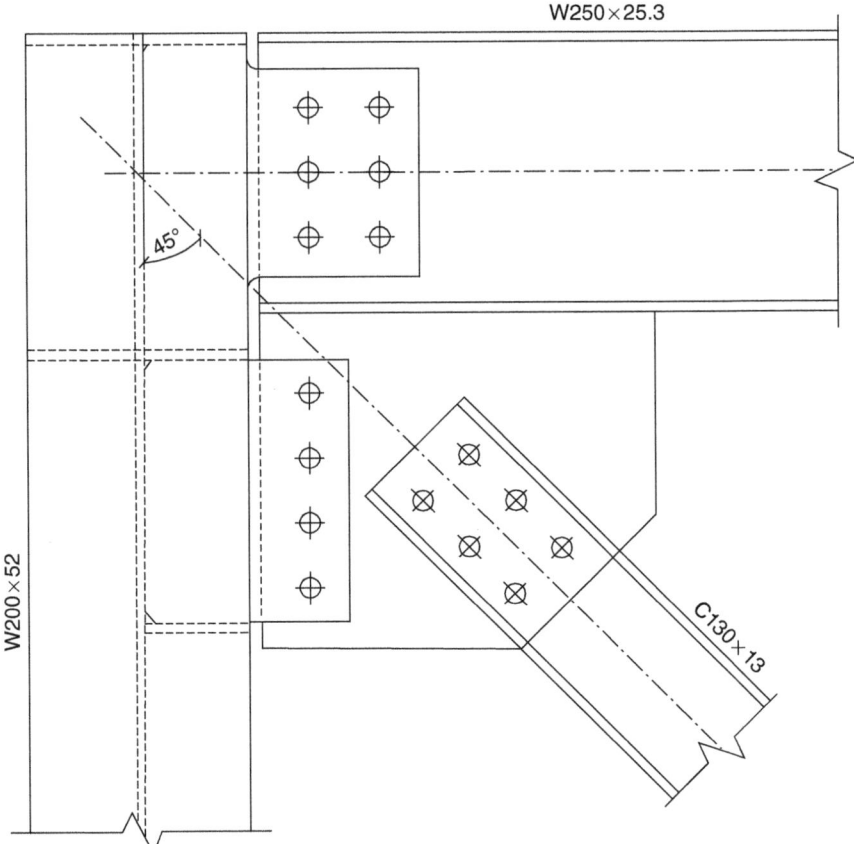

Figure 4.90 Brace connection; the geometry of the gusset is designed to minimize eccentricities.

$0.53 \times 1 \times 0.33 \times 0.82 \times 827 \times 201 = 23.8$ kN. Hence, six bolts are necessary and they resist about 143 kN.

4.14.3.2 Brace and Gusset Bearing

The channel web is 8.26 mm so we choose a little bigger gusset plate thickness (there might be local instabilities), say, 10 mm. Considering an edge distance that is twice the bolt diameter (suggested minimum by the author for braces), the bearing check becomes $0.8 \times 3 \times 16 \times 10 \times 450 = 173$ kN, that is, a $(150/6)/173 = 0.14$ exploitation ratio.

Similarly, for the brace we get (see SCS) 18%.

4.14.3.3 Block Shear

As it can be anticipated for the limited bearing ratios, the block shear is also widely verified. From SCS we read values around 29% for the brace and 22% for the gusset.

4.14.3.4 Channel Shear Lag

Since the channel is only connected by the web there is a 0.75 shear lag coefficient to apply. The tension resistance is then $0.75(=\Phi) \times 0.75 \times 450 \times (1664 - 8.26 \times 44) = 329$ kN (the gross yield area does not govern), which is acceptable.

4.14.3.5 Whitmore Section for Tension Resistance and Buckling of Gusset Plate

The Whitmore section in the gusset is $10(120 \times \tan 30° \times 2 + 60) = 10 \times 198.6 = 1986$ mm². The tension resistance for the net area (governing) is then $0.75 \times 10(198.6 - 2 \times 22) \times 450 = 521$ kN (29% ratio).

For buckling, let us recall the discussion in Section 3.21.1. If we take [26] as reference, the limit thickness to define the plate as "compact" is, taking $c = 120$ (conservative), $t_\beta = 1.5\sqrt{(300 \times 120^3/(200\,000 \times 120/\sin 45°))} = 5.9$ mm and thus the plate can be considered compact and no buckling check is necessary (slenderness $= 0$) as per [26]. Taking instead the advice in [27] and assuming a 0.65 coefficient with $l_{avg} = \sim(225 + 140 + 140)/3 = 170$, we get (the weak-axis radius of gyration of the plate Whitmore section is $10/\sqrt{12} = 2.9$ mm) $Fe = \pi^2 \times 200\,000/(170 \times 0.65/2.9)^2 = 1359$ MPa and, therefore, a resistance (see Section 13.3.1 in [25]) $0.9 \times 1986 \times 300(1 + (\sqrt{(300/1359)})^{\wedge}(2 \times 1.34))^{\wedge}(-1/1.34) = 488$ kN, which is acceptable.

4.14.3.6 UFM Forces

The forces distributed to the column and beam connections of the gusset are computed according to the UFM. If we design the gusset to be 360 mm wide, geometrically b (see Figure 4.87) is about 135 mm and a (again in Figure 4.87) is about 163 mm: This is done on purpose to have the centerlines meet on the brace axis so that moments are zero. The vertical connection being bolted is in fact correct to take its axis on the bolt group centerline instead of the end of the gusset. Also, the center of the connection (where the distance b is measured, see Figure 4.91) should be taken from the middle of the bolt group. We calculate $r = \sqrt{((135 + 158)^2 + (128.5 + 163)^2} = 413$, and hence $V_c = 163/413 \times 150 = 59$ kN, $H_c = 157/413 \times 150 = 57$ kN, $V_b = 128.5/413 \times 150 = 47$ kN, and $H_b = 135/413 \times 150 = 49$ kN. The values match very well with the ones given by SCS (UFM shared moments option). The remaining moment is just negligible (eccentricities < 0.3 mm) as confirmed by SCS.

4.14.3.7 Gusset-to-Column Shear Tab

See Section 4.6.6 for a detailed calculation example. SCS sets this up automatically: The governing limit state is the bolts in shear (37%).

4.14.3.8 Gusset-to-Beam Weld

We design a double fillet of 6 mm nominal size (throat about 4 mm), which is largely acceptable (8% ratio by SCS).

4.14.3.9 Beam-to-Column Shear Tab

Even in this case no detailed calculations are shown since they are similar to previously studied situations. A 10-mm-thick plate gives acceptable results (see

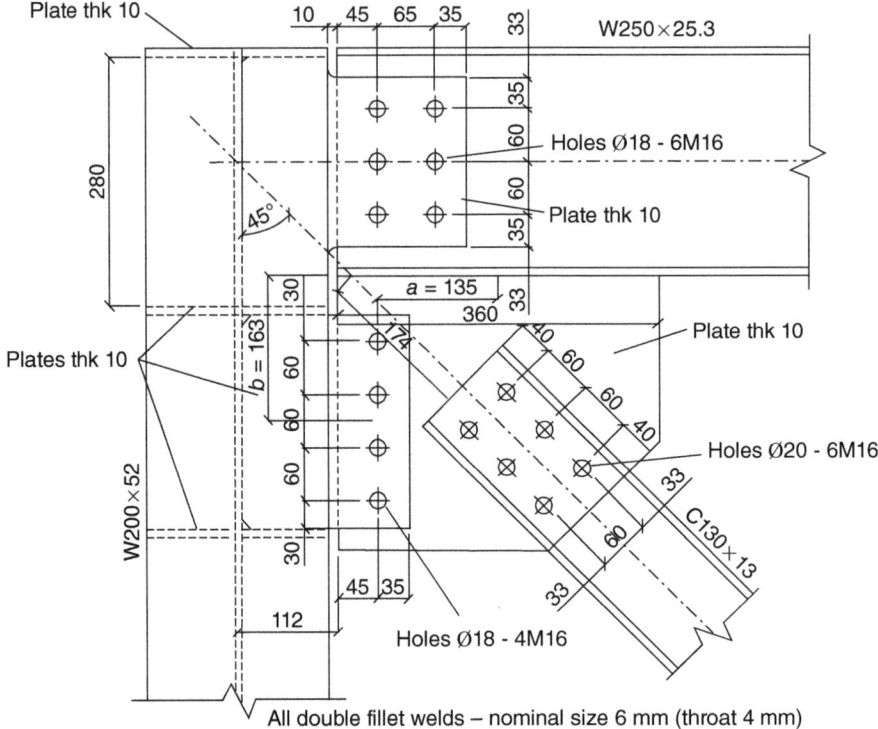

Figure 4.91 Dimensional details.

Figure 4.91). By positioning, as rigorously correct, the connection axis on the column axis, the bolts have a remarkable eccentricity (about 190 mm), which makes their shear limit state work at 88% (by SCS; governing with the case SLUlat). If we do not consider it (the gusset-to-column shear tab could help a lot in resisting it with a combined action), the bolt shear is at 22% (still governing).

4.14.3.10 Ductility and Structural Integrity

Not considering the beam-to-column shear tab value at 88%, which, as we just mentioned, has very conservative hypotheses, the brace-to-gusset bolts in friction govern, which gives plenty of additional potential resistance (the same bolts will then react in pure shear).

The connection seems therefore positively redundant and also satisfies the initial request of easing erection operations by providing oversized holes.

4.15 Seated Connection

Seated connections as in Figure 4.92 are currently not very popular, at least for H-shaped and I-shaped profiles.

One of the issues is that, although they have always been traditionally taken as pin connections, the EC (if this is the basis for the design) considers them as

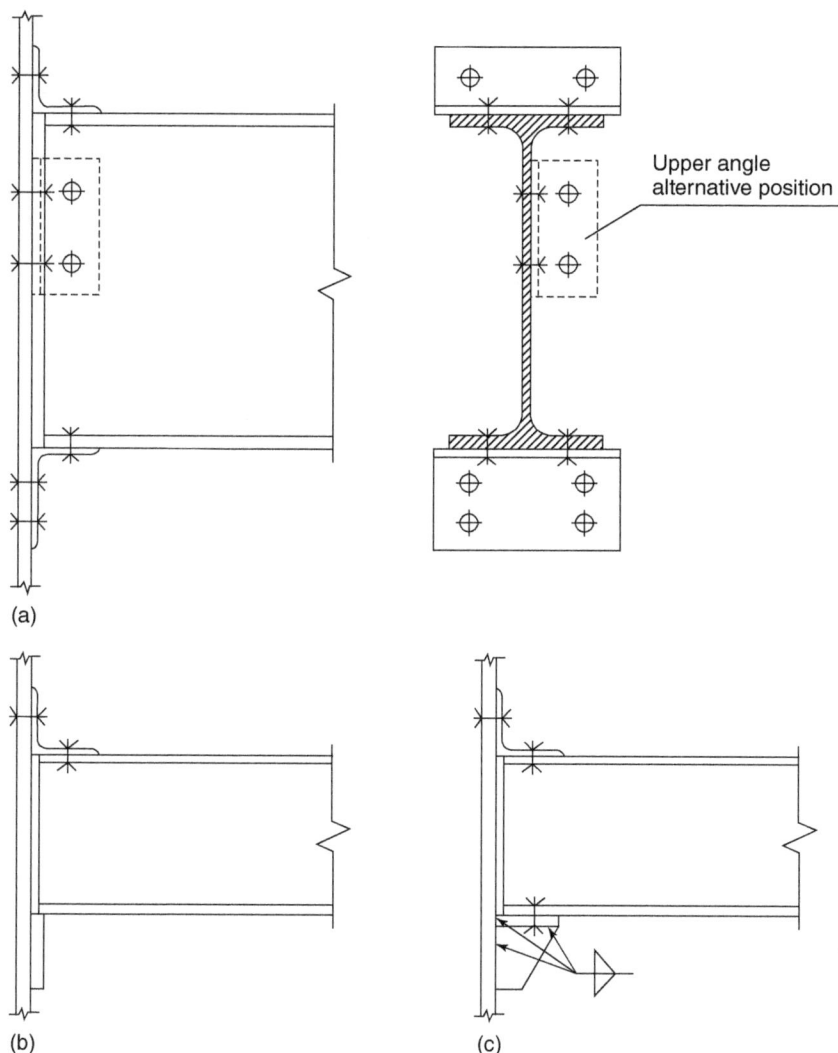

Figure 4.92 Seated supports: unstiffened (a), bearing pad (b), and stiffened (c).

semirigid joints, with consequent analytical complications. If, for other types of connections, for example, end plates and fin plates, there might be "loopholes" to still assume them as pins (considerations by [13], mentioned in Sections 4.6.1 and 4.8.1), for seated connections with an angle on top (which is recommended to provide stability) it would be necessary to evaluate the rotational stiffness (see Ref. [5]) for a design that strictly suits the EC. However, Australian and US design codes consider, with some necessary details, this kind of joint as a simple pin connection. For example, bolt pitches must not be tight or, if realized by welding, the angle connected to the beam top flange (minimum angle depth is 100 mm) must be welded to the column only on its top edge, leaving the angle free to deform on its sides. For additional details the reader is referred to [1, 28].

Similar supports with only the bottom angle are sometimes adopted to ease erection (temporary supports) or are used for C-shaped profiles, channels, and tubes when they are secondary frames such as girts and purlins (see also Section 4.16). Since the shape of the profiles provides stability or, in girts, the load is essentially horizontal being given by the wind, it is not strictly necessary to put the top angle and this helps in considering the connection as "simple," with no bending moments at the supports.

The vertical-load check for this kind of connection is done for the support plate (often the leg of an angle, see Figure 4.93) by considering it a cantilever. The bending moment will be equal to the shear by the distance of the assumed point of support (to be defined) from the point where the support ends or increases its dimensions (e.g. if we have an angle, at the start of the root radius with the other leg). The effective width of the equivalent section will be obtained by distributing the load at 45° on the horizontal plane. Another check to be done is for the beam web, which might be unable to oppose the concentrated reaction at the support. Even here, the force can be spread (now vertically) at 45° to calculate the effective width of the web that can resist the force. Some authors (e.g. Ref. [3]) use this consideration to locate the point of reaction for the shear (necessary to find the bending moment as just seen); that is, they calculate the web width necessary to resist the concentrated force and, from this, with an α distribution at 45° (or at 60° as in some sources, e.g. DIN), they go backward in the computations.

Using formulas (and symbols) as in [3], we have (see also Figure 4.93)

$$b = \frac{R}{\frac{1.3f_y}{\gamma_M} t_w} - c$$

where $c = (tf + r) \tan \alpha$ and having taken, as per [3], 1.3 times the yield strength for the contact pressure. From b we can thus get the force eccentricity and check the leg of the support angle as a cantilever.

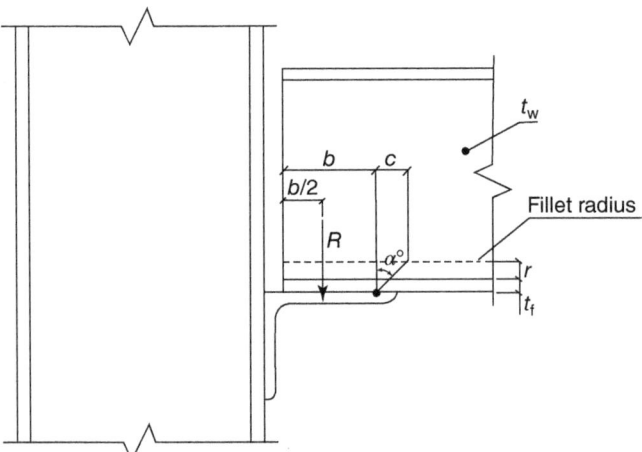

Figure 4.93 Seat, parameters to calculate the supporting leg resistance.

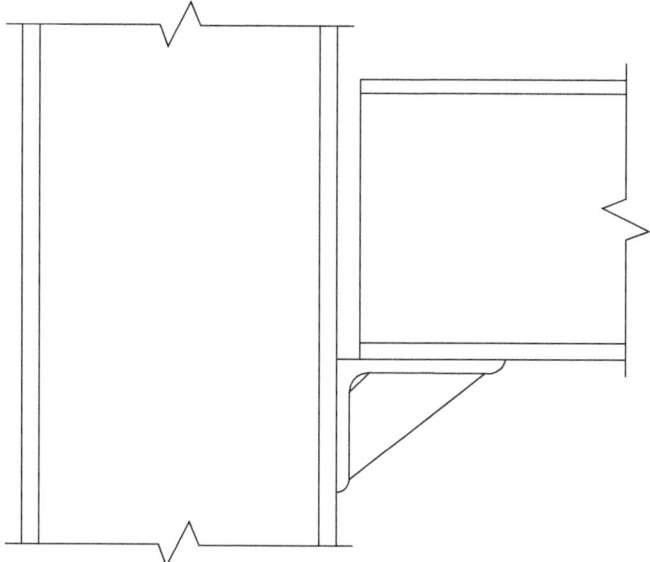

Figure 4.94 Stiffened seat detail.

Adding a local stiffener in the beam might help in reducing the eccentricity to its minimum. Stiffening the seat (Figure 4.94) makes the cantilever check useless. In this case it is preferrable to use a stiffener thickness that is similar to the beam web. Depending on the kind of connection between the seat and the primary member, the engineer will proceed to check welds and/or bolts according to the methods already shown (see Section 4.3). If necessary, the primary member (usually a column) will also be checked for local limit states.

4.16 Connections for Girts and Purlins

Secondary frames in roofs are often made with light cold-formed sections of small thickness (3 mm, i.e. $1/8$ in., maximum 4 mm usually). Those profiles might be rectangular hollow sections (RHSs) but they are more frequently C, Z, or Ω (omega), as shown in Figure 4.95. Similarly, the same types (most of all Cs and RHSs) are used as girts, that is, to support lateral panels or the like. One of the main advantages in choosing thin cold-formed sections is that the panels can normally be fixed by self-drilling or self-threading screws.

The purlins lie on beams and are fixed in several possible ways:

- With angles bolted to the beam top flange and the purlin web if this is, for example, a channel or RHS (Figure 4.96).
- With systems similar to an upside-down U (actually many others are possible) welded to the beam and bolted through both sides (webs) of Ω purlins; this connection eases the erection and can also take small bending moments by

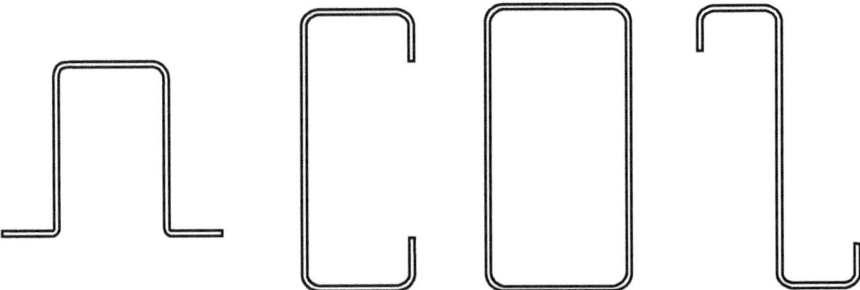

Figure 4.95 Typical profiles for purlins: Ω, C, RHS, Z.

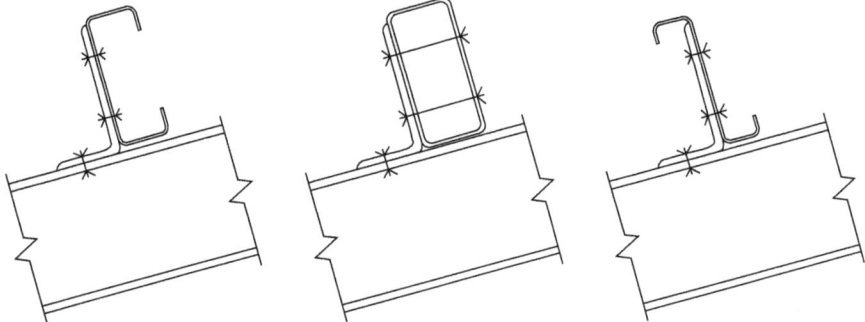

Figure 4.96 Connection with bolted angle, suitable for C, Z, or RHS profiles.

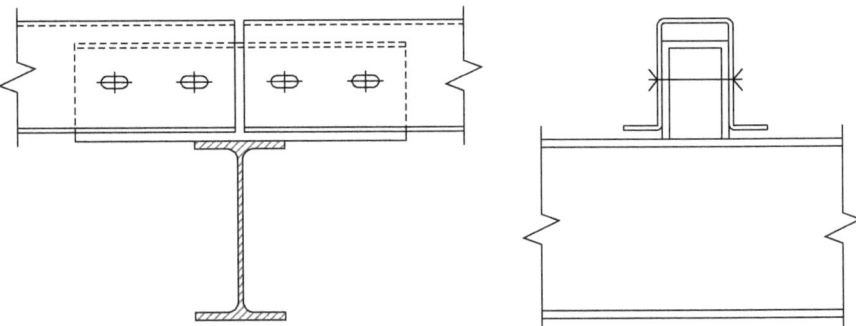

Figure 4.97 Possible connection of Ω purlins with an upside-down U-shaped bent plate.

making the two (Figure 4.97) bolts work together (which, notice, work on two resistant sections) on each end.
- By bolting the lower part of the purlin directly to the beam top flange if the purlin is Ω shaped or by bolting to the beam a plate that is welded to the bottom of the purlin if this is in RHS; this joint is not recommended for the same reasons discussed in Section 4.11.

The various limit states to be considered are easily derivable by the engineer, case by case, and they are mostly related to bolt shear and bearing.

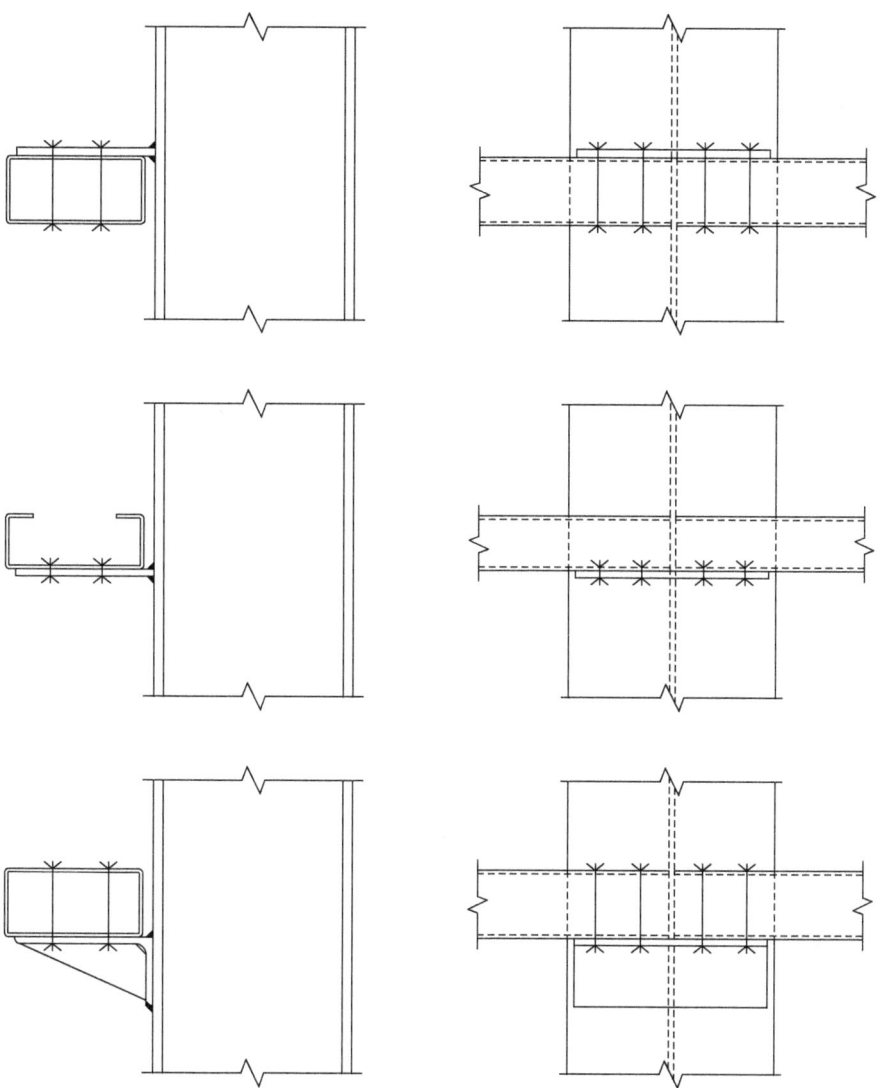

Figure 4.98 Possible connections bolted to the girt and welded to the column.

For RHS- and C-shaped profiles used as girts, support angles (or equivalent cantilevers made of plates) welded (or, more rarely, bolted) to columns are commonly designed. The girts are then bolted to the supports. Some possibilities are shown in Figures 4.98 and 4.99.

The limit states to check are analogous to the ones for purlins, with the important addition that the supporting cantilever is stressed by both horizontal and vertical loads, unless the vertical wall directly leans on the foundation (see Section 4.15).

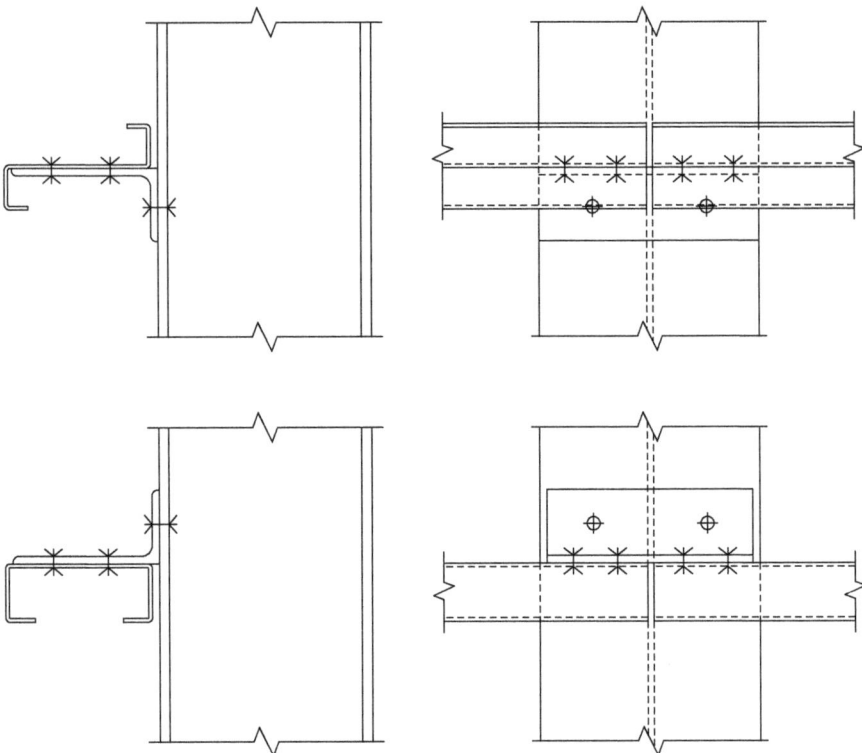

Figure 4.99 Possible completely bolted connections for girt-to-column joints.

4.17 Welded Hollow-Section Joints

This kind of joint, being fully welded, is not in the scope of this text.

Detailed guidance on this point is provided in [5] (in Chapter 7) and to a lesser extent in [1] (in Chapter K of the provisions) (there are however several errors in the original version of [5] and hence the reader should check the list of errata-corrigenda).

4.18 Connections in Composite (Steel–Concrete) Structures

Composite steel–concrete joints are not in the scope of this text and the reader is referred to [29–31].

4.19 Joints with Bolts and Welds Working in Parallel

As seen in Section 2.3, bolts and welds should not work in parallel, in the sense that they should not share the resistance to the same force.

Recent studies at the University of Alberta [32], already included in the latest AISC provisions, show that it is possible to make welds and bolts work together if the resistance of the bolts is taken as 50% of its maximum capacity.

Another way to make bolts and welds share the same loads is to design the bolted connection as slip resistant to the ultimate limit states, as requested by the EC, with class 8.8 (or above) bolts in category C, pretensioned after the welds are completed.

An interesting situation occurs when an existing structure is modified to take new loads (and/or members) and a quick solution would be to add welding to existing bolted connections. Concerning this, the AISC provisions suggest that the welds can be designed only for the additional load if the existing bolts are pretensioned and well designed for the previous actions. Caution and engineering judgment must always be used though.

4.20 Expansion Joints

Thermal variations in steel structures usually do not require special expansion joints because the structural dimensions that would require them are quite large and the problem is commonly solved by "dividing" the structure, that is, by designing a double line of columns that separates the parts.

Figure 4.100 provides some interesting general indications in the hypotheses of a structure internally heated and with pinned column bases.

The design variables are several and [33] suggests, generally speaking, the following adjustments:

- If the structure also has summer temperature control (air conditioning), the maximum distance can be increased by approximately 15%.

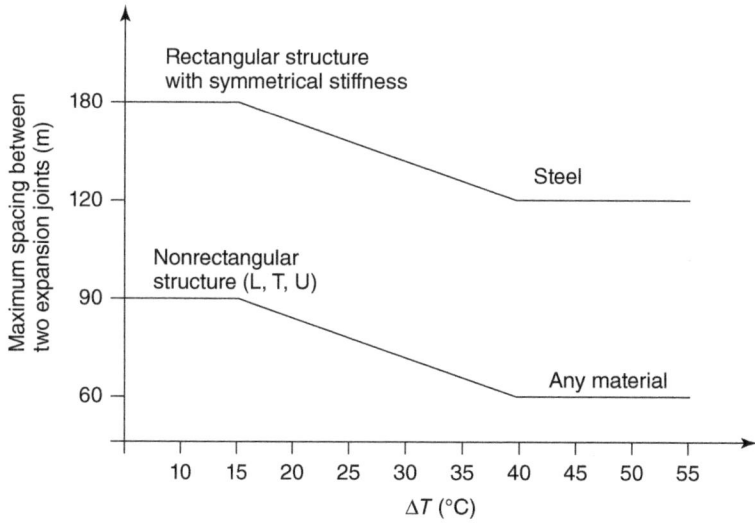

Figure 4.100 Maximum suggested reference distance for expansion joints. Source: From Ref. [33].

- If there is no temperature control (no heating either), the distances must be reduced by one-third.
- If the column bases are fixed, the distances should be reduced by approximately 15%.

The bracing configuration is also important: only one central brace, if sufficient, allows the structure to expand (and shrink) while two pairs of braces in the outer spans block the lateral deflection with consequent internal stresses that become a serious design issue in long buildings.

Going back to the main topic of the discussion, that is, the connection, it is possible to realize some "roller" constraints (see Section 4.22) by using Teflon (PTFE) or a similar material (that does not eliminate the problem completely since there will be some friction left and/or maintenance required) but the best choice is, as anticipated and if possible, to separate the parts of the structure by introducing double frames where needed.

4.21 Perfect Hinges

There are special applications (e.g. details of bridges or parts that must rotate) where a "real" pin is needed.

As reference formulas to study the problem we take the ones (probably the most recent and clear among standards) in EC that divide the case where the thickness is given from the case where the geometry is assigned.

If the thickness is given, the following (see symbols in Figure 4.101) must be verified:

$$a \geq \frac{F_{Ed}\gamma_{M0}}{2tf_y} + \frac{2d_0}{3}$$

$$c \geq \frac{F_{Ed}\gamma_{M0}}{2tf_y} + \frac{d_0}{3}$$

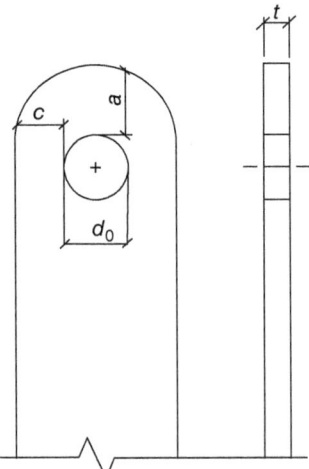

Figure 4.101 Given thickness.

Figure 4.102 Given geometry.

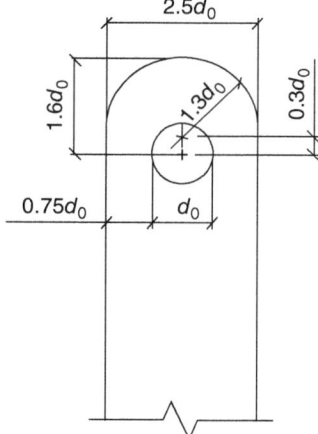

where f_y is the lowest yield value among the connected parts and F_{Ed} is the design axial force.

If the engineer wants to adopt a suggested "standard" geometry (Figure 4.102), the thickness must verify the following:

$$t \geq 0.7\sqrt{\frac{F_{Ed}\gamma_{M0}}{f_y}}$$

$$t \geq \frac{d_0}{2.5}$$

The additional limit states to check are (d is the pin diameter) as follows:

Pin shear: $\quad F_{v,Rd} = \dfrac{0.6Af_{up}}{\gamma_{M2}} \geq F_{v,Ed}$

Bearing: $\quad F_{b,Rd} = \dfrac{1.5tdf_y}{\gamma_{M0}} \geq F_{b,Ed}$

Pin bending moment $\quad M_{Rd} = \dfrac{1.5W_{el}f_{yp}}{\gamma_{M0}} \geq M_{Ed}$

Combined shear and bending $\quad \left[\dfrac{M_{Ed}}{M_{Rd}}\right]^2 + \left[\dfrac{F_{v,Ed}}{F_{v,Rd}}\right]^2 \leq 1$

It must be noted that it is necessary to also check the bending moment and not only the shear as in bolts.

More information is provided in [5] if the pin has to be replaceable.

4.22 Rollers

As seen earlier, in special applications such as bridges where the design might need special details like rollers, Teflon (PTFE) joints with a very low friction coefficient are sometimes adopted (Figure 4.103).

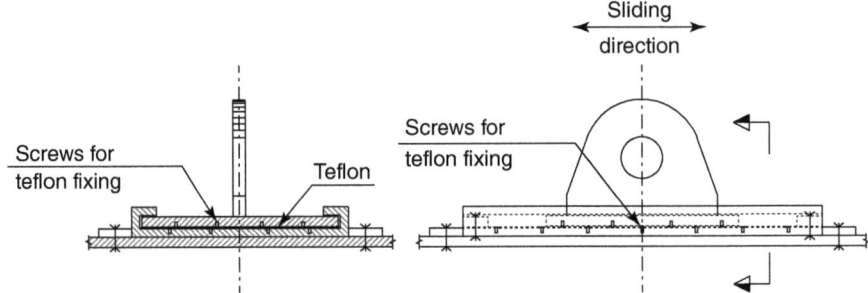

Figure 4.103 Possible joint scheme that allows translation and rotation; stainless steel and rubber parts might be required to be part of the system.

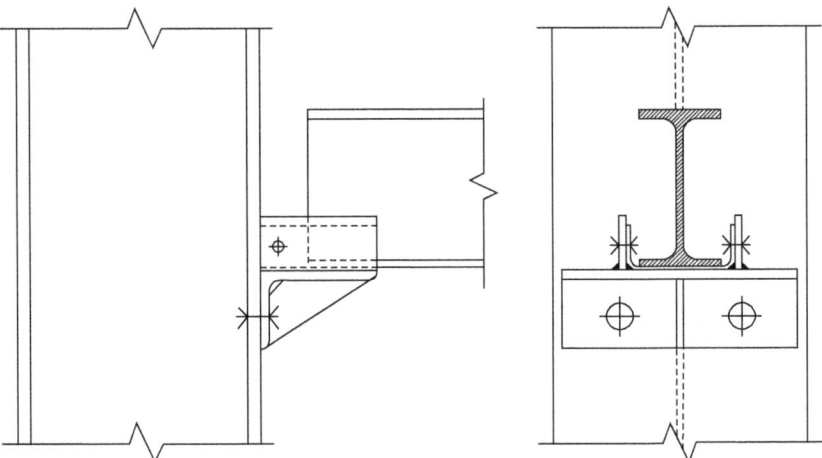

Figure 4.104 Alternative solution that allows the beam to slide.

An alternative to Teflon (less efficient from a friction point of view but with better characteristics such as durability) is to create contact using (lubricated) stainless steel (an example is given in Figure 4.104).

The designer has to check that the sliding area does not get uplift or overturning forces, in which case suitable measures must be enforced.

Recently, a new reference standard was introduced in Europe, EN 1337, and, depending on the geographical location of the project, the engineer might have to carefully consider it (local check for the PTFE stress might also be required). The standard requires that the bearing systems be certified in order to guarantee durability and maintenance. Other local requirements might apply.

4.23 Rivets

Rivets were used in the past, before high-strength bolts and when welding processes were not automated or large laminated sections were not available, to create for example deep beams composed of plates.

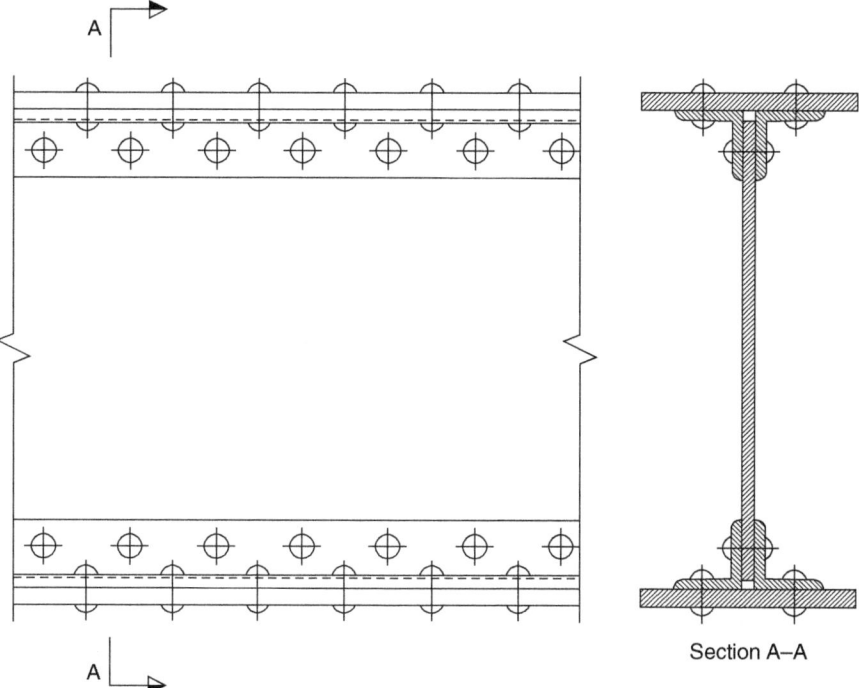

Figure 4.105 "Old" beams made by riveted plates.

Currently, the usage of rivets in steel structures is extremely limited and hence, because the goal here is not to cover the subject in detail but to concentrate on the practical cases of steel structures, it is not in the scope of this book. Current standards do provide formulas if the engineer needs to use rivets so it will be easy to find reference equations.

Remember that, if the forces in the rivets of Figure 4.105 (or, similarly, in the welds if that is the case) need to be calculated, they have to absorb the shear between the web and flange, that is, the difference in axial action on the flange, which happens to be the beam shear in that point if the axial action is uniform (see Ref. [3] for more insights).

4.24 Seismic Connections

A detailed discussion of joints for seismic applications is not within the scope of this text and consideration of "special" seismic connections is excluded (for example, isolators or dissipators, maybe even patented, that reduce the quake energy applied to the structure by exploiting various phenomena such as fluid deformation, solid viscoelasticity, friction, yielding, and spring dynamics).

In this section we discuss how to modify or improve the "classical" joints that we have studied (say moment end plates or brace connections) to guarantee an efficient seismic response and to be able to define a certain structure as ductile so that higher behavior factors (>1 for EC) or response modification coefficients (>3 essentially with US standards) can be used in the structural analysis.

Seismic engineering experience shows that the connections are a focal point in the resistance to earthquakes.

The basic concept is that the joint must not be a weak link in the structure but must conversely guarantee, due to its resistance and/or deformation capacity, that the connected members can yield and thus absorb some energy of the quake. The material yield value being, by definition, a statistically guaranteed minimum, the effective yield limits are higher than is necessary to apply some overstrength factors to make sure that the connection resistance is higher than the connected parts. It is important to remember that some provisions allow to take as design actions forces that are lower than the connected part strength if the analysis forces are smaller than the full strength when considering a unit (or more, see reference standards) behavior factor (or its equivalent).

The joint deformation is essential too because it allows to dissipate seismic energy and in addition allows some redistribution of forces to other parts of the structure (ductility).

4.24.1 Rigid End Plate

A rigid end plate must be able to transmit the full bending strength of the connected beam, amplified by an overstrength factor as per the applicable building code.

To realize a full-strength bolted end plate, the guidelines illustrated in Section 4.12 can be followed once the required actions are imposed.

It seems useful to mention a possible "loophole" that could be taken to have an end plate that fully recovers the beam strength without excessive calculations and details: The joint might be welded on-site. This method was common practice for years in the United States in order to achieve full-strength portals in highly seismic areas. However, the 1994 Northridge earthquake in California made evident the limits of this practice, which consisted in bolting the web (to ease erection and the initial positioning of the beams) and then site weld the beam flanges to the column flanges. The quake recorded many fragile failures, most of all near the bottom flange of the beam (later tests identified that the main problem was in the limited rotation capacity of the joint). The corrective actions taken after the 1994 experience were as follows (technically, the topic is very complex and cannot be summarized in a few lines):

- Some tests and checks (among which the plastic rotation, which has to be between 0.01 and 0.04 rad depending on some variables) are now required for some welded (and nonwelded) beam-to-column connections.
- The joint needs to be reinforced locally.
- Reducing (hence weakening) the beam section near the connection will cause a plastic hinge to form there and let the beam rotate as necessary.

This latest method [reduced beam section (RBS)], is also known as "dog bone" because the shape of the beam flanges after the intervention looks like a bone.

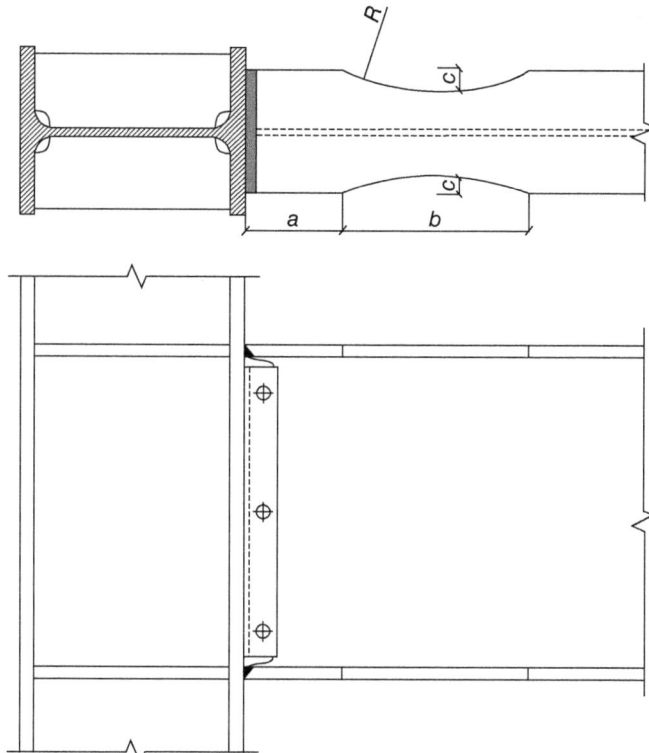

Figure 4.106 Dog bone.

As approximate values (for details look at [34]) for a, b, c, and R shown in Figure 4.106 (h_b is the beam depth and w_b the width) consider

$$0.5w_b \leq a \leq 0.75w_b$$
$$0.65h_b \leq b \leq 0.85h_b$$
$$0.1w_b \leq c \leq 0.25w_b$$
$$R = \frac{4c^2 + b^2}{8c}$$

AISC provisions have often been ahead of the rest of the world (Japan excluded) if the topic is seismicity and, therefore, it is suggested to study the subject (that quickly updates to be refined) in [34], which is also quite rich in design examples of seismic connections.

4.24.2 Braces

Even here it is essential (if we want to use consistent response modification factors) that the connection is not the weakest link but rather allows the brace to go over the elastic limit and enter plasticization. To achieve a full-strength connection, the engineer must, first, design the connection so as to avoid shear lag issues (see Section 3.19.1) and, second, not penalize too much the brace section with holes.

If in fact we connect the brace only on a part of the profile, the entire force that the member can transmit cannot go through the joint. This concept was discussed in Section 3.19 with the applicable design formulas.

The holes must also be designed in order to avoid excessive weakening of the net section: The holes should not decrease the gross section of a factor larger than the ratio of the yield and ultimate tensile strength of the material (including in the computation the partial safety factors). The yield of the gross section (ductile event) must prevail over the net section failure (fragile). The Italian NTC rules ([35], where f_{tk} is the ultimate tensile strength of the material) define this by the equation (which includes an additional 1.1 coefficient)

$$\frac{A_{res}}{A} \geq 1.1 \frac{f_{yk}/\gamma_{M0}}{f_{tk}/\gamma_{M2}}$$

If the brace is an H-shaped profile, a possible solution consists in welding to the column a stub of the same size of the brace and then bolting the parts by using a splice (see Section 4.13) that connects both the flanges and the web.

Similar considerations apply if the brace is made by L- or U-shaped profiles. Even in this case it is important to connect both the flanges (L and U) and the web (U braces) to avoid fragile failure. If the connection, as in Figure 3.40 (lug angle, see Section 3.19.1, it can also be applied in a similar fashion to U-shaped profiles) is not enough (even staggering the bolts), some local reinforcing plates can be welded.

There are recent engineering approaches aimed at designing the gusset plate so that it buckles on purpose in order to dissipate energy in the "postbuckling" phase. In the United States this solution is an alternative to dimensioning the connection as full strength and it can be reached as shown by the joint in Figure 4.107. For additional information the reader is referred to [34]. Other, more recent, studies (Ref. [36] being one of them) consider a modified distance ($8t_p$ instead of $2t_p$, see Figure 4.108) on an ellipse, letting the engineer design thinner and more compact plates.

4.24.3 Eccentric Braces and "Links"

An interesting joint for seismic applications is the one in eccentric braces (Figure 4.109), where there is a part of the beam (the link) intended for absorbing the quake energy through the cyclic deforming applied by the eccentric braces.

The reader is referred to the various regulations to dimension (space and thickness of the plates) the link.

4.24.4 Base Plate

The base plate is a crucial detail from a seismic point of view, though the provisions do not always clearly define the performance requirements.

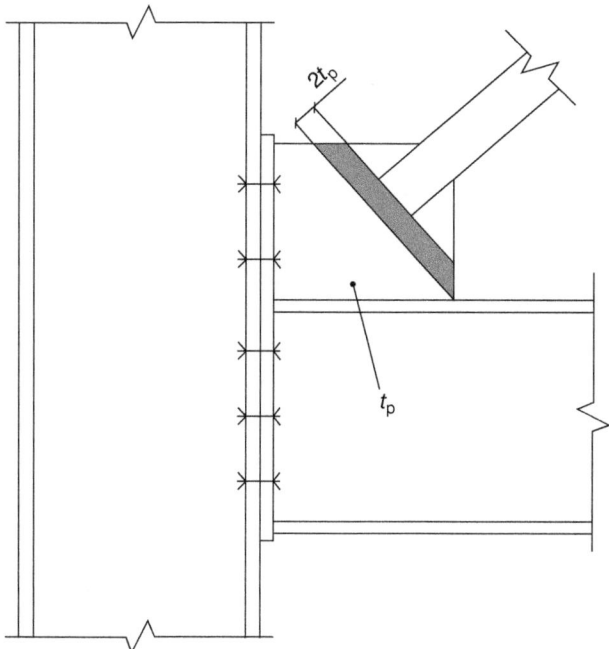

Figure 4.107 Brace connection that can buckle out of plane, thus allowing a plastic hinge to develop. Source: From Ref. [34].

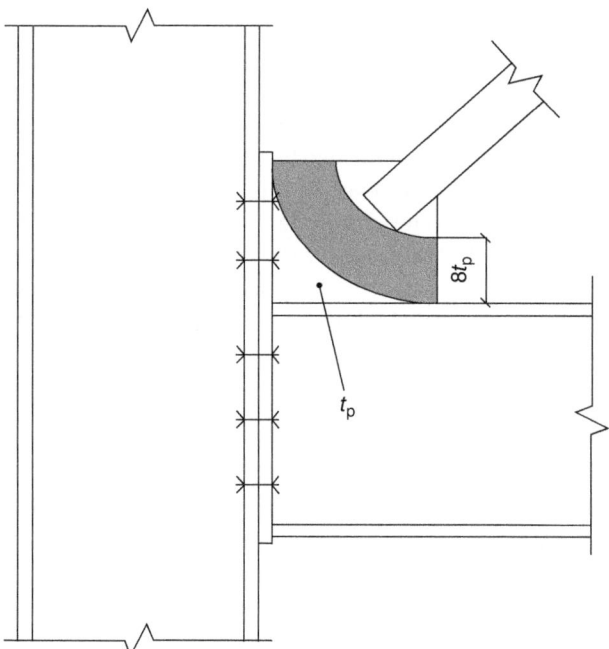

Figure 4.108 Modified distance considering an ellipse. Source: From Ref. [36].

Figure 4.109 Links in eccentric braces.

As a general concept, it must be kept in mind that the column-to-foundation joint must be able to resist all the actions transferred to all the connected elements, both as shear and as axial action or bending. This means that if a base plate also connects a brace, the requirements for the brace (likely the full strength with even an overstrength factor) will have to be combined with the ones for the column.

If the applicable code allows this, it could be economic to evaluate the maximum action that can be elastically transferred by the system (i.e. with a behavior factor of 1 or, where allowed, e.g. in some cases in the US provisions, by amplifying the results with an Ω_0 factor) since it can be considered a maximum limit to the strength request. An interpretation could also be that if a ground restraint is a pin from an engineering point of view, even amplifying the results leaves the bending moment equal to 0. If the reference code does not request the full strength of the column to be restored at the base for the applicable lateral resisting system, this could be a viable approach, but the general advice is not to be cheap with this fundamental detail, most of all where the seismicity is medium to high.

References

1 American Institute of Steen Construction (AISC) (2011). *Steel Construction Manual*, 14e. Chicago, IL: AISC.
2 Crawford, S.F. and Kulak, G.L. (1968). *Behaviour of Eccentrically Loaded Bolted Connections*, Studies in Structural Engineering, vol. 4. Halifax, NS: ASTM.
3 Ballio, G. and Mazzolani, F. (1983). *Theory and Design of Steel Structures*. London: Taylor & Francis.

4 Fisher, J.M. and Kloiber, L.A. (2010). *Base Plate and Anchor Rod Design, AISC Design Guide 1*, 2e. Chicago, IL: American Institute of Steel Construction.

5 CEN (2005). *Eurocode 3: design of steel structures – Part 1–8: Design of joints*, EN 1993-1-8: 2005. Brussels: CEN.

6 Wald, F., and others (1999). *Column Bases in Steel Building Frames*Brussels: COST C1.

7 American Concrete Institute (ACI) (2008). *Building code requirements for structural concrete and commentary*, ACI 318-08. Farmington Hills, MI: ACI.

8 DeWolf, J.T. and Bicker, D.T. (2003). *Column Base Plates, AISC Design Guide 1*, 1e. Chicago, IL: American Institute of Steel Construction.

9 CEN (2005). *Design of concrete structures, general rules and rules for buildings*, EN 1992-1-1: 2005. Brussels: CEN.

10 The Steel Construction Institute (SCI), The British Constructional Steelwork Association (BCSA) (1995). *Joints in Steel Construction: Moment Connections*. Ascot, UK: SCI, BCSA.

11 Fisher, J.M. (1981). Structural details in industrial buildings. *Eng. J. AISC* 18: 83–89, Quarter 3.

12 Goldman, C. (1983). Design of column base plates and anchor bolts for uplift and shear. *Struct. Eng. Pract.* 2 (2): 3–12.

13 Jaspart, J.P., Demonceau, J.F., Renkin, S., Guillaume, M.L., and ECCS Technical Committee 10 (2009). *Structural connections, european recommendations for the design of simple joints in steel structures*, Eurocode 3, Parts 1–8, ECCS Guide No. 126. Portugal: ECCS.

14 Jaspart, J.P., Renkin, S., and Guillame, M.L. (2003). *European Recommendations for the Design of Simple Joints in Steel Structures*: Université de Liège.

15 The Steel Construction Institute (SCI), The British Constructional Steelwork Association (BCSA) (2002). *Joints in Steel Construction: Simple Connections*. Ascot, UK: SCI, BCSA.

16 German Institute for Standardization (DIN) (2011). *Steel structures – Part 1: Design and construction*, DIN 18800-1:2008-11. Berlin: Beuth Verlag GmbH.

17 Muir, L.S. and Thornton, W.A. (2014). *Vertical Bracing Connections—Analysis and Design, AISC Design Guide 29*. Chicago, IL: American Institute of Steel Construction.

18 Bureau of Indian Standards (BIS) (2007). *General Construction in Steel – Code of Practice*, 3rd rev. New Delhi, India: IS 800: 2007, BIS.

19 UNI – Ente Nazionale Italiano di Unificazione (1988). *Costruzioni di Acciaio – Istruzioni per il calcolo, l'esecuzione, il collaudo e la manutenzione*, CNR-UNI 10011. Milan: UNI.

20 Consiglio Superiore dei Lavori Pubblici (2009). *Istruzioni per l'applicazione delle 'Norme tecniche per le costruzioni' di cui al D.M., January 14, 2008*. Circolare, February 2, 2009, No. 617, Gazzetta Ufficiale 47, February 26, 2009.

21 Carter, C.J. (1999). *Stiffening of Wide Flange Columns at Moment Connections, AISC Design Guide 13*. Chicago, IL: American Institute of Steel Construction.

22 Murray, T.M. and Shoemaker, W.L. (2002). *Flush and Extended Multiple Row Moment End Plate Connections, AISC Design Guide 16*. Chicago, IL: American Institute of Steel Construction.
23 Murray, T.M. and Sumner, E.A. (2003). *Extended End Plate Moment Connections – Seismic and Wind Applications, AISC Design Guide 4*, 2e. Chicago, IL: American Institute of Steel Construction.
24 Australian Building Codes Board (1998). *Australian standard – steel structures*, AS 4100: 1998. Sydney: Standards Australia.
25 Canadian Standards Association (CSA) (2014). *Limit states design of steel structures*, CSA S16-14. ON, Canada: Canadian Standards Association.
26 Dowswell, B. (2006). Effective length factors for gusset plate buckling. *Eng. J. AISC* 43: 91–102, Quarter 2.
27 Thornton, W.A. (1984). Bracing connections for heavy constructions. *Eng. J. AISC* 21: 139–148, Quarter 3.
28 Hogan, T.J. and Thomas, I.R. (1994). *Design of Structural Connections*, 4e. Sydney: Australian Institute of Steel Construction.
29 Tamboli, A.R. (1999). *Handbook of Structural Steel Connection Design and Details*. New York: McGraw Hill.
30 The Steel Construction Institute, The British Constructional Steelwork Association (BCSA) (1998). *Joints in Steel Construction: Composite Connections*. Ascot, UK: SCI, BCSA.
31 European Convention for Constructional Steelwork (ECCS) Technical Committee 11 (1999). *Composite structures, design of composite joints for buildings*, Guide No.109. Brussels: ECCS.
32 Kulak, G.L. and Grondin, G.Y. (2003). Strength of joints that combine bolts and welds. *Eng. J. AISC* 40: 89–98, Quarter 2.
33 Federal Construction Council (1974). Expansion Joints in Buildings. *Tech. Rep.* 65. National Research Council: Washinton, DC.
34 American Institute of Steen Construction (AISC) (2012). *Seismic Design Manual*, 2e. USA: American Institute of Steel Construction.
35 Ministero delle Infrastrutture e dei Trasporti (2008). *NTC – Norme tecniche per le costruzioni*, Decreto Ministeriale (D.M.) January 14, 2008. Rome: Gazzetta Ufficiale.
36 Lehman, D.E., Roeder, C.W., Herman, D. et al. (2008). Improved seismic performance of gusset plate connections. *J. Struct. Eng. ASCE* 134 (6): 890–901.

5
Choosing the Type of Connection

Once a joint is a pin connection (or a rigid one) in a calculation model, how should one choose the pin (or rigid) connection that fits best?

The engineer should remember that, even though the initial problem of resisting the model forces is achieved (which is the minimum target), the quality of his or her work is measured against other important benchmarks, for example, the simplicity of fabrication and erection, the performance (e.g. the ductility to provide a strength surplus), and economic competitiveness.

5.1 Priority to Fabricator and Erector

The first advice is to talk and, at the end of the day, let the fabricator choose the preferred connections. If the fabricator does not erect the structure, it is advisable to have a meeting including the erectors.

We can discuss the economy of each joint and weigh the ease of fabrication and erection (which is discussed in the next section) but it is more relevant and important to listen to the habits of the fabricator: past experiences, shop organization, machinery, personnel, company experience, and culture will likely make the choice among types of joints quite convenient to each fabricator. It may come as a surprise that different fabricators will probably choose different solutions for the same project.

It is therefore crucial that the engineer discusses the project with people involved in the erection and production to understand the preferred types of connections before designing them. It is arguably a good idea to ask for past successful projects by the fabricator in order to try to design similar joints wherever possible.

5.2 Considerations of Pros and Cons of Some Types of Connections

Here we look at the advantages and disadvantages of some types of joints without forgetting that priority should go to the fabricators and erectors. The topic

Design and Analysis of Connections in Steel Structures: Fundamentals and Examples,
First Edition. Alfredo Boracchini.
© 2018 Ernst & Sohn Verlag GmbH & Co. KG. Published 2018 by Ernst & Sohn Verlag GmbH & Co. KG.

here focuses on "simple" (pin) connections, the ones that do not transfer bending moments and that are commonly the most used in a project.

Table 5.1 contains only some general considerations, largely the opinion of the author.

5.3 Shop Organization

5.3.1 Plates or Sheets

Some shops make connection plates starting from sheets, others from "plates."

What we just called plates here refer to a steel sheet with a commercial well-defined width usually about 6 m (20 ft) long while the sheets commonly are 1 m × 2 m or 1.25 m × 2.5 m or 1.5 m × 3 m (similarly with imperial units).

Generally speaking, small shops would start from plates, while larger fabrication facilities would buy sheets, but, as in many other situations, the engineer should talk with the fabricator directly (e.g. a small shop winning a large job might have an interest in having a third part produce laser plates from sheets while, in contrast, a large fabricator might subcontract small jobs to small fabricators that use plates or because there is something stored to be finished).

This difference is relevant for detailers preparing shop drawings because one of the two dimensions of the plate should be a commercial one if the fabricator uses plates. This avoids expensive double cuts. If, say, a plate is 113 mm × 243 mm, the shop worker must take a 120-mm plate and cut it twice, once to 243 mm and then from 120 to 113 mm. It would be better if the detailer (and the engineer earlier) used a 120-mm plate when possible, to save some labor. On the other hand, if plates are laser cut (or similar technique), the smallest dimension is recommended for plate nesting and weight savings (plates will cost accordingly).

5.3.2 Concept of "Handling" One Piece

Medium large fabrication facilities frequently have lines for punching and cutting separated from welding lines. In other words, one beam is first cut and (holes) drilled, then, in another line and by different personnel, the necessary plates are welded to it. Sometimes this happens in different shops in distinct locations. Plates are even normally made by external subcontractors specialized in this and beams sometimes come from a steel producer already cut and with holes.

When a beam or a column must have at least a plate welded to it, it is said that the member must be handled, which means that we are not only talking about welding parts but that we must, most importantly, transport and manage this piece. Whether we weld 1 plate or 10 plates, we still have to "handle" and move the piece, and therefore the difference in cost is not proportional to the number of plates to be welded; the main difference is related to the fact that at least one welding operation has to be made.

If we do not have to weld any plate to the member, one complete phase of production comes off because we do not have to "handle" the part. This means that some parts could even be shipped from the steel producer directly to the site.

Table 5.1 Pros and cons of most common "simple" joint types.

	Fin plate/shear tab	Double-bolted shear plate	Double angles	Flexible end plate
Joint shear capacity in relation to beam shear capacity	Up to 50% with single line of bolts; up to 75% with more bolt columns	Up to 50%	Up to 50% with single line of bolts; up to 75% with more bolt columns	Up to 75%; up to 100% with plate as deep as beam
Structural integrity	Good	Average	Good	Good
Adaptability to skewed joints	Good	Average	Not possible	Average
Adaptability to beams eccentric to columns	Good	Not possible	Poor	Average
Connection to column webs	Good	Average	Average	Good
Fabrication	Very good (no welded plates for secondary beams)	Very good (few welded plates)	Very good (no welded plates)	Good
Surface treatment economy	Good	Average (some components to be treated separately)	Average (some components to be treated separately)	Good
Ease of erection	Good	Good	Average (difficult for two-sided connections)	Good (difficult for two-sided connections)
Site adjustment	Average	Very good	Good	Average
Erection temporary stability	Average	Poor	Average	Good

Source: Adapted from Ref. [1].

It is economical (but, as always, to verify with the fabricator, who may have potentially different habits) to totally avoid welding any plate to some pieces and only design for them connections directly bolted to the piece itself. Typical examples where this concept can be applied are as follows:

- *Secondary beams*: If fin plates or double angles are used, the secondary beam might only need to have the web holes punched.
- *Truss diagonals or vertical members*: In particular, if realized with angles (L) or channels (U), the bolts can directly be connected to the piece.
- Braces (as mentioned in *Truss diagonals or vertical members*).

5.4 Culture

Different geographic areas often favor different kinds of connections. This is an indication that maximum efficiency (in other words, the ratio of quality over price) is not clearly defined and demonstrable and, at least to some extent, culture takes a relevant role in choosing joints.

For example, consider the common habit of welding some parts in the field in the United States, typical for some kinds of connections. The same proposal, say to weld on-site columns of tall multistory buildings to complete strength in order to reach some kind of performance, will likely be rejected by a European fabricator.

Once again, the cautious engineer will exchange opinions with the fabricator to understand the "culture" about the connections, which might even only be the local "shop culture" but that is the hardly debatable, at least initially, consolidated experience.

Reference

1 The Steel Construction Institute (SCI), The British Constructional Steelwork Association (BCSA) (2002). *Joints in Steel Construction: Simple Connections*. Ascot, UK: SCI, BCSA.

6

Practical Notes on Fabrication

6.1 Design Standardizations

It is good practice to try to reduce the range of design materials of connections in a project, such as:

- Material quality
- Plate thickness
- Bolt diameters.

6.1.1 Materials

With regard to the material, it is suggested that a standard grade for plates be defined and, for larger jobs, an additional grade with a higher resistance for certain connections and/or thicknesses that are heavily loaded. If a certain thickness must be of higher grade, this grade should be maintained for all the connections having similar thickness. For example, if a 30 mm thickness is necessary to be S355 for one special connection, it is good practice to use the same grade for all connections involving 30-mm-thick plates.

6.1.2 Thicknesses

The thicknesses should be chosen to be as similar as possible, for example, deciding to use (metric unit) plates every 5 mm. It is uneconomical to have many different thicknesses (especially for medium to low thicknesses) for the same job since it does not allow good optimization during shop fabrication.

6.1.3 Bolt Diameters

Metric diameters such as M14, M18, and M22 are currently not recommended since their production and availability are consistently declining (again, this should be checked with the fabricator).

It is a good general rule to limit the kinds of bolts in order to:

- limit the different holes.
- save when buying bolts.

Design and Analysis of Connections in Steel Structures: Fundamentals and Examples,
First Edition. Alfredo Boracchini.
© 2018 Ernst & Sohn Verlag GmbH & Co. KG. Published 2018 by Ernst & Sohn Verlag GmbH & Co. KG.

- limit the stockpile (bolts are always purchased in excess so it is common to have some left at the end of a job).
- help erectors to avoid taking with them an excessive number of bolt sizes in uncomfortable erection positions.
- avoid possible dangerous errors during erection: it is easy to recognize that an M12 bolt is not in the right place in an 18 mm hole created for an M16 but it might be trick to recognize the same for an M14. This might occur if the design has been realized using both M14 and M16, but if the engineer prudently used only M12 and M16 (dropping the M14 size), this does not happen.

6.2 Dimension of Bolt Holes

Some standards (e.g. the Italian NTC) are quite rigid about bolt hole tolerances, prescribing (Italian NTC) only 1 mm of tolerance until size M20, then 1.5 mm, which can be derogated by the statement "when possible settlements under service loads do not go over acceptable limits." This recalls the discussion in Section 3.2.

Eurocode requests 1 mm until M14 included, 2 mm from M16 to M24, then 3 mm. The 2 mm is acceptable also for M12 and M14 if bearing resistance is less than shear resistance (which means that bolt shear must not control design since it is a nonductile limit state) and if the engineer takes into account a reduced shear resistance of bolts (85% of the full value).

Internationally renowned publications (as [1]) give 2 mm until M24 as a reference standard and 3 mm when over it.

Some fabrication shops might ask for larger tolerances to help erection, but it is not recommended to go over the mentioned limits unless connections are designed by friction. EN 1090 allows standard clearances (here considered as the difference between the hole dimension and bolt nominal diameter) as in Table 6.1 for oversize holes and slots. Note that slot width is by rule the same as a regular (standard) hole size.

Table 6.2 gives the metric tolerances for holes according to AISC [2].

Using oversize holes means designing bolts by friction ("slip-critical connection") also according to AISC. Slots (short or long) can be used even without the slip-critical condition if the slot is perpendicular to the load according to the AISC practice. Furthermore, [2] requests long slots to be used in only one of the elements making the connection.

Table 6.1 Maximum bolt hole clearance (in mm) according to EN 1090.

Nominal diameter (mm)	12	14	16	18	20	22	24	27 and beyond
Round standard hole	1 (2)		2					3
Oversize standard hole	3		4				6	8
Short slot (length)	4		6				8	10
Long slot (length)	1.5d							

Table 6.2 Maximum bolt hole clearance (in mm) according to AISC.

Nominal diameter (mm)	16	20	22	24	27 and beyond
Round standard hole		2			3
Oversize standard hole	4		6		8
Short slot (length)	6		8		10
Long slot (length)			1.5d		

It is to be noted that slots or oversize holes might not be compatible with some seismic prescriptions. According to AISC 341-10, for example, only standard holes and short slots (perpendicular to force) can be used in lateral load resisting systems, with the exception of oversize holes on one part only (plate or profile) for brace connections.

6.2.1 Bolt Hole Clearance in Base Plates

Base plates are usually fabricated with a larger clearance since it is very common that anchor bolts are not correctly positioned. Anchors are in fact laid by workers used to centimeters instead of millimeters (masons instead of steel workers) and this kind of personnel external to the steel fabricator is not sensible to the problems erectors will face because of poor anchor bolt positioning.

Even a few millimeters laid wrong in anchors (which means wrong distance between columns) will cause problems when erecting braces and may cause the structure to be out of plumb since beam bolting will "push" columns out of the vertical position (or alternatively some on-site adjustments will be necessary, drilling and welding steel pieces, with consequent economical damage). This is the reason why many fabricators go to the field and survey the anchor bolts before sending steel to the site so that, if needed, some adjustments can be adopted (e.g. welding a base plate asymmetrically to make up for an error). This adds cost but it is by far cheaper than the problems faced during erection.

Base plates need larger bolt hole clearances as explained. For example, Table 6.3 provides AISC standards (imperial values with metric translation) with

Table 6.3 Limit dimensions for base plate holes and washers according to AISC.

Anchor diameter, in. (mm)	Maximum hole diameter, in. (mm)	Minimum external washer dimension, in. (mm)	Minimum washer thickness, in. (mm)
3/4 (19.1)	1 5/16 (33.3)	2 (50.8)	1/4 (6.4)
7/8 (22.2)	1 9/16 (39.7)	2 1/2 (63.5)	5/16 (7.9)
1 (25.4)	1 13/16 (46.0)	3 (76.2)	3/8 (9.5)
1 1/4 (31.8)	2 1/16 (52.4)	3 (76.2)	1/2 (12.7)
1 1/2 (38.1)	2 5/16 (58.7)	3 1/2 (88.9)	1/2 (12.7)
1 3/4 (44.5)	2 3/4 (69.9)	4 (102)	5/8 (15.9)
2 (50.8)	3 1/4 (82.6)	5 (127)	3/4 (19.1)
2 1/2 (63.5)	3 3/4 (95.3)	5 1/2 (140)	7/8 (22.2)

maximum allowed values being even larger than oversize holes. Washers with minimum dimensions and thicknesses are also given.

It is recommended that large washers be used in order to completely cover holes even when the anchor is on one side of the hole. If necessary, plates (circular or square) can be used in column base details since, according to [3], it is possible to use nonhardened material (i.e. without thermal treatment to better resist pre-tensioning). Plate thickness of at least one-third of the anchor bolt diameter is commonly adopted.

It is good and consolidated practice to cast grout or the equivalent after the structure is erected (see Section 6.17 for additional recommendations).

From a design point of view, if the anchor bolt hole is larger than standard, it is appropriate not to have bolts in shear but to verify it by considering friction or a shear lug. Concerning this, the reader is referred to the information on base plates in Chapter 4.

6.3 Erection

The engineer must dedicate time and attention thinking about erection because during this phase some critical problems might arise, for example, structure lability or space clearance issues that do not allow some bolts to be inserted or even some parts to be positioned.

6.3.1 Structure Lability

A braced structure is temporarily composed only by columns and beams before braces are inserted. It is an assignment for the engineer to evaluate the stiffness and resistance of the column "pins" at the base because they must provide the partial rigid restraint that is necessary during erection. If this is not sufficient, some temporary stabilizing structures are needed.

It is therefore common to start erection with braced columns (and it is good practice to underline this in erection drawings too). This also helps the structure to be "plumbed" and "squared."

The engineer must anyway be conscious of the potential lability during erection in order to take some countermeasures when necessary.

6.3.2 Erection Sequence and Clearances

The connection designer must consider how beams can be inserted into their final location because in some connection types the beam could only be positioned rotating it. In other cases it could be required to add notches (to be considered in connection design since the beams are obviously weakened) as shown in Figures 6.1 and 6.2.

Sometimes even the erection sequence can be essential to assemble a structure since the obstruction of a beam might impede the other to being inserted.

Figure 6.1 Notches can ease erection.

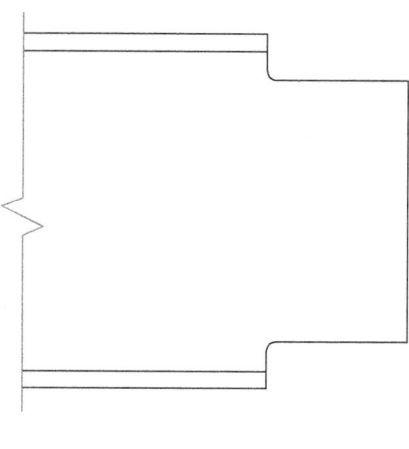

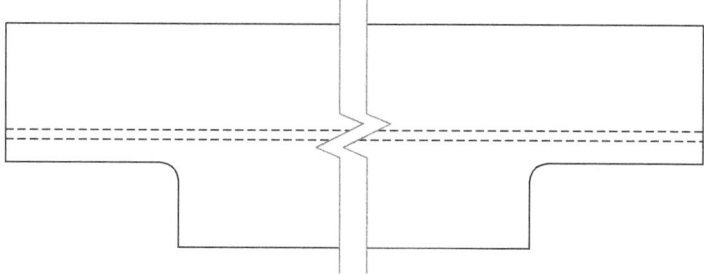

Figure 6.2 Example of a beam partially notched on flanges to help its insertion during erection.

6.3.3 Bolt Spacing and Interferences

Also for inserting and tightening bolts it is necessary to pay attention to clearances and erection sequences in order to find admissible solutions. Mostly in industrial plants where some special situations of machinery support or concentration of bracings step in, it is necessary to increase care to avoid erection issues (see, for example, Figure 6.3).

6.3.4 Positioning and Supports

It is good design habit to appoint some support for the beams before they are bolted in order to facilitate the erection operations. If it is true that in multistory structures the parts are mainly bolted when on the ground and then lifted, it is difficult in tall structures and in general to keep the profiles in position with cranes when bolts are inserted through holes (the operation is not simple at all if the beams are not supported). It is therefore helpful to provide angles or any sort of "seated" details to ease tightening.

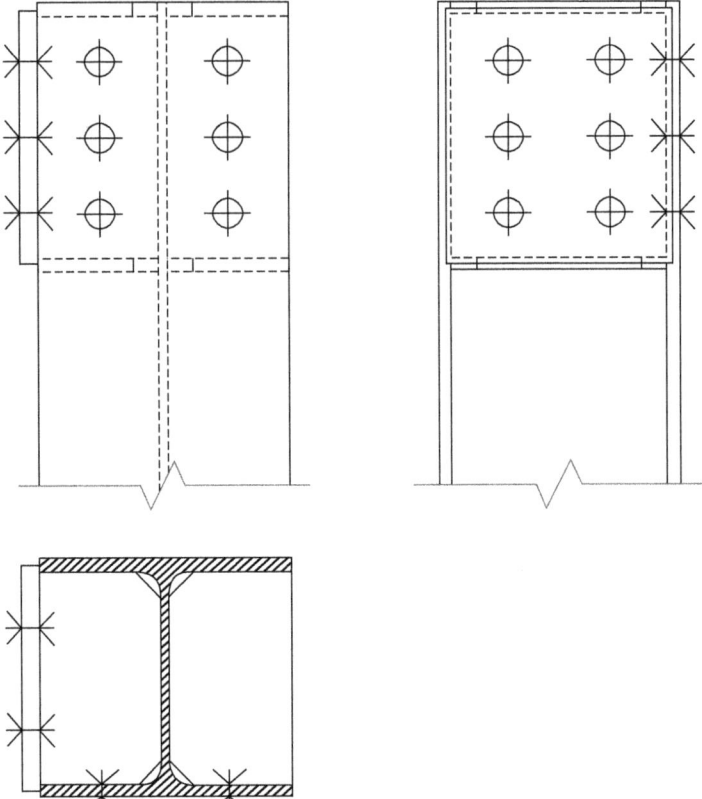

Figure 6.3 Situations where some parts cannot be reached during erection.

6.3.5 Holes or Welded Plates for Handling and Lifting

Profiles need to have some "areas" where the steel can be fixed for crane lifting or handling. Erectors often use the holes of connections, but this might not always be safe (the erection load has not been tested for those holes).

It would generally be good practice to make some calculations about those erection loads, but this control is meaningful only if there is a close relationship with erectors, which can indicate how they will handle materials.

This reaffirms the importance of having good communication with fabricators and erectors in order to make projects successful in terms of quality, safety, and economy.

6.4 Clearance Needed to Operate Tightening Wrenches

In addition to standard wrenches, some other types might become useful in certain situations, like polygonal or "pipe" (locally known also with other names) or socket wrenches, that might be hand operated or in motor-driven screwdrivers. The engineer must consider the necessary space to use tightening

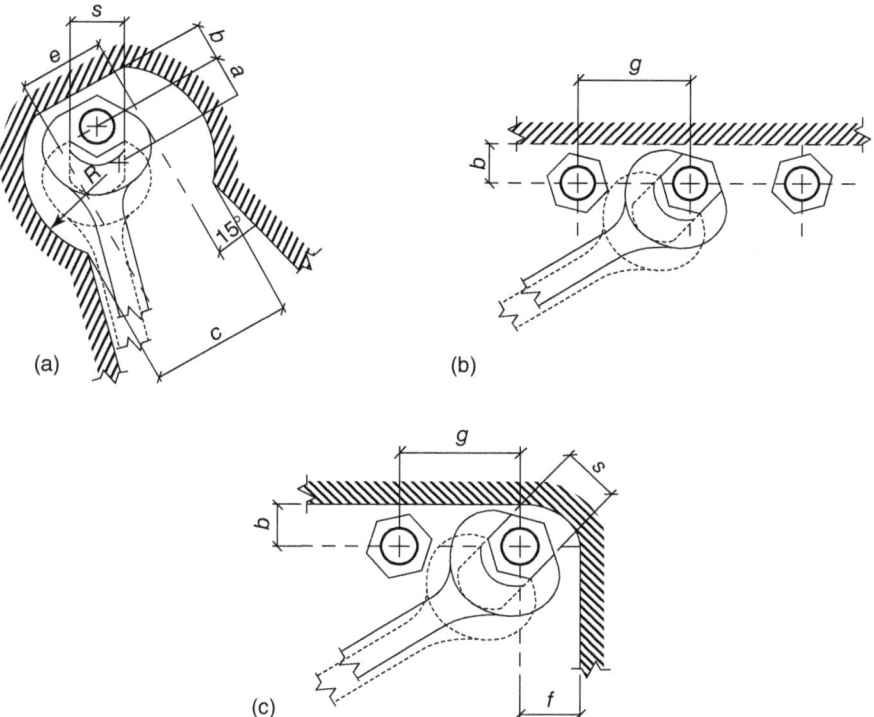

Figure 6.4 Standard wrenches.

Figure 6.5 Polygonal wrench.

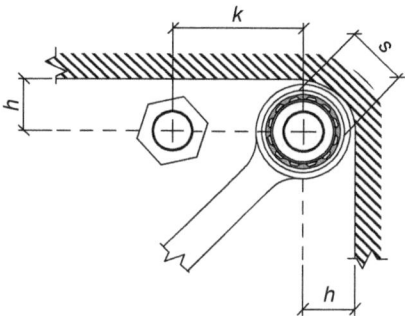

tools and the reader is referred to Figures 6.4–6.6 and Tables 6.4–6.9. Actually screw-driving tools can have different sizes depending on the producer and sometimes hydraulic wrenches are utilized and they need even larger clearances that are different depending on the brand (SKF being the most common).

6.4.1 Double Angles in Connections

Double-angle bolting in joints needs correct assessment of the necessary tolerances to tight the bolts. As in Figure 6.7, if the distance x is not enough, tightening bolts might become difficult (if not impossible, since bolts might collide).

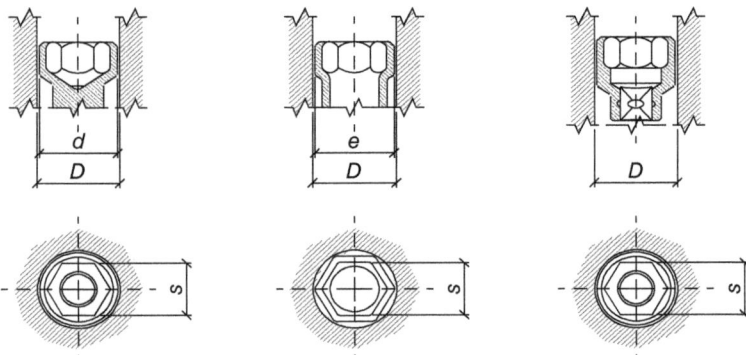

Figure 6.6 Pipe and socket wrenches.

Tamboli [4] gives helpful insights on how to avoid the problem when using imperial bolts and those data are given in Table 6.10. It should be noted that, depending on the tensioning tools (see also previous tables), the required tolerances might even be larger (to be considered when inserting eccentricities in connection design).

6.5 Bolt Spacing and Edge Distances

Spacing between bolts and, most of all, between bolts and free edges is a key aspect when optimizing connection design. Therefore, the engineer should provide the detailer with the necessary information in the connection sketches. It is recommended that the drafter be educated about those basic notions of spacing and edge distance in order to avoid problems when detailed information about bolting is not provided to them.

The schemes in the pages that follow may be used to compare instructions provided by the different international standards. The examples in this section are EC (Figure 6.8), DIN (Figure 6.9), and AS 4100 (Figure 6.10).

Simply put, the minimal notions that a steel detailer should have are that the standard spacing between bolts is normally around 3 times the diameter while the edge distance is about 1.5 times the bolt diameter. When this geometry cannot be respected, it can be roughly lowered to 2.4 times for bolt spacing and 1.2 times for edge distance but the drafter should inform the connection designer and ask for authorization (bearing can become a relevant design issue).

Standards also give maximum values for bolt spacing and edge distance that avoid local buckling issues when the bolts locally compress the plate and prevent corrosion in exposed elements. Eurocode clearly states that there are no limitations when the above conditions do not apply.

6.6 Root Radius Encroachment

It is important when designing connections to always take into account the root radius that fillets web and flange in laminated profiles because connecting plates

Table 6.4 Characteristic dimensions needed for standard wrenches.

Wrench s	a	b	c	e	f	g	R
17	15	12.5	52	31	19	35.5	32
19	16	14	56.5	32	21.5	40	35.5
22	18	16	64	36	23.5	45	40.5
24	19	17.5	69	38.5	25	48	43
27	21	19	76.5	43	28	55	48.5
30	23	20.5	84	48	30	60	53
32	24	22	90	51	31.5	62.5	55.5
36	26.5	23.5	99	54	37	73	63.5
41	30	26	112	62	41.5	82.5	71
46	35	28.5	122	67	45	90	77.5
50	40	31	134	74	47	96.5	83
55	45	32.5	142	77	49	102.5	88.5
60	50	35.5	150	82	51.5	109.5	94.5
65	57	39	164	88	57.5	121	104
70	60	44	180	94	64	134	115
75	65	46	198	108	69	144	124
80	63	47	209	122	78	152	122
85	68	50	220	125	82	162	130
90	74	53	240	138	90	176	144
95	76	57	244	138	90	180	146
100	80	64	268	155	100	195	158
105	83	64	270	155	100	197	160
110	90	65	280	155	104	206	168
115	93	72	300	168	114	221	180
120	93	72	300	168	114	225	180
130	105	77	328	184	123	244	197
135	110	80	340	188	127	254	205
145	120	86	368	212	132	269	218
150	123	90	382	216	142	284	230
155	127	92	392	216	142	284	232
165	135	98	410	230	150	305	246
170	140	103	425	244	160	316	260
175	143	104	432	244	160	323	262
180	146	106	440	248	164	332	267
185	148	108	468	264	175	348	280
190	152	112	470	264	175	350	282
200	160	117	500	276	182	368	296
210	170	124	512	280	190	388	312
220	178	128	556	306	200	405	328
230	182	132	560	310	210	410	335

Table 6.5 Characteristic dimensions needed for polygonal wrenches.

Wrench s	h	k
17	14.25	27
19	15.75	30
22	18.25	35
24	19.75	38
27	21.75	42
30	23.75	46
32	25.25	49
36	28.25	55
41	32.25	63
46	36.25	71
50	39.25	77
55	45	88
60	48	93
65	50	95
70	60	110
75	62.5	115
80	68	122
85	66	130
90	77	152
95	77	152
100	78.5	155
105	87	172
110	87	172
115	87	172
120	98	194
130	103.5	205
135	103.5	205
140	116	230
150	116	230
155	116	230
165	136	276
170	136	276
175	136	270
180	136	270
190	148.5	295
200	148.5	295
210	163.5	235

Table 6.6 Characteristic dimensions needed for hand-driven socket wrenches.

Wrench dimensions	Socket				
	6.3	10	12.5	20	25
	D min	D min	D min	D min	D min
17		26	27		
19		28	29		
22		32	32.5	42	
24			35	42	
27			39	42	
30			43.5	46.5	
32			46	49	
36				54	
41				60	
46				66.5	69.5
50				71.5	74.5
55				77.5	80.5
60					87

Table 6.7 Characteristic dimensions needed for motor-driven socket wrenches.

Wrench dimensions	Socket						
	6.3	10	12.5	16	20	25	40
	D min	D min	D min	D min	D min	D min	D min
17		30	39	37	50		
19		30	39	37	50		
22			39	37	50		
24			39	39.5	50		
27			42	45	51	61	
30				48	51	61	
32				50.5	61	61	
36				55.5	61	61	89
41					64	67	89
46					70.5	73.5	89
50						78.5	89
55						85	91
60						91	97

or double angles might interfere with it. Actually, some encroachment is possible and [2] gives permissible values depending on some root radius ranges (radii vary overall between 5/16 and 1 3/8 in., that is, between 8 and 35 mm). Toward interpolating a linear function to have a metric tool that gives us approximately sound encroachments, Tables 6.11–6.13 have been compiled for commonly used EC

Table 6.8 Characteristic dimensions needed for pipe wrenches.

Wrench dimension s	Light series		Heavy series	
	d max	D min	e max	D min
17	25.5	27.5	26	28
19	28	30	28.5	30.5
22	31.5	33.5	32	34
24	34	36	34.5	36.5
27	38	41	38.5	41.5
30	41.5	44.5	42	45
32	44.5	47.5	45	48
36	50	53	—	—
41	57	60	—	—

style profiles. The encroachment is roughly equivalent (it would be less for small radii, slightly more for larger radii) to have the plate 1 mm from the profile web assuming perfect rigidity (in reality there is some local deformation). This 1 mm distance is approved also by EC [5], which exactly defines 1 mm as acceptable in preloaded connections and 2 mm if there is no preload (however to be limited when corrosion is an issue).

Tables 6.11–6.13 show the maximum space wa and wb (see Figure 6.11) that can be used in connections on the flange and web, respectively. The values wa and wb are the maximum allowed including encroachment but, if possible, it is better to avoid (or lower) the interference (therefore deduct encroachment from wa and two times encroachment from wb).

The tables also report two additional dimensions that can be useful: when a plate is bolted internally to a flange, as in the case of a splice with double plates on flanges, the maximum hole dimension depends on the maximum plate dimension, itself depending on the allowable encroachment. The tables then provide the maximum hole (the bolt size will consequently come) that can be drilled in the two cases of a hole that is 1.5 times distant from the edge (standard dimension) or 1.2 times the bolt diameter (rough lower limits, not allowed by some standards). If the connection is realized with two rows of bolts on each side of the flange, the maximum value of d_0 is obtained by halving the data in the table.

6.7 Notches

Notches (Figure 6.13) should have some minimum dimensions for technological reasons (size of the machinery tools) so the designer should take this into account (see Figure 6.12). Manual operations, say by torch, would allow any dimension but the approach should be cautious.

For flange notches, the recommendation is to not cut over the start of the root radius.

Table 6.9 Minimum and maximum wrench dimensions depending on bolt size.

Bolt	s min	s max
M10	17	—
M12	19	22
M14	22	24
M16	24	27
M18	27	30
M20	30	32
M22	32	36
M24	35	41
M27	40	46
M30	45	50
M33	50	55
M36	55	60
M39	60	65
M42	65	70
M45	70	75
M48	75	80
M52	80	85
M56	85	90
M60	90	95
M64	95	100
M68	100	105
M72	105	110
M76	110	115
M80	115	120
M90	130	135
M100	145	150
M110	155	165
M125	180	185
M140	200	210

6.8 Bolt Tightening and Pretensioning

AISC norms clearly divide bolt categories and consequently the requested pretension:

- *Bearing bolts*: This kind of joint (e.g. simple shear) must have the parts in firm contact, and therefore a preload has to be applied, but it is empirical and not defined in magnitude; it corresponds to the force applied by a worker with a standard wrench.

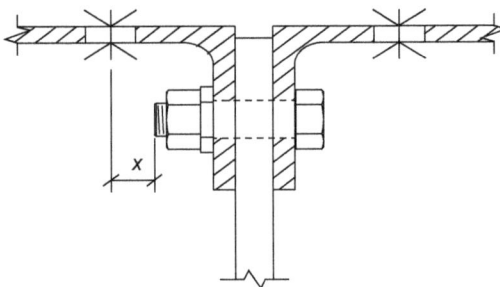

Figure 6.7 Distance *x* for Table 6.10.

Table 6.10 Distance *x* as in Figure 6.7 suggested by [4] for imperial bolts.

Imperial bolt (in.)	5/8	3/4	7/8	1	1 1/8	1 1/4
Nut depth, in. (mm)	5/8 (15.9)	3/4 (19.1)	7/8 (22.2)	1 (25.4)	1 1/8 (28.6)	1 1/4 (31.8)
x min, in. (mm)	1 (25.4)	1 1/4 (31.8)	1 3/8 (34.9)	1 7/16 (36.5)	1 9/16 (39.9)	1 11/16 (42.9)

- *Pretensioned joints*: The bolts work in tension and pretensioning is defined and must be applied.
- *Slip-critical connections*: Shear loads are resisted by friction; pretensioning is defined (in addition to some requirement for surfaces in contact) and has to be applied.

The tightening values to be applied according to AISC (when necessary) are given in Table 6.14.

Table 6.15 gives the EC values [5] for the pretension to reach in preloaded connections (derived using formula 0.7 $f_{ub}A_s$ as defined in Chapter 3).

The torsion to apply to reach the correct preload can be obtained [5] as $M_r = kdF_{p,C}$, k depending on the kind of tensioning method and on the bolts (k values are from 0.10 to 0.23; precise instructions are in the EN 14399 series). Alternatively, the instruments can be calibrated in the field (see Annex H of [5]).

6.8.1 Calibrated Wrench

According to [5], once the moment to be applied to the bolt is defined, the operation must be executed in two phases: an initial one that brings all the bolts in the connection to 75% of the final desired value and a second one that reaches 110% of the necessary torsion.

The case made in [6] that the applied torsion is dependent on the condition of the surfaces demands that the calibration between applied torsion and final tension be executed daily.

6.8.2 Turn of the Nut

The old system of controlling tensioning by a certain rotation (turn) of the nut is, contrary to appearances, scientifically sound and, for many aspects, better than the others.

6.8 Bolt Tightening and Pretensioning

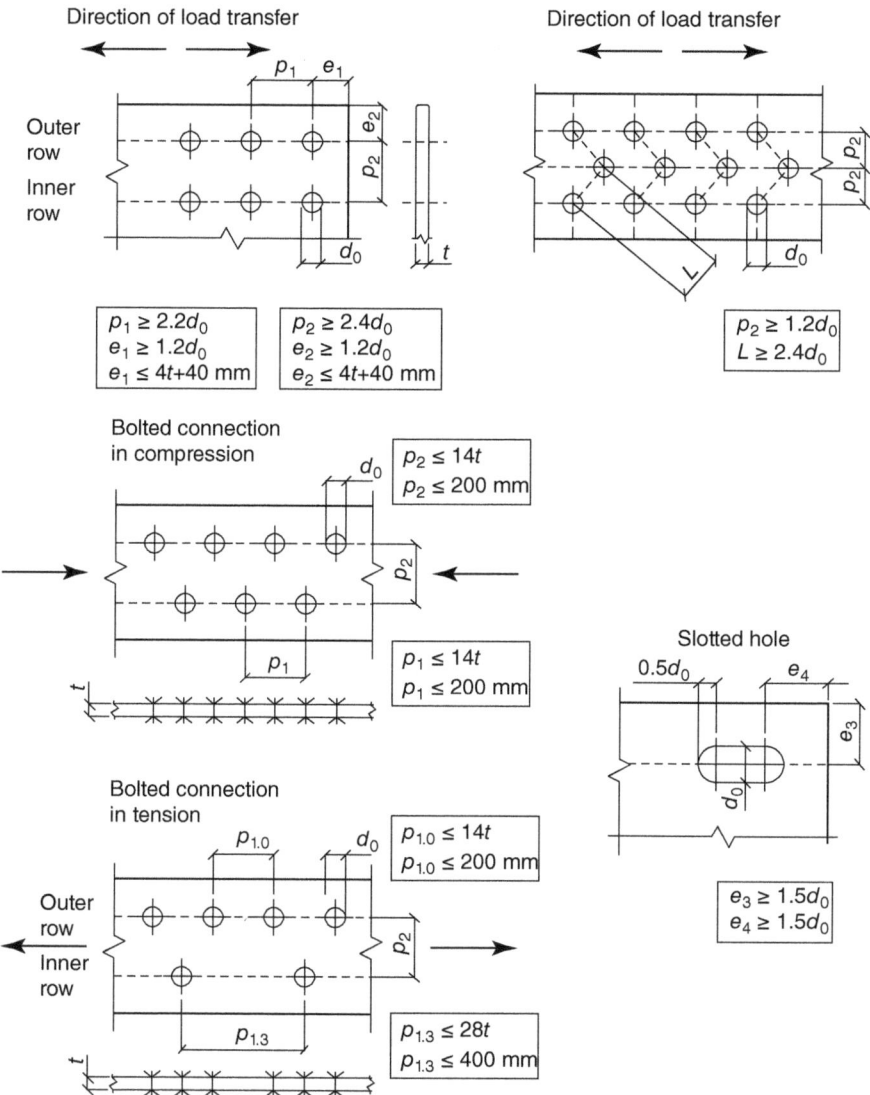

Figure 6.8 Eurocode limits.

Once the parts are in firm contact, an additional rotation of a half turn allows a correct pretensioning. If it is true that the definition of "firm contact" is empirical and not really precise, the tolerance on the rotation is ample (60° more or less is still fine) and the method is surely valid [1, 7]. The statistical results are better than the others since the deformation (bolt elongation) controls and it is several

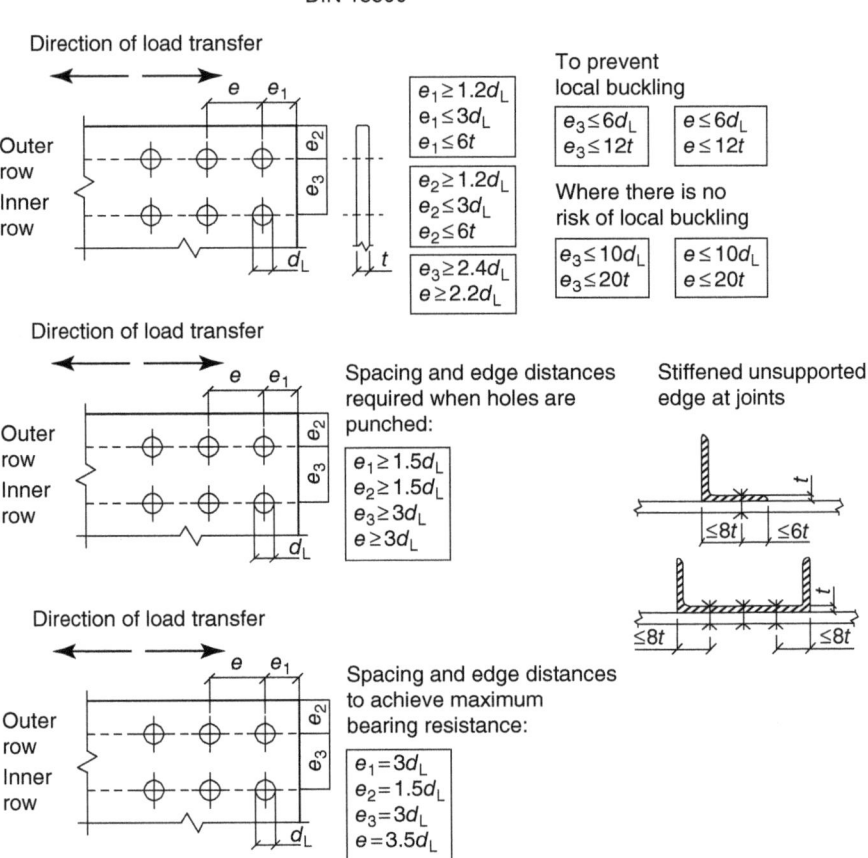

Figure 6.9 DIN 18800 indications.

times less than the collapse deformation; conversely a method like the calibrated wrench, for example, is controlled by the elastic strength of the material, which is only 20–40% less than the ultimate strength.

The rotation to be applied in accordance with AISC is a half turn beyond the firm contact for bolts with standard length, that is, included between four and eight times the diameter. It is requested to rotate only one-third turn when the bolt length is less than 4 diameters but two-thirds turn for long bolts, which means with a length between 8 and 12 times the diameter (see Table 6.16).

Instead, [5] calls this system a "combined method" since the "firm contact" phase is supposed to reach the tightening stage with a calibrated wrench (or the equivalent) until it reaches 75% of the M_r value previously defined. For the exact values of the turn of the nut according to [5], they should be evaluated after a formal mark (by crayon or paint) of the first step and they are a function of the "total

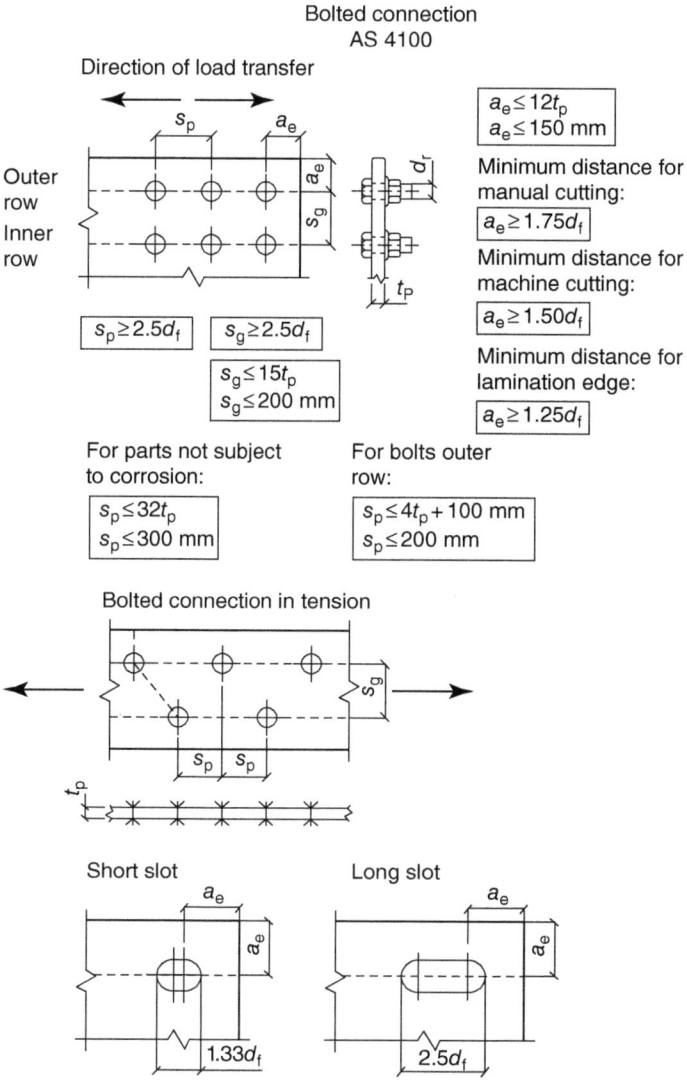

Figure 6.10 Australian standard AS 4100 indications.

bolted ply," which is the total thickness including plates and washers (everything between bolt head and nut). Table 6.17 provides the exact requirements.

The method allows reaching a pretension that is bigger than the minimum and, as discussed in Chapter 2, this is not a concern. Let us also mention that the force required to break the bolt is 1.5 turns from the firm contact (and it is unlikely to be applied by a regular wrench for large bolts) so the method is reasonably safe when workers are correctly educated.

It is to be emphasized that if the connected plies include material that is not steel (e.g. gaskets or thermal insulation), then the turn of the nut relations are

Table 6.11 Useful dimensions for IPE profiles.

Profile	Depth	Width	Web thickness	Flange thickness	Radius	Encroachment	wa	Flange Max $\Phi - 1.5d_0$	Flange Max $\Phi - 1.2d_0$	Web wb
IPE 80	80	46	3.8	5.2	5	2	18	6	7	63
IPE 100	100	55	4.1	5.7	7	3	21	7	9	80
IPE 120	120	64	4.4	6.3	7	3	25	8	10	99
IPE 140	140	73	4.7	6.9	7	3	30	10	12	118
IPE 160	160	82	5	7.4	9	4	33	11	14	135
IPE 180	180	91	5.3	8	9	4	38	12	15	154
IPE 200	200	100	5.6	8.5	12	5	40	13	16	169
IPE 220	220	110	5.9	9.2	12	5	45	15	18	187
IPE 240	240	120	6.2	9.8	15	6	48	16	20	202
IPE 270	270	135	6.6	10.2	15	6	55	18	23	231
IPE 300	300	150	7.1	10.7	15	6	62	20	26	260
IPE 330	330	160	7.5	11.5	18	7	65	21	27	285
IPE 360	360	170	8	12.7	18	7	70	23	29	312
IPE 400	400	180	8.6	13.5	21	7	71	24	30	345
IPE 450	450	190	9.4	14.6	21	7	76	25	31	392
IPE 500	500	200	10.2	16	21	7	81	27	33	440
IPE 550	550	210	11.1	17.2	24	8	83	27	34	483
IPE 600	600	220	12	19	24	8	88	29	36	530

no more valid and therefore Research Council on Structural Connections (RCSC [8]) specifications demand that the connected material be steel only.

The method can be considered simple and reliable if erectors have been instructed and the necessary supervision is performed. However, [5] requirements make it a little more complex when demanding calibrated wrenches to reach M_r in the first step.

6.8.3 Direct Tension Indicators

As evident from the widespread use in the United States, direct tension indicators (Figure 6.14) look like washers and they are positioned under the bolt head or the nut (in case it is coupled with a real washer); they are produced with small embossed parts that will be pressed down when the preload is correct. The pretension is therefore controlled by the deformation of the embossed parts (no gap must be present at the end of the installation). Tension indicators must be chosen depending on the bolt size and the resistance class.

Annex J of [5] gives good guidance for a successful use on the field.

Table 6.12 Useful dimensions for HEA profiles.

Profile	Depth	Width	Web thickness	Flange thickness	Radius	Encroachment	Flange wa	Flange Max Φ – 1.5d_0	Flange Max Φ – 1.2d_0	Web wb
HEA 100	96	100	5	8	12	5	40	13	17	66
HEA 120	114	120	5	8	12	5	50	16	21	84
HEA 140	133	140	5.5	8.5	12	5	60	20	25	102
HEA 160	152	160	6	9	15	6	68	22	28	116
HEA 180	171	180	6	9.5	15	6	78	26	32	134
HEA 200	190	200	6.5	10	18	7	85	28	35	148
HEA 220	210	220	7	11	18	7	95	31	39	166
HEA 240	230	240	7.5	12	21	7	102	34	42	178
HEA 260	250	260	7.5	12.5	24	8	110	36	46	193
HEA 280	270	280	8	13	24	8	120	40	50	212
HEA 300	290	300	8.5	14	27	9	127	42	53	226
HEA 320	310	300	9	15.5	27	9	127	42	53	243
HEA 340	330	300	9.5	16.5	27	9	127	42	53	261
HEA 360	350	300	10	17.5	27	9	127	42	53	279
HEA 400	390	300	11	19	27	9	126	42	52	316
HEA 450	440	300	11.5	21	27	9	·126	42	52	362
HEA 500	490	300	12	23	27	9	126	42	52	408
HEA 550	540	300	12.5	24	27	9	125	42	52	456
HEA 600	590	300	13	25	27	9	125	41	52	504
HEA 650	640	300	13.5	26	27	9	125	41	52	552
HEA 700	690	300	14.5	27	27	9	124	41	52	600
HEA 800	790	300	15	28	30	9	121	40	50	692
HEA 900	890	300	16	30	30	9	121	40	50	788
HEA 1000	990	300	16.5	31	30	9	120	40	50	886

Some of those tension indicators are now sold with a colored resin under the embossment that will spread out when the correct preload is reached. This allows faster inspection because the resin (usually orange) is readily visible and easier to inspect than the gap.

6.8.4 Twist-Off Type Bolts

These bolts (Figure 6.15), called HRC in Europe, in accordance with EN 14399-10, have a splined end that extends beyond the threaded portion to allow a specially

Table 6.13 Useful dimensions for HEB profiles.

Profile	Depth	Width	Web thickness	Flange thickness	Radius	Encroachment	wa	Flange Max $\Phi - 1.5d_0$	Max $\Phi - 1.2d_0$	Web wb
HEB 100	100	100	6	10	12	5	40	13	16	66
HEB 120	120	120	6.5	11	12	5	49	16	20	84
HEB 140	140	140	7	12	12	5	59	19	24	102
HEB 160	160	160	8	13	15	6	67	22	28	116
HEB 180	180	180	8.5	14	15	6	76	25	32	134
HEB 200	200	200	9	15	18	7	84	28	35	148
HEB 220	220	220	9.5	16	18	7	94	31	39	166
HEB 240	240	240	10	17	21	7	101	33	42	178
HEB 260	260	260	10	17.5	24	8	109	36	45	193
HEB 280	280	280	10.5	18	24	8	118	39	49	212
HEB 300	300	300	11	19	27	9	126	42	52	226
HEB 320	320	300	11.5	20.5	27	9	126	42	52	243
HEB 340	340	300	12	21.5	27	9	126	42	52	261
HEB 360	360	300	12.5	22.5	27	9	125	42	52	279
HEB 400	400	300	13.5	24	27	9	125	41	52	316
HEB 450	450	300	14	26	27	9	125	41	52	362
HEB 500	500	300	14.5	28	27	9	124	41	52	408
HEB 550	550	300	15	29	27	9	124	41	52	456
HEB 600	600	300	15.5	30	27	9	124	41	51	504
HEB 650	650	300	16	31	27	9	124	41	51	552
HEB 700	700	300	17	32	27	9	123	41	51	600
HEB 800	800	300	17.5	33	30	9	120	40	50	692
HEB 900	900	300	18.5	35	30	9	119	40	50	788
HEB 1000	1000	300	19	36	30	9	119	39	49	886

designed tightening wrench to handle it and apply the necessary torque: when the correct preload is reached, the splined part, conveniently weaker, will break.

For those kinds of bolts it is suggested to complete the operation, for each group of bolts, in two phases: an initial phase for the firm contact and a final phase where the splined part is broken.

One disadvantage is that special wrenches are needed (to be summed with the cost of those bolts) but a noteworthy advantage is that there is no need for calibration (lubricants must however be used and surfaces cleaned to guarantee the correct friction). Another advantage is that they can be tightened from one side by only one person after the bolt has been inserted from the other side.

Figure 6.11 Allowable encroachment.

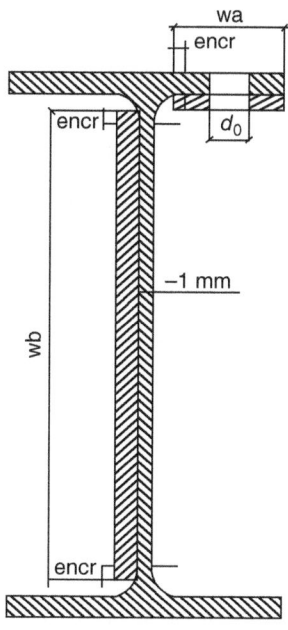

Figure 6.12 Possible instructions (in mm) for notches.

6.8.5 Hydraulic Wrenches

Hydraulic tools to assist pretensioning have evolved from the design of calibrated wrenches, with roughly the same pros and cons but likely a better control (in particular if the system takes into account the bolt elongation). However, a disadvantage of this kind of tool is the ample clearances that are needed for operating them: this might become a sensitive design issue.

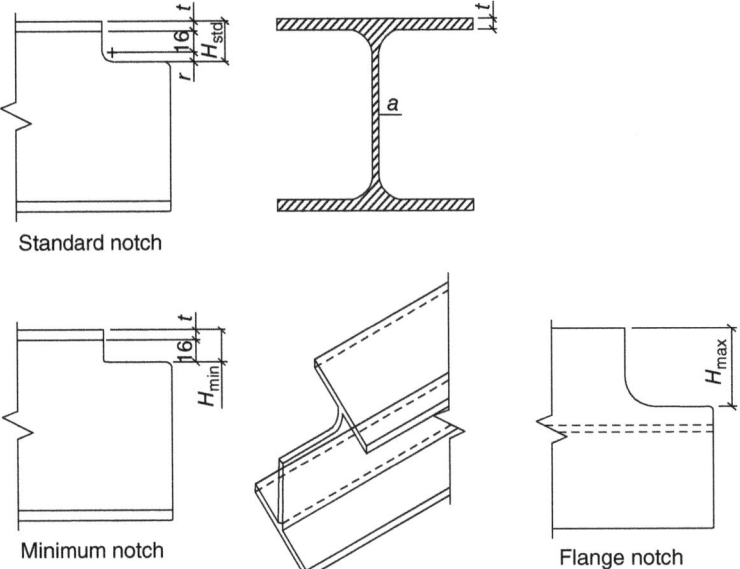

Figure 6.13 Top flange notch example.

Table 6.14 AISC tightening values.

Nominal bolt diameter d_b (in.)	Specific minimum bolt pretension T_m (kips)	
	ASTM A325, F1852	ASTM A490
1/2	12	15
5/8	19	24
3/4	28	35
7/8	39	49
1	51	64
1 1/8	56	80
1 1/4	71	102
1 3/8	85	121
1 1/2	103	148

6.9 Washers

Washers are needed to extend the contact area between the bolt head (or nut) and the plate in order to make the tightening elastic (thus decreasing the risk of unscrewing).

Washers (characterized by hardening) should be located under the part that is tightened (usually the nut).

Eurocode demands two washers (under the nut and the head) when there are single lap joints with only one bolt of rows, as shown in Figure 6.16.

Table 6.15 Tightening values according to EC [5].

Nominal bolt diameter (mm)	Specific minimum bolt pretension $F_{p,C}$ (kN)	
	Class 8.8	Class 10.9
12	47	59
16	88	110
20	137	172
22	170	212
24	198	247
27	257	321
30	314	393
36	458	572

Table 6.16 Tightening according to [2].

	Disposition of outer faces of bolted parts		
Bolt length	Both faces normal to bolt axis	One face normal to bolt axis, other sloped not more than 1 : 20	Both faces sloped not more than 1 : 20 from normal to bolt axis
Not more than $4d_b$	⅓ turn	½ turn	⅔ turn
More than $4d_b$ but not more than $8d_b$	½ turn	⅔ turn	⅚ turn
More than $8d_b$ but not more than $12d_b$	⅔ turn	⅚ turn	1 turn

Table 6.17 Tightening according to [5] in hypothesis of perpendicular surfaces.

t = Total thickness of connected parts, washers included, d = bolt diameter	Rotation to be applied after first step
$t < 2d$	⅙ turn
$2d \leq t < 6d$	¼ turn
$6d \leq t < 10d$	⅓ turn

According to [5], the same requirements of two washers apply with class 10.9 bolts. Again, [5] tells us that if holes have standard dimensions and connections are not pretensioned, washers are not essential.

6.9.1 Tapered (Beveled) Washers

There are commercial washers with a tapered surface (also known as beveled washers, see Figure 6.17) to be used with channels, I- and T-shaped profiles,

Figure 6.14 Direct tension indicator.

Figure 6.15 Twist-off round head bolt.

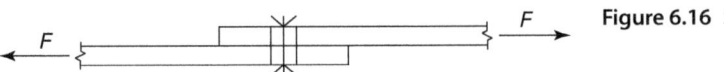

Figure 6.16 Single lap joint.

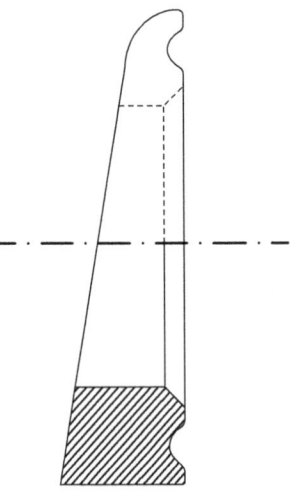

Figure 6.17 Tapered (beveled) washers.

or anything else having surfaces lacking parallelism (8%, 14%, or other slopes depending on the profile type).

6.9.2 Vibrations

Elastic or locking washers (combined to bolt pretensioning) are an option to evaluate if vibrations are expected, unless other systems (e.g. double nuts) are implemented. In similar situations of constant vibrations, frequent and periodic inspections are vital too.

6.10 Dimensions of Screws, Nuts, and Washers

In Europe the most recent standard for preloaded high-resistance bolts is EN 14399, made of several parts. Commercially speaking, though, other bolts that fit different designations can still be found, such as EN ISO, UNI 5712 (those actually surpassed), DIN 6914, and BS 4190. The dimensions of heads and nuts are the most important geometric data that vary, making the bolts with larger head surfaces the ones that work better for preloaded and slip-resistant connections. For their values the engineer can check bolt catalogs or Section 6.4.

6.10.1 Depth of Bolt Heads and Nuts

To help design (e.g. when calculating the bolt elongation causing prying action according to EC), Table 6.18 gives some general dimensions about the depth of bolt heads and nuts. The tabulated values are approximated and only indicative since, as just mentioned, the exact number depends on each reference standard; they can however be useful for a quick reference (roughly, the nut depth can be taken to be about 0.8 times the diameter while the head is about 0.65). Let us also mention that nuts for high-resistance bolts (e.g. those following EN 14399) are on average about 2 mm deeper than the listed values.

6.10.2 Washer Width and Thickness

Similar to what is just discussed, it is not in the scope of this book to provide precise washer dimensions since there are several bolt standards; this means that to set up, say, shop drawings, it is first necessary to check what is commercially available in local resellers' catalogs (e.g. there are washers with larger dimensions than the values given in Table 6.19). Nevertheless, it seems useful to give some general values of the external diameters of the most common washers so that the connection designer can quickly verify if the correct spaces are taken into account.

Table 6.19 also lists the thickness nominal value (some tolerance to be applied) of common washers. Once again, as for any bolt components, these are just general values (the values in the range between M12 and M36 are supposed to be for EN 14399 washers).

Table 6.18 "Average" dimensions for depth of common bolt heads and nuts.

Bolt	Nut depth	Head depth
M10	8	7
M12	10	8
M14	11	9
M16	13	10
M18	15	12
M20	16	13
M22	18	14
M24	19	15
M27	22	17
M30	24	19
M33	26	21
M36	29	23
M39	31	25
M42	34	26
M45	36	28
M48	38	30
M52	42	33
M56	45	35
M60	48	38
M64	51	40
M68	54	43
M72	58	45
M76	61	48
M80	64	50

6.11 Reuse of Bolts

Eurocode says that preloaded high-resistance bolts must not be reused since the preload brings plasticization in the bolt (in the shank or in the thread depending on the HR or HV bolt system).

Also in the United States, (check in particular [7, 8]), galvanized high-resistance bolts must not be reused. It might be allowed, if approved by the engineer of record, that "black" bolts (i.e. plain or no-finish bolts) are reused. The decision should be based on such criteria as ease in inserting the nut (yielding or elongation during the life of the bolts would show in this operation and the reuse would not be suggested at all).

Untightening and retightening bolts during erection to help insert other parts should not be considered as reuse, as explained in [8].

Table 6.19 External diameter values for common washers.

Bolt	External D	Thickness
M10	21	2
M12	24	3
M14	28	4
M16	30	4
M18	34	4
M20	37	4
M22	39	4
M24	44	4
M27	50	5
M30	56	5
M33	60	5
M36	66	6
M39	72	6
M42	78	7
M45	85	7
M48	92	8
M52	98	8
M56	105	9
M60	110	9
M64	115	9
M68	120	10
M72	125	10
M76	135	10
M80	140	12

6.12 Bolt Classes

Section 3.3 can be referred for bolt classes. For nuts, the corresponding resistance classes according to the International Organization for Standardization (ISO) are mainly classified with only the first digit, that is, a class 8 nut goes with a 8.8 screw, a 10 goes with a 10.9 screw, and so on. Other commercial names exist (some of them actually outdated), for example, 6S (it might initially look like a mechanical tolerance), which means it is a class 8 nut.

The approach in the recent EN 14399 is quite interesting: a single vendor must supply the bolt system (that is screw + nut + washers) so that it can be held responsible for the performance of the bolt group. The producer must also mark the bolts with a characteristic symbol and guarantee the finish. In the medium to long term, this kind of approach can hopefully solve the problem of using different names just mentioned earlier.

It is noted that, generally speaking, steel structures are more often designed with bolts exclusively in classes 8.8 and 10.9 (or equivalent like ASTM A325 and A490); therefore, it is suggested not to prescribe bolts with lower resistances.

6.13 Shims

The shims (Figure 6.18) are thin plates used during erection and they can allow compensating fabrication tolerances.

"Finger shims" are fabricated in various thicknesses (e.g. 2, 3, and 5 mm) and they are installed and inserted among bolts as "combs." It should be remembered that for jobs according to EC, Ref. [5] demands are that there are maximum three plates.

Shims can be used in many types of connections. For example, a base plate that does not have locknuts on the bottom needs prefabricated shims, to be installed on-site as necessary.

End-plate connections are routinely fabricated with a tolerance on each side in order to ease erection (and to compensate for galvanization thickness). It is therefore good practice to ship some finger shims to the erection site to be inserted by erectors when they are needed (Figure 6.19).

(a)

(b)

Figure 6.18 Shims prebolted in the shop for shipping.

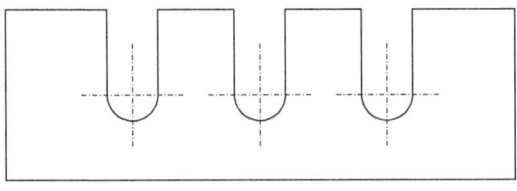

Figure 6.19 Finger shims.

6.14 Galvanization

Galvanization (Figure 6.20) by a molten zinc bath to profiles after they have been cut, drilled, and assembled with plates is, alternatively to painting or in combination with it in "Duplex" processes, a typical finish for steel parts. It is needed in order to improve the resistance against corrosion and consequently the life of the structure.

Galvanization has to be considered also when preparing drawings of connections and steel parts because some special details are often needed.

6.14.1 Tubes

It happens when using tubes that there are plates closing and sealing both ends. Thinking indeed about a tubular column (classical solution in indoor mezzanines), it is easy to find a plate on the bottom (base plate) and a horizontal plate on top, where the primary beam will lean and be bolted.

In such a situation it is essential to drill holes on both the end plates so that, during galvanization, the zinc can go inside (and air come out) and then flow out when the tube is pulled from the tub.

If holes are missing, the air inside the tube will be trapped. Zinc high temperatures will make the air expand and, since there is no way out, one of the plates might actually explode (usually breaking welds).

Holes on just one side would avoid explosion but they would not allow the zinc to go in and out as explained during the bathing process.

Figure 6.21 shows some examples of holes but better solutions taken from [9] are represented in Figure 6.22.

6.14.2 Plate Welded over Profiles as Reinforcement

A profile can be reinforced by welding a plate over it. A typical example is welding a web doubler plate over a column web in a moment connection in order to help web panel shear. This kind of detail might create problems if the column

Figure 6.20 Graphical representation of hot dip galvanization of a steel profile.

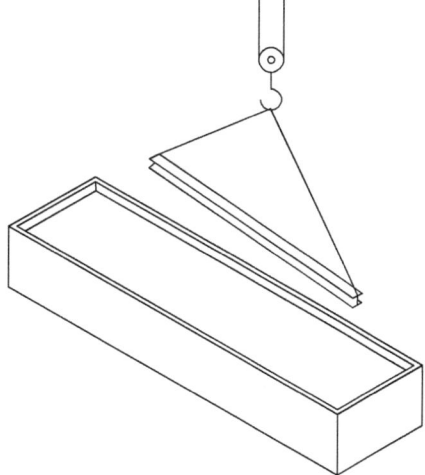

Figure 6.21 Minimum hole (not optimal) for zinc drainage.

is later galvanized. If the plate is welded by an all-around fillet weld, air bubbles trapped inside the plate and the web might cause explosion and detach the plate (as already explained the heat will make this happen). On the other hand, if the weld is intermittent, a different situation might develop: the acid in the zinc bath will penetrate in the interstices between the plate and the web while the zinc, which is more viscous, will not. The acid though, evaporating, will prevent the zinc from gripping steel in the adjacent areas. The final result is that the galvanization near the plate will be of poor quality, and likely absent.

How can this be prevented? Apart from avoiding these types of situations (e.g. to reinforce the column web, a diagonal stiffener might suffice, if this does not interfere with other joints on the weak side of the column), Ref. [10] proposes central holes in the reinforcing plates (the size depending on the plate dimensions) to vent air and at least concentrate the defects in those areas (to be locally treated later). More details are available in [10].

6.14.3 Base Plates

The advice for base plates or similar situations in geometry is to adopt appropriate chamfers (or holes or similar as in Figure 6.23) in the reinforcing plates in order to have the necessary clearances not only for welds but also to allow the zinc bath to flow out, most of all from corners, where it could pile up.

6.15 Other Finishes After Fabrication

When designing connections the engineer is supposed to check that they will not be in areas that will later be "treated": in some projects, for example, columns might be later covered by fire-resistant material and therefore it might be advised (in order to allow maintenance operations like disassembling) to make the bolted connection at an appropriate distance from the column, through beam stubs as in Figure 6.24.

Similar situations happen with composite constructions where columns or beams are partially or completely encased by concrete.

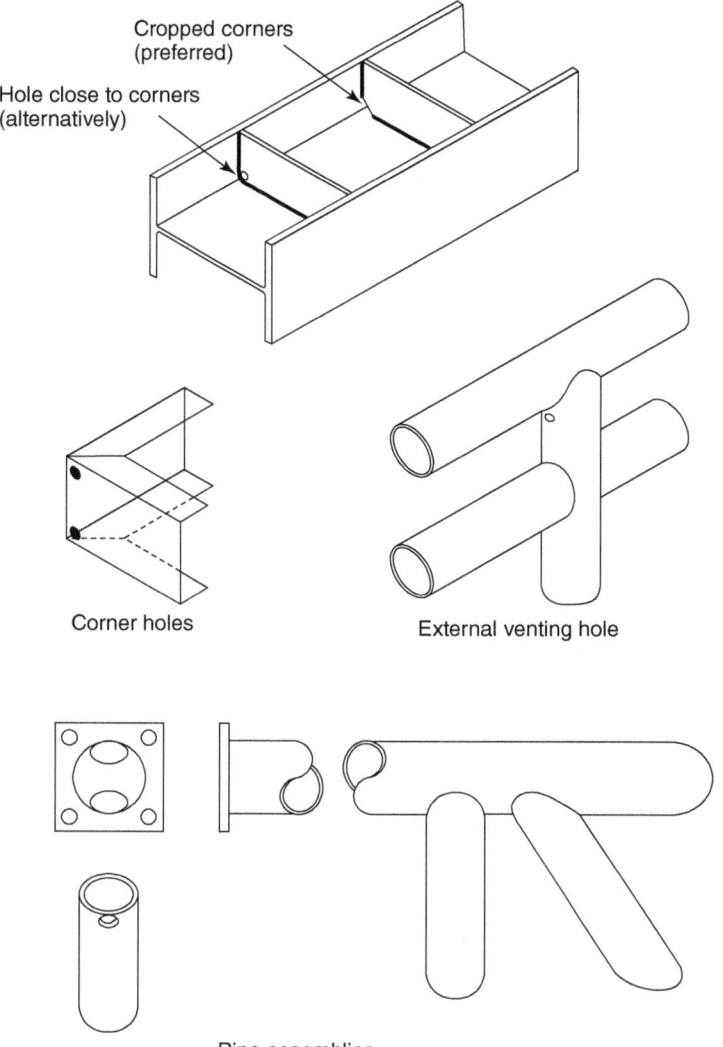

Figure 6.22 Optimal details (various situations are represented) to avoid galvanizing problems).

6.16 Camber

Cambering a profile means fabricating it as curved in a vertical plane. This is usually done to compensate for some vertical deflection that is expected after erection, normally in long spans. The deformation might indeed be unacceptable for some reasons (structural but more frequently architectural and esthetic).

The camber can be given by cold (e.g. press) or hot (e.g. heat source) treatments but this is outside the scope of this text.

From a connection design perspective, which is what the book is about, some problems might arise from the fact that the hypothetical horizontal position

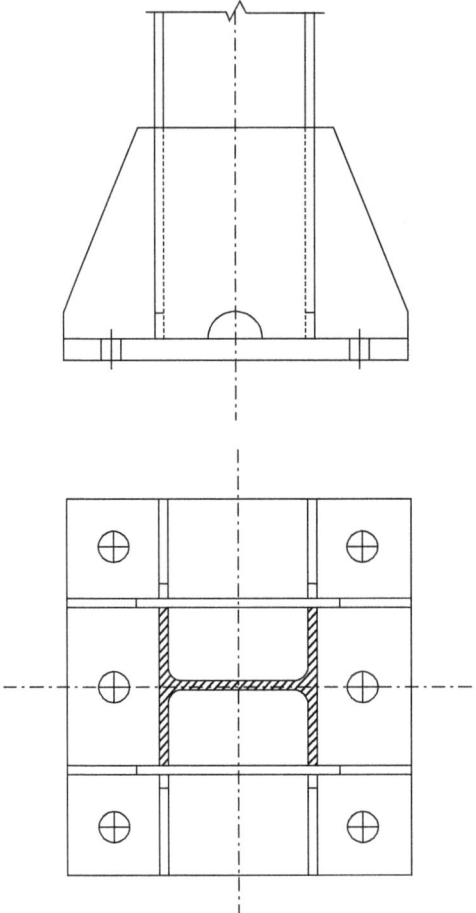

Figure 6.23 Reinforced base plate: note the half-moon-shaped cut to drain zinc.

is usually reached after the application of the final dead loads (therefore, after erection) so, sometimes difficult geometric situations like the ones shown in Figure 6.25 might occur.

It is good practice for the engineer to anticipate the issue and inform the detailer so that the necessary geometric countermeasures are taken into account when generating shop drawings.

6.17 Grout in Base Plates

As explained in Section 4.4, base plates commonly have grout (also called expansive grout or nonshrinkage grout or by the commercial name Emaco) cast after erection and this is very important to create the correct design contact pressure at the base necessary to resist compression loads.

Sometimes the quality of the grout is quite poor (or it is even totally absent!) and the reasons for this might be as follows:

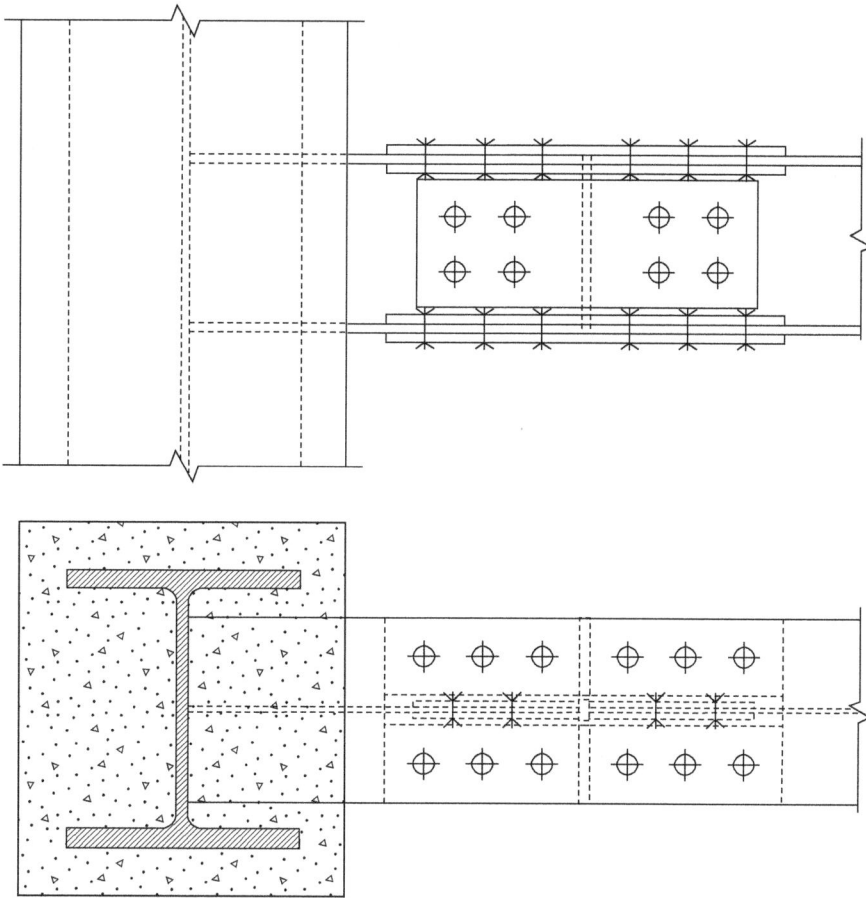

Figure 6.24 Composite column – connection designed outside the encased part.

- It is done after the structure is assembled by masons that are likely not in the erector crew and therefore the significance of the detail is not clear.
- The customer might not understand the importance and might even forget calling the masons to cast the grout.
- It is easy for the pit under the base plate to get dirty with soil or other material that is not removed before casting.
- The base plates often do not have holes of the correct size to allow masons to stir and compact the grout during casting.

This is the reason some designers only take into consideration the steel area of anchor bolts even to resist compression. This approach is not recommended, but it is necessary to make sure that correct holes are drilled in the base plates to help the casting operation, that the right emphasis on the process is shown in the drawings, and that the correct execution (e.g. the proper fluidity) is performed.

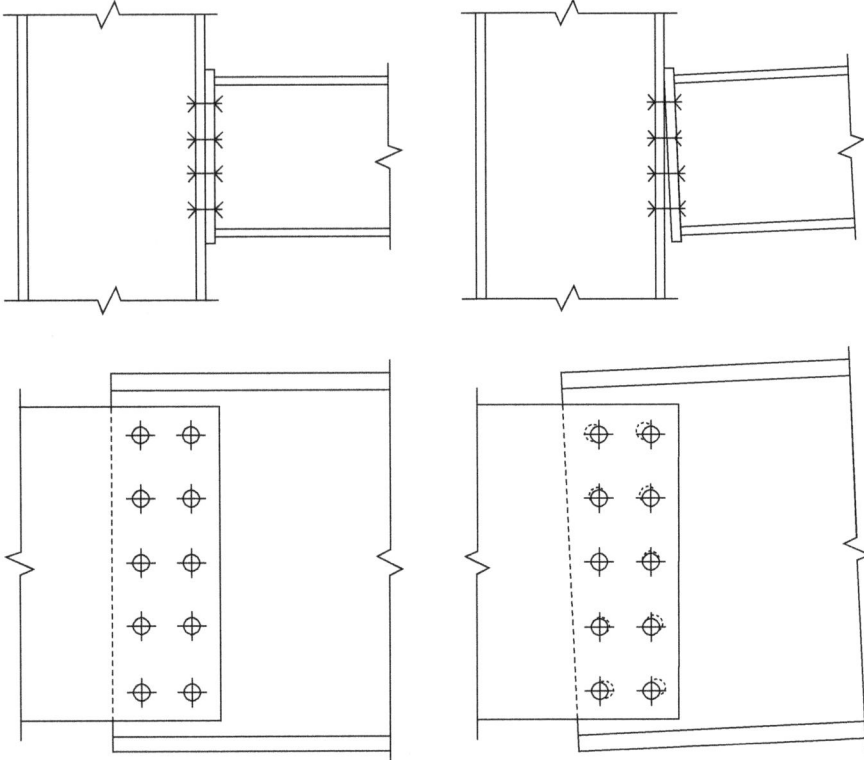

Figure 6.25 Possible troubles due to camber – the geometry at erection (on the right-hand side) does not fit with design geometry (on the left).

6.18 Graphical Representation of Bolts and Connections

A typical standard in the industry is to represent with a triangle the connections that must withstand a moment in the schematic drawings (each single line indicates a profile) with the design members of a structure.

It could also be convenient to show holes and/or bolt sizes with symbols like, for example, the ones in Figure 6.26 (this is just a proposal, different symbols are also used and, therefore, a legend is necessary).

Symbology for holes and related bolts

Symbol	⊗	●	⌀	⊕	●	⊕	●	⊕	⊕
Bolt	M10	M12	M14	M16	M18	M20	M22	M24	M27	M30

Figure 6.26 Possible symbols for holes and corresponding bolt sizes.

6.19 Field Welds

In some countries (e.g. most of Europe), field welds are avoided whenever possible, while in other countries (e.g. the United States), they are adopted more commonly and are even used to reach specific seismic performances (for more details see Section 4.24).

It is true that field welding has many disadvantages:

- It might be difficult to keep pieces in the correct position.
- While in the shop the steel pieces can be rotated and the welds executed horizontally, this is not likely on-site and it is common to have difficult welds (say, overhead welds, where "the fight" is against gravity).
- The wind can disturb the gas protecting the weld, lowering final quality.
- Gas metal arc welding (GMAW) (see Section 3.9.5) as commonly done in the shop is not possible (the equipment cannot easily be moved around) and therefore shielded metal arc welding (SMAW) is needed.
- Highly qualified personnel are needed to deal with the issues.
- There are problems in welding hot-dip galvanized parts since the zinc will deteriorate the weld and the weld will deteriorate the galvanization; a (partial) remedy is to apply by hand (there are special sprays) some "cold" zinc after the weld is done to recover some protection from corrosion.

6.20 Skewed Joints

Figures 6.27–6.38 represent some examples of skewed joints (among the examples some are obtained from [11, 12]).

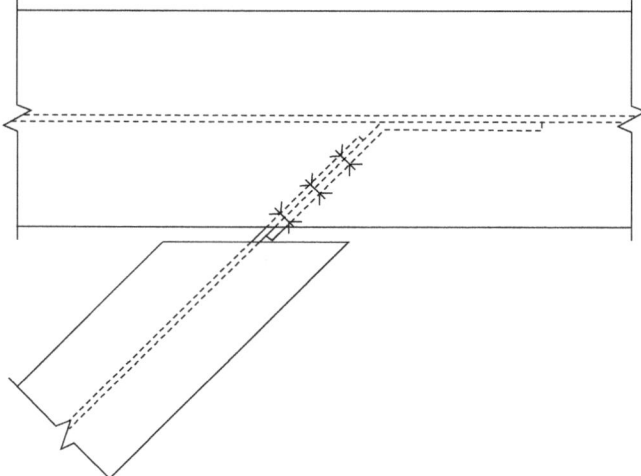

Figure 6.27 Bent plate welded over the main member web and bolted to the notched secondary member similarly to a fin plate.

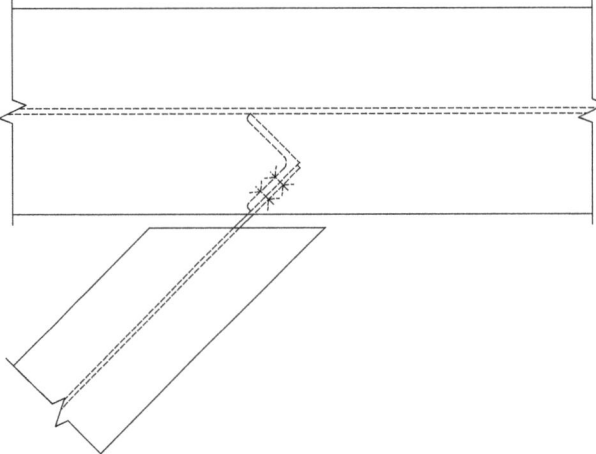

Figure 6.28 Double angle welded to the primary web (stiffener on the back side possible) and bolted to the secondary member (notched).

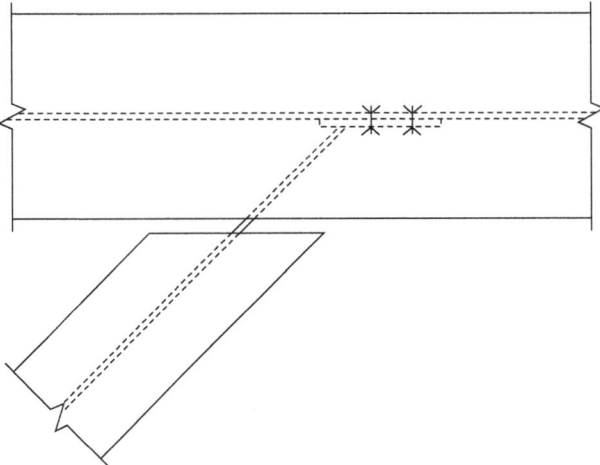

Figure 6.29 Plate butt welded to the secondary web and bolted to the primary web.

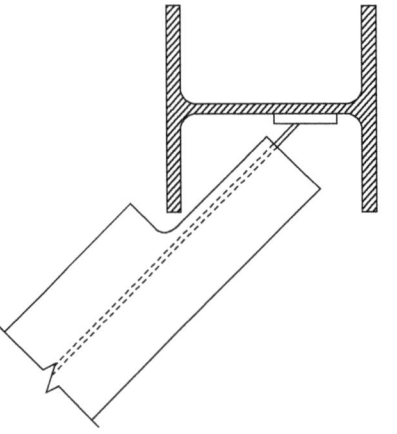

Figure 6.30 Plate welded to the secondary member (notched on flanges on one side) and field welded to the column web.

Figure 6.31 Fin plate (shear tab) inclined as necessary on column web and bolted to the beam.

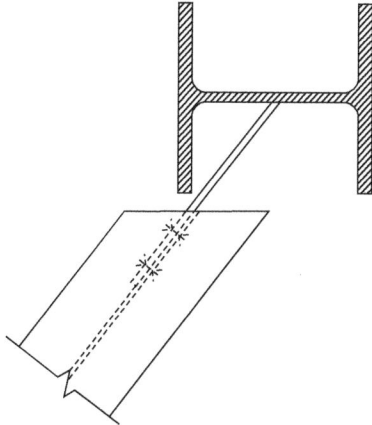

Figure 6.32 Fin plate (shear tab) inclined as necessary on column flange and bolted to the beam.

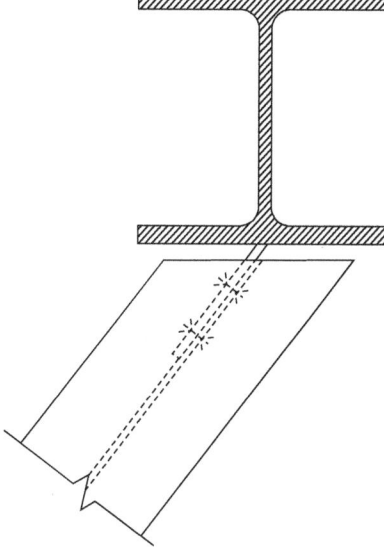

Figure 6.33 Fin plate inclined as in Figure 6.32 but laterally; stiffeners on the column are likely necessary.

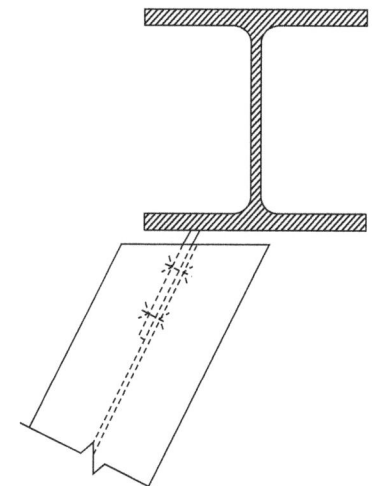

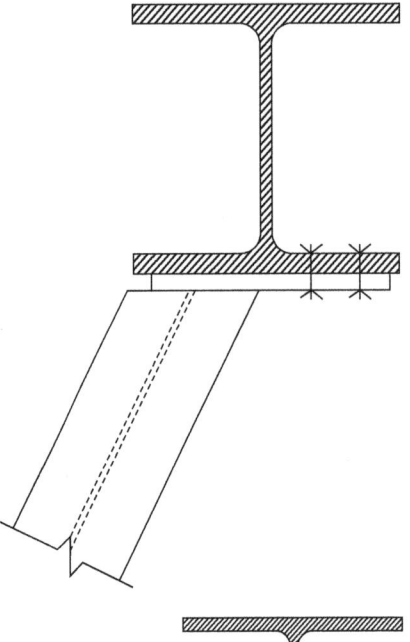

Figure 6.34 End plate welded to secondary and bolted to only half flange of the column; solution to be calculated carefully because of the position of the bolts.

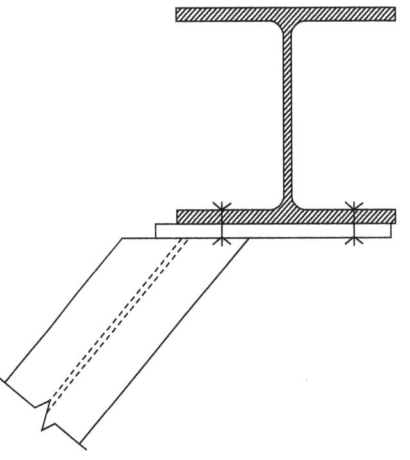

Figure 6.35 Representation of Figure 6.34 but the bolt position, whenever possible, will allow a better performance.

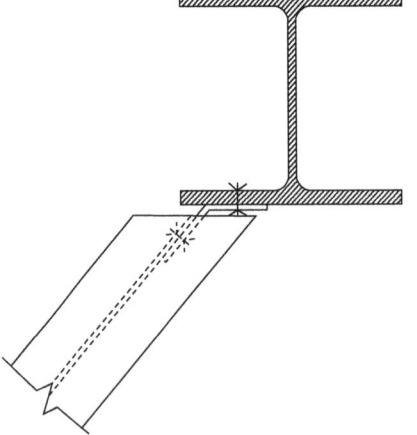

Figure 6.36 Bent plate bolted to the web secondary and to the primary flange (half of it).

Figure 6.37 Bent plate welded (or bolted) to the column flange and bolted to the secondary beam (or brace) on the left; two double angles bolted to the column flange (bolts through the other plate) and welded (or bolted) to the web of the beam.

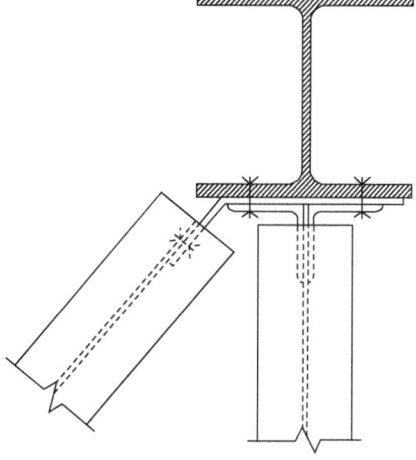

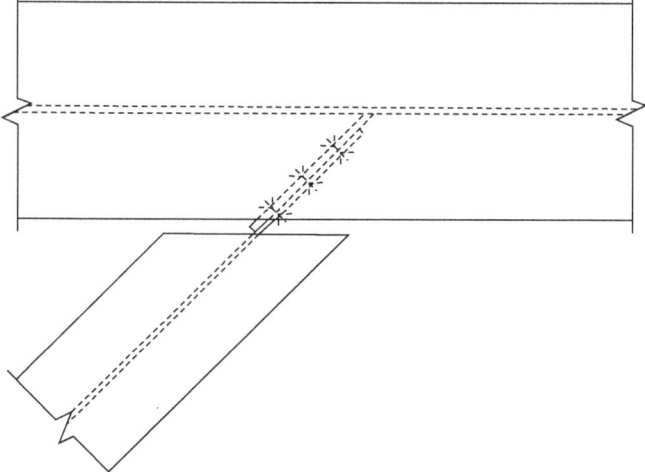

Figure 6.38 Shear-tab-like plate skewed and welded over the main beam and bolted to the notched secondary.

References

1 Ballio, G. and Mazzolani, F. (1983). *Theory and Design of Steel Structures*. London: Taylor & Francis.
2 American Institute of Steen Construction (AISC) (2011). *Steel Construction Manual*, 14e. Chicago, IL: AISC.
3 Fisher, J.M. and Kloiber, L.A. (2010). *Base Plate and Anchor Rod Design, AISC Design Guide 1*, 2e. Chicago, IL: American Institute of Steel Construction.
4 Tamboli, A.R. (1999). *Handbook of Structural Steel Connection Design and Details*. New York: McGraw Hill.

5 CEN (2008). *Execution of steel structures and aluminium structures – Part 2: technical requirements for the execution of steel structures*, EN 1090-2:2008. Brussels: CEN.
6 Kulak, G.L., Fisher, J.W., and Struik, J.H.A. (2001). *Guide to Design Criteria for Bolted and Riveted Joints*, 2e. Chicago, IL: RSCS – American Institute of Steel Construction.
7 Kulak, G. (2002). *High strength bolts, a primer for engineers*, AISC Design Guide 17. Chicago, IL: American Institute of Steel Construction.
8 Research Council on Structural Connections (RCSC) Committee A.1 (2009). *Specification for Structural Joints Using ASTM A325 or A490 Bolts*: RCSC www.boltcouncil.org (accessed 25 January 2018).
9 American Galvanizers Association (AGA) (2005). *The Design of Products to Be Hot Dip Galvanized After Fabrication*. Centennial, OK: AGA.
10 American Society for Testing and Materials (ASTM) (2003). *Providing high quality zinc coatings (hot dip)*, Recommended Practice A385-03. West Conshohocken, PA.
11 Kloiber, L. and Thornton, W. (2001). Design of skewed connections. *Eng. J. AISC* 38: 140–147, Quarter 3.
12 The Steel Construction Institute (SCI), The British Constructional Steelwork Association (BCSA) (2002). *Joints in Steel Construction: Simple Connections*. Ascot, UK: SCI, BCSA.

7

Connection Examples

This chapter presents some connection examples (without any dimensions) to help the designer in finding the right solutions. The cases are actually innumerable and it is hoped that the realizations given here (Figures 7.1–7.112) will inspire the engineer by visualizing some real situations that worked in other projects.

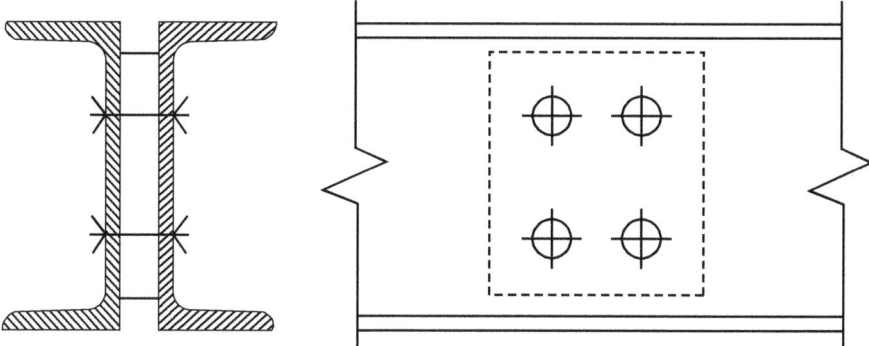

Figure 7.1 Tie plates for large channels.

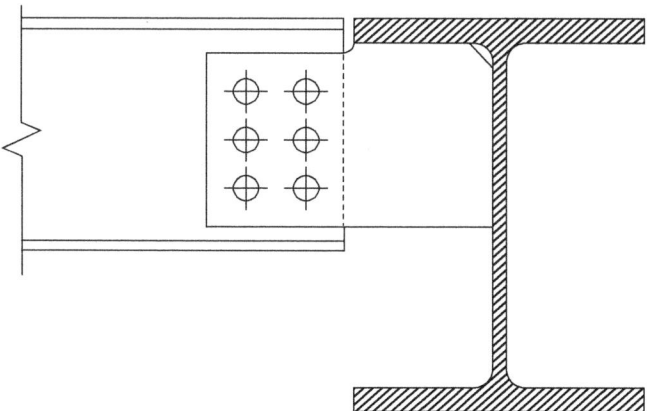

Figure 7.2 Beam-to-beam fin plate (extended shear tab) welded to the primary top flange and half web.

Design and Analysis of Connections in Steel Structures: Fundamentals and Examples,
First Edition. Alfredo Boracchini.
© 2018 Ernst & Sohn Verlag GmbH & Co. KG. Published 2018 by Ernst & Sohn Verlag GmbH & Co. KG.

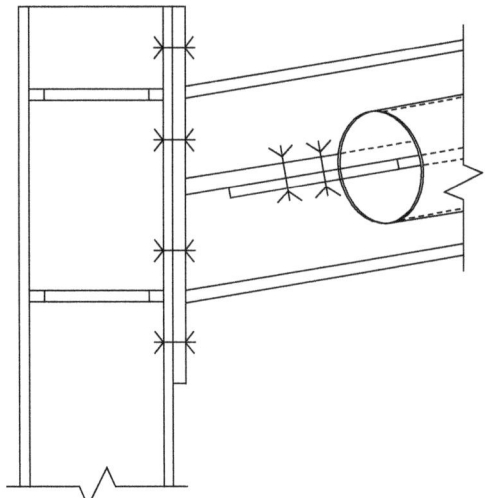

Figure 7.3 Beam-to-column moment connection with horizontal pipe brace also framing into it.

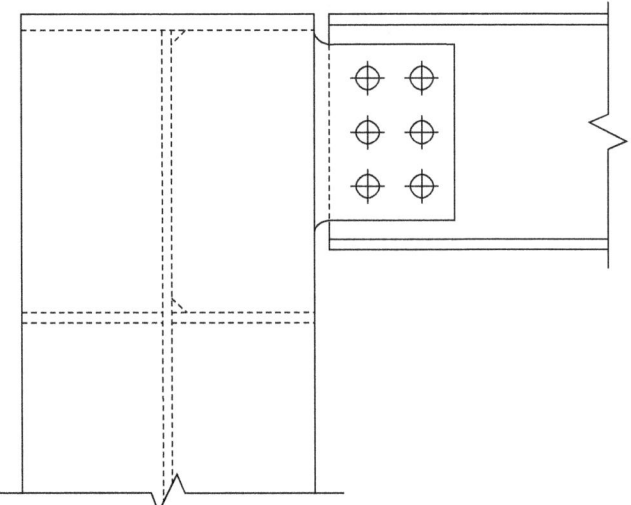

Figure 7.4 Extended shear tab (fin plate) welded to horizontal stiffeners and column web.

7 Connection Examples | 295

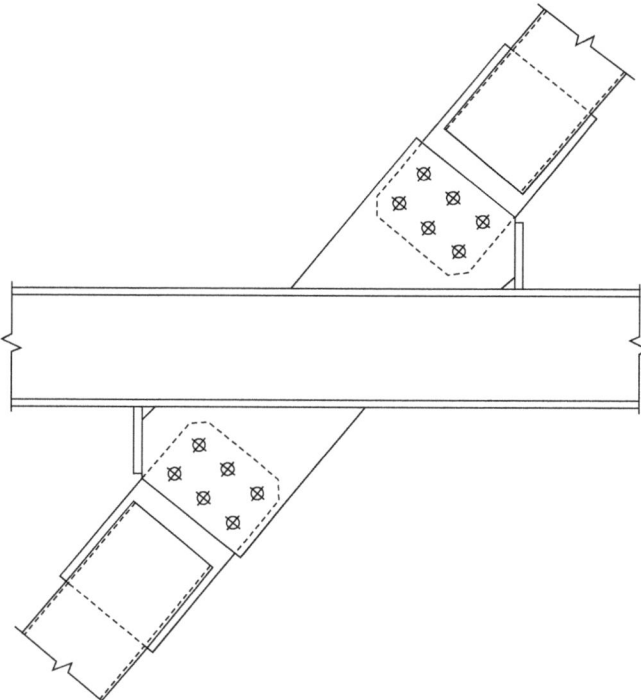

Figure 7.5 Tube brace (or, generally speaking, a diagonal) going through a beam.

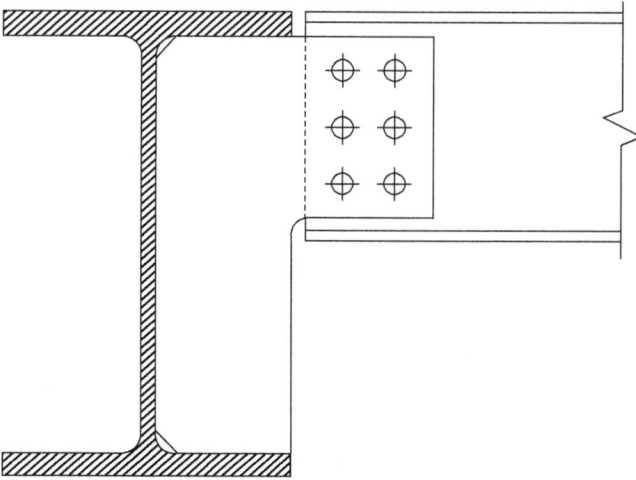

Figure 7.6 Full-depth shear tab (fin plate) welded to primary beam and bolted to secondary beam.

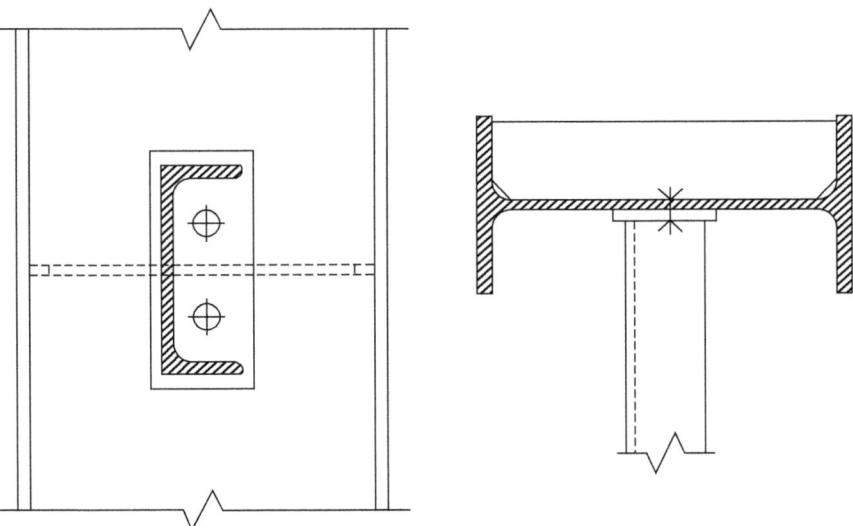

Figure 7.7 Channel (likely a stair stringer) bolted to the column web (stiffener on the back).

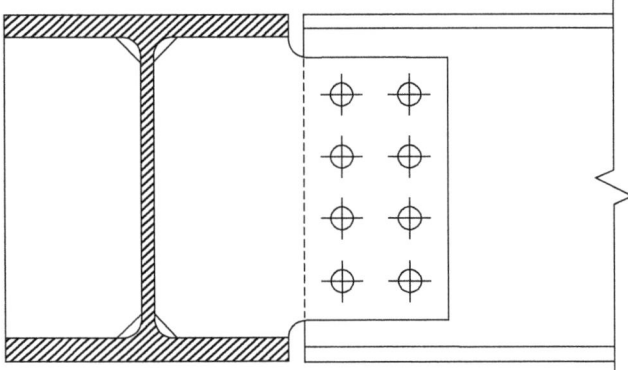

Figure 7.8 Fin plate with a stiffener on the opposite side.

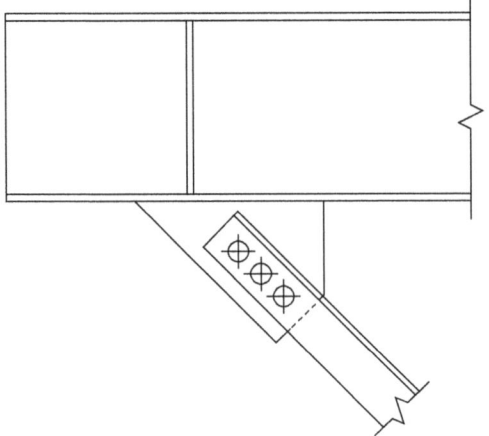

Figure 7.9 Diagonal angle(s) supporting a beam (likely a cantilever).

7 Connection Examples | 297

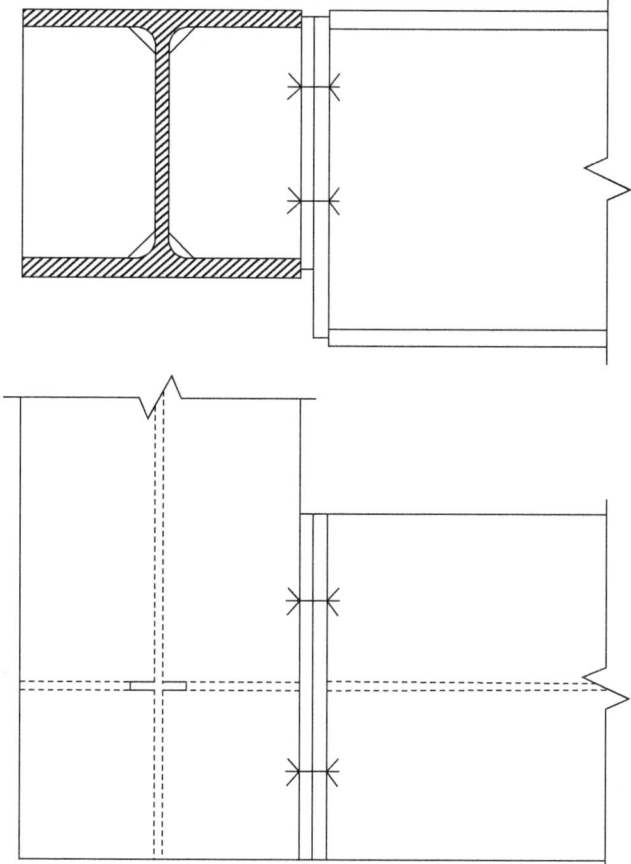

Figure 7.10 End plate in a corner.

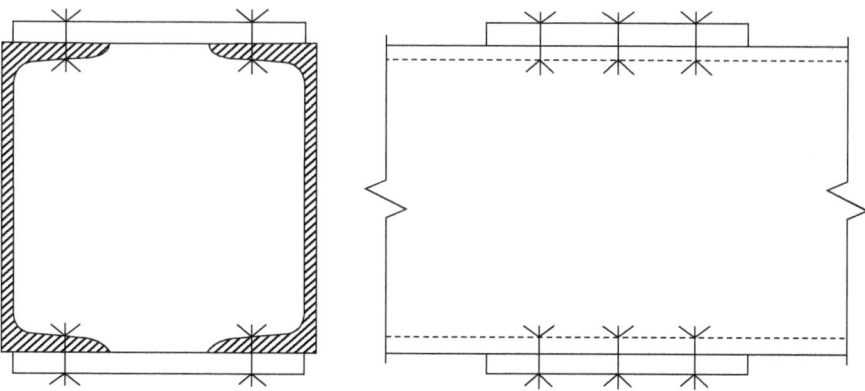

Figure 7.11 Channels connected on flanges.

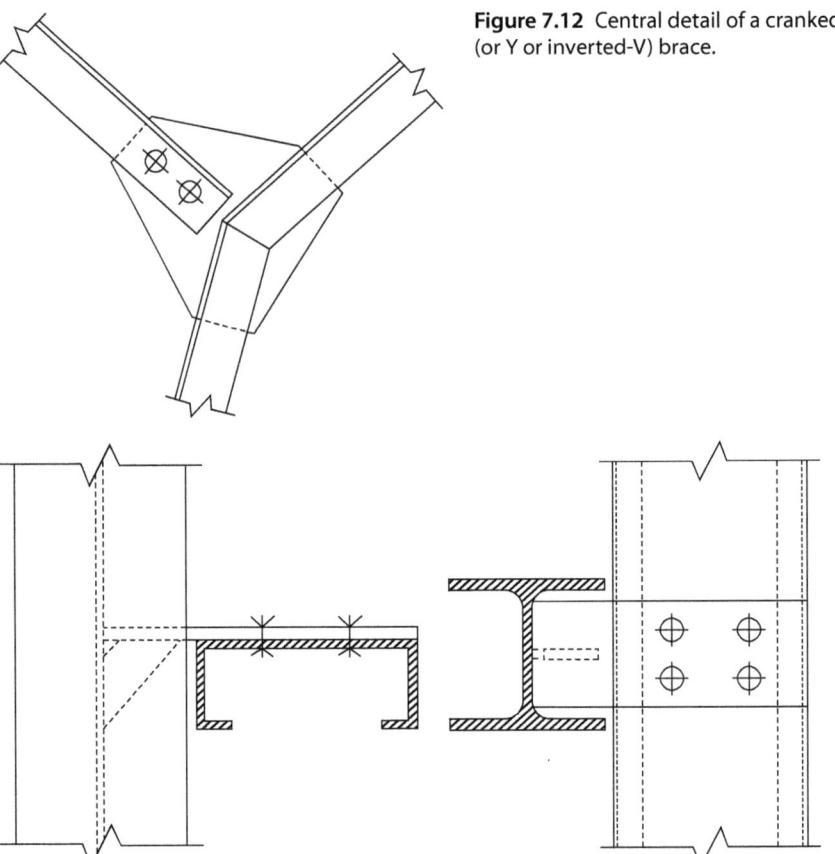

Figure 7.12 Central detail of a cranked K (or Y or inverted-V) brace.

Figure 7.13 Connection to column of a girt realized with a channel.

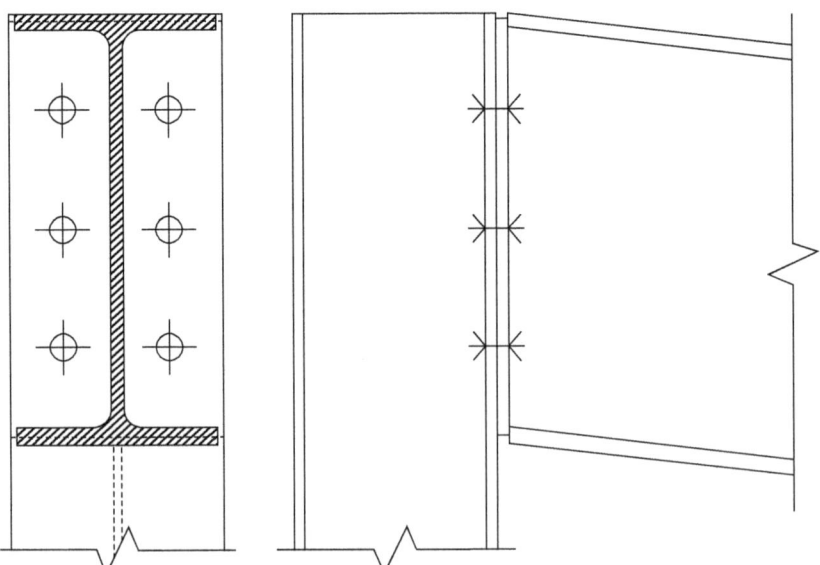

Figure 7.14 Sloped beam end plate framing into a column.

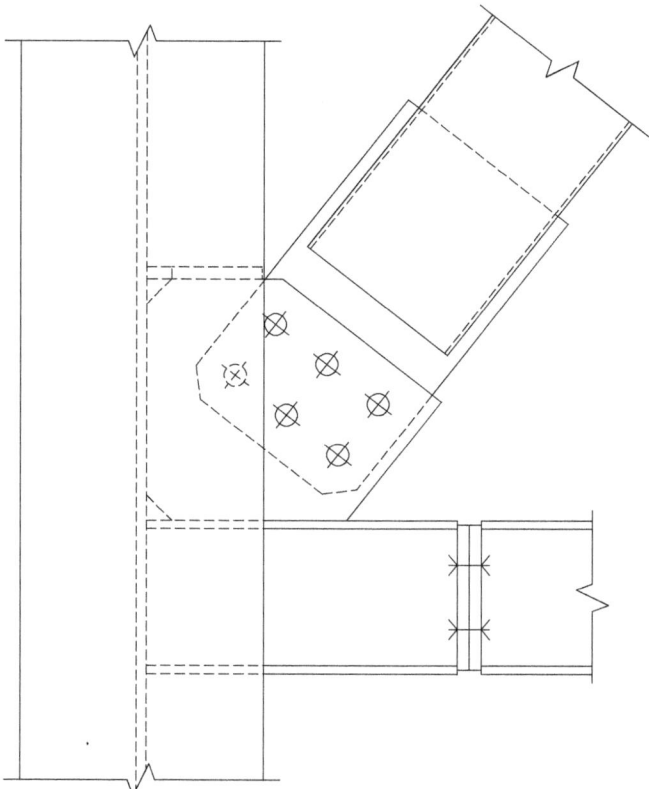

Figure 7.15 Brace connection of a pipe into column and beam.

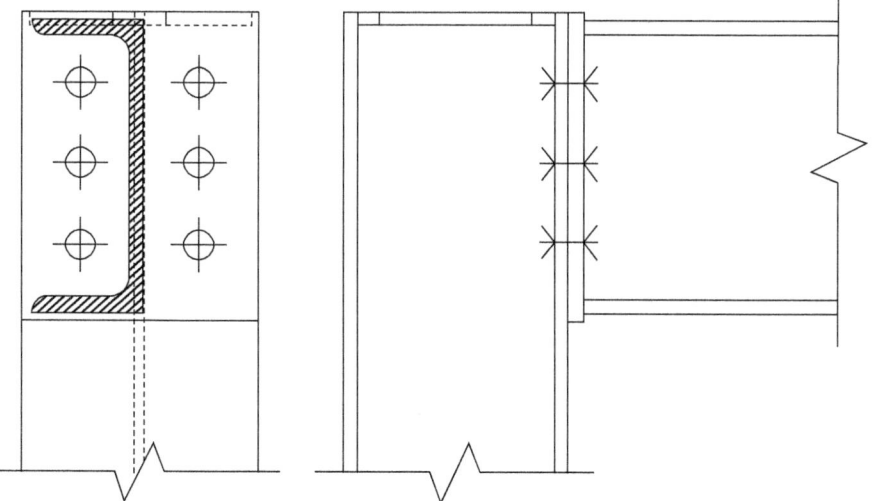

Figure 7.16 Joint of a channel (possibly a stair stringer) into a column.

300 | 7 Connection Examples

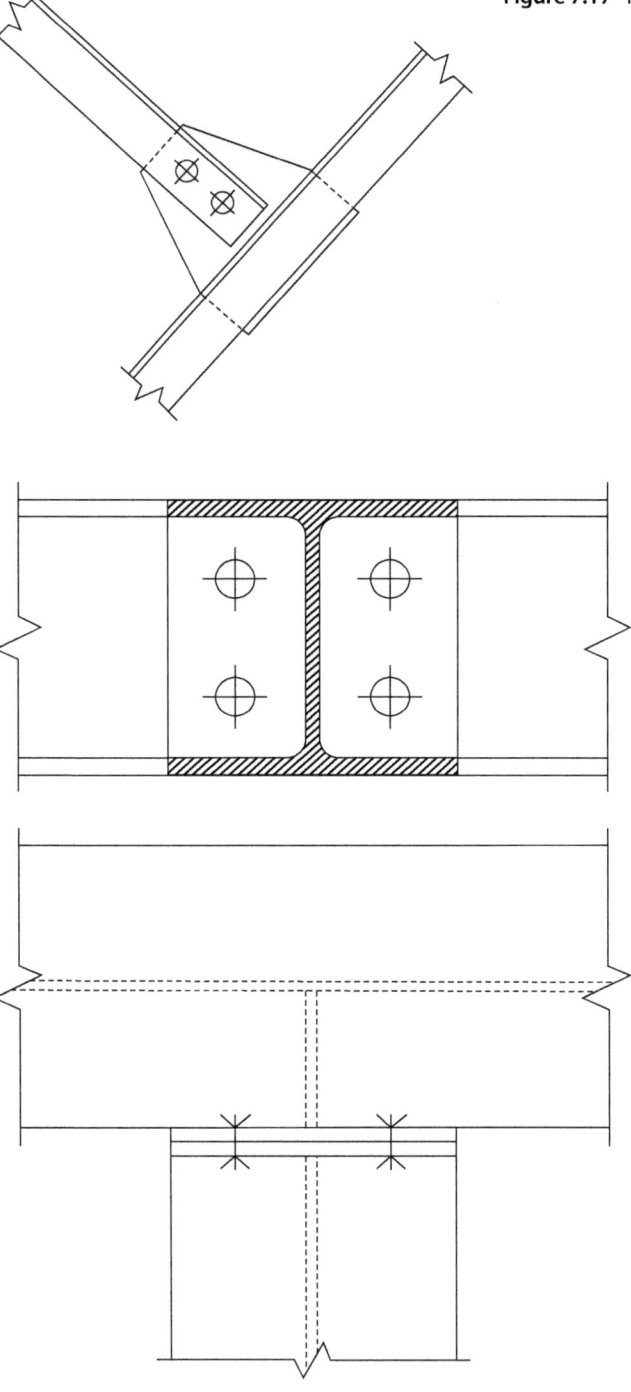

Figure 7.17 Truss detail.

Figure 7.18 End plate with central stiffener between wide flange type I beams.

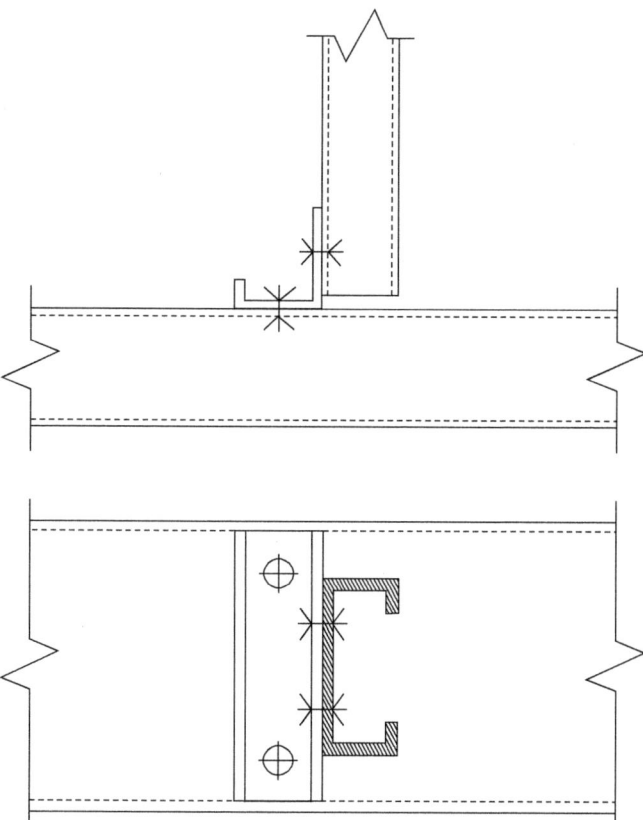

Figure 7.19 Fully bolted detail to connect C-shaped purlins with a stabilizing profile.

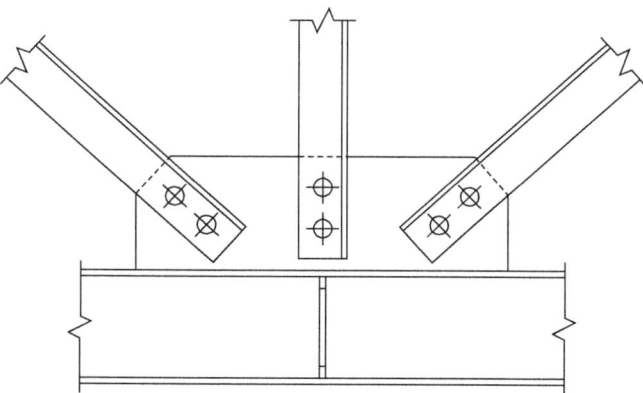

Figure 7.20 Angles framing into a beam.

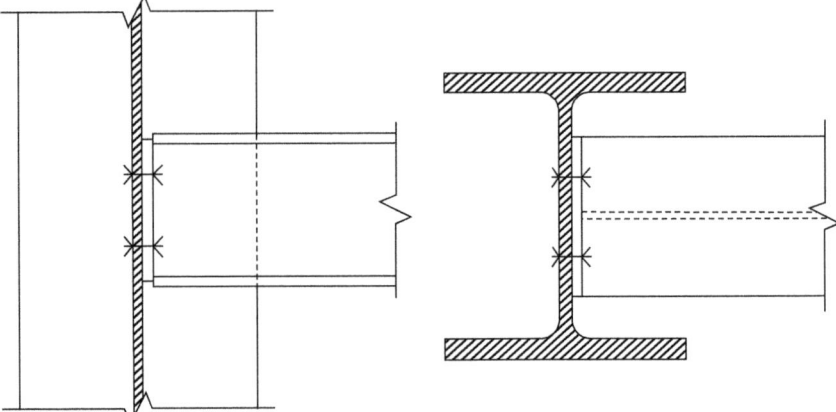

Figure 7.21 End plate on column web.

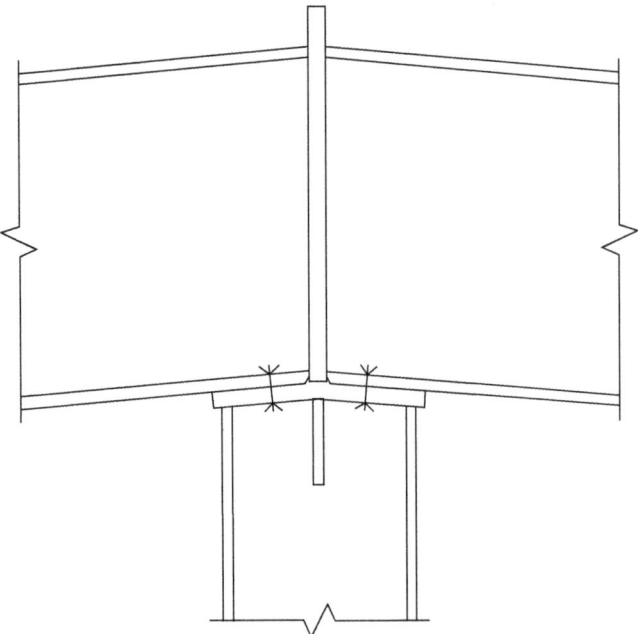

Figure 7.22 Welded apex between beams leaning on a central column.

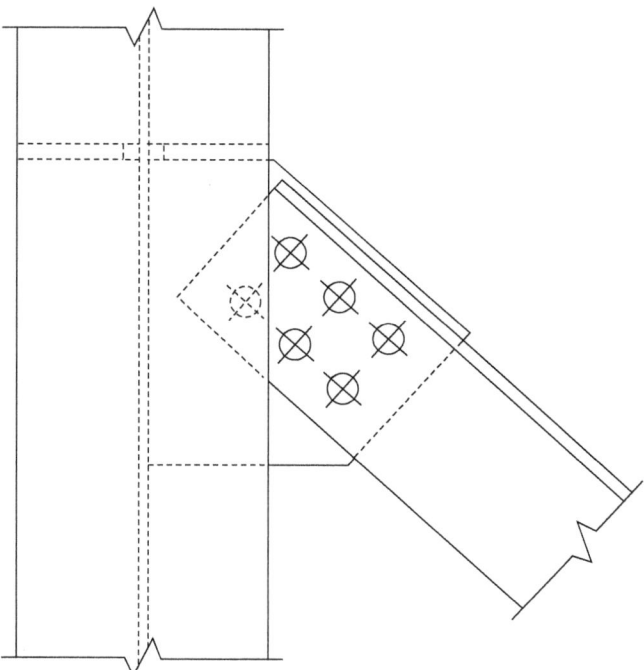

Figure 7.23 Heavy angle (probably a brace) bolted to plate welded to column weak side.

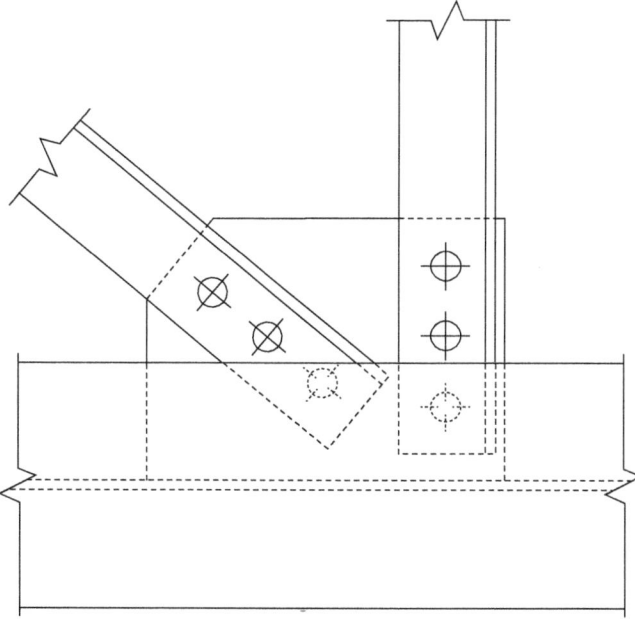

Figure 7.24 Bolted connections in a truss having a rotated I beam as lower chord (stiffeners actually suggested opposite to the plate).

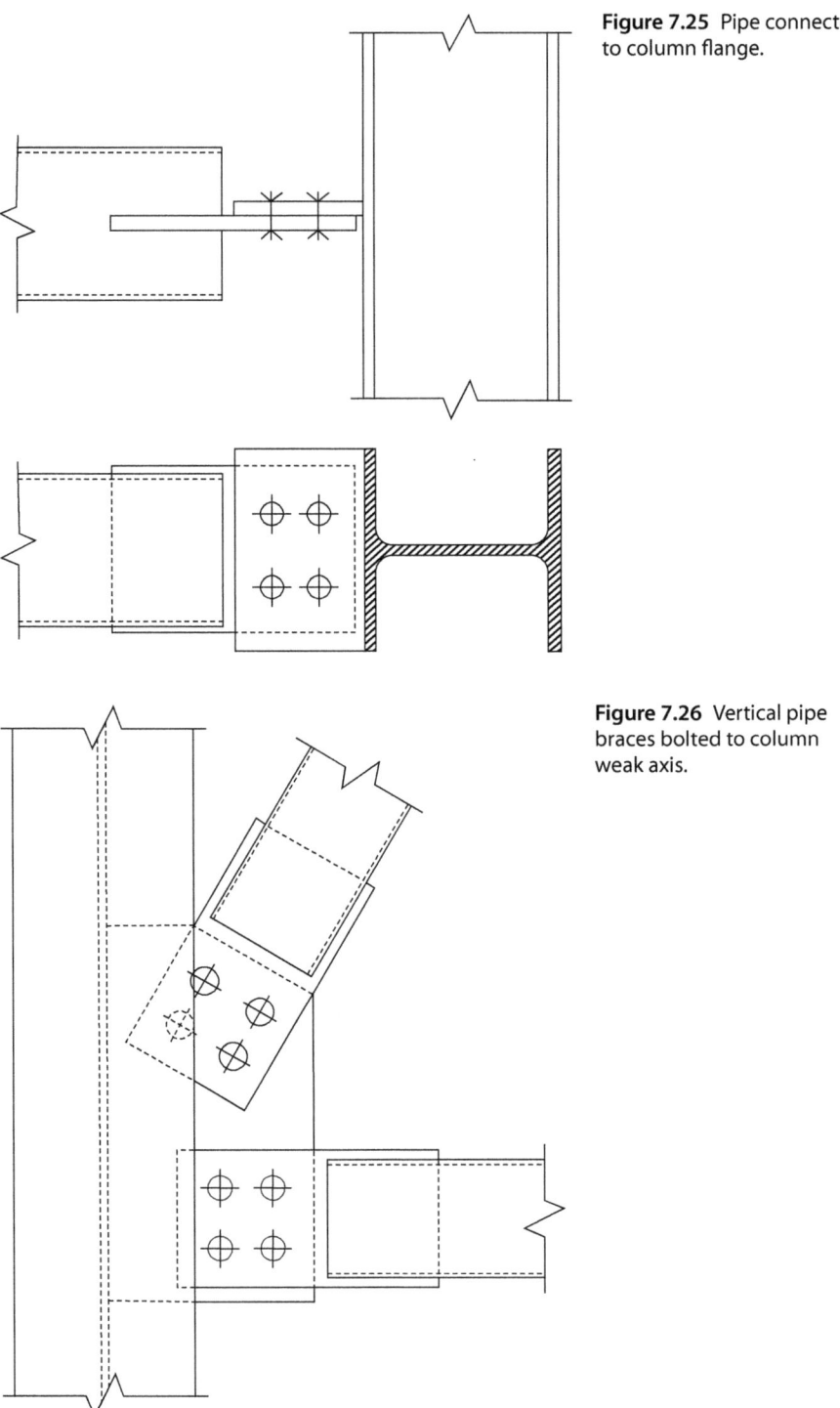

Figure 7.25 Pipe connected to column flange.

Figure 7.26 Vertical pipe braces bolted to column weak axis.

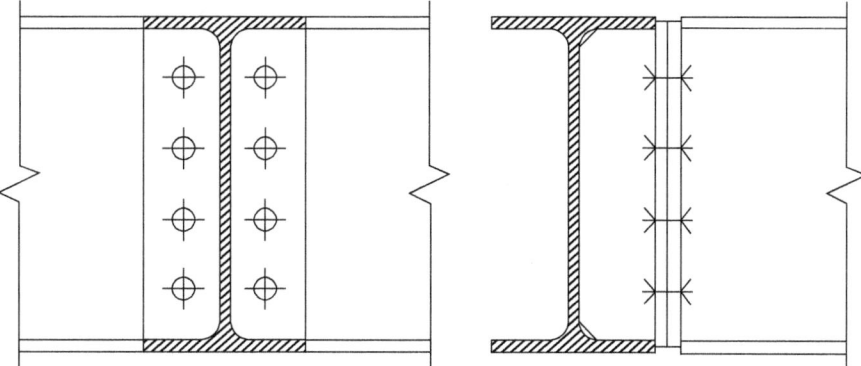

Figure 7.27 End plate between beams of the same size.

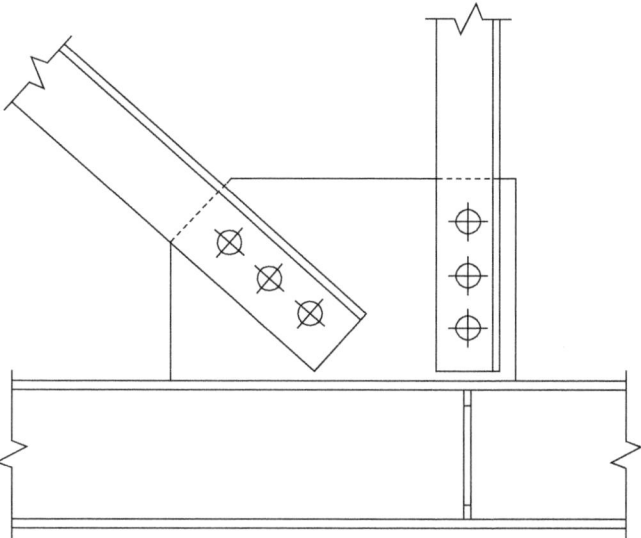

Figure 7.28 Angles connected to a beam (stiffener to be noted).

Figure 7.29 Omega-shaped purlins connected to supporting beam.

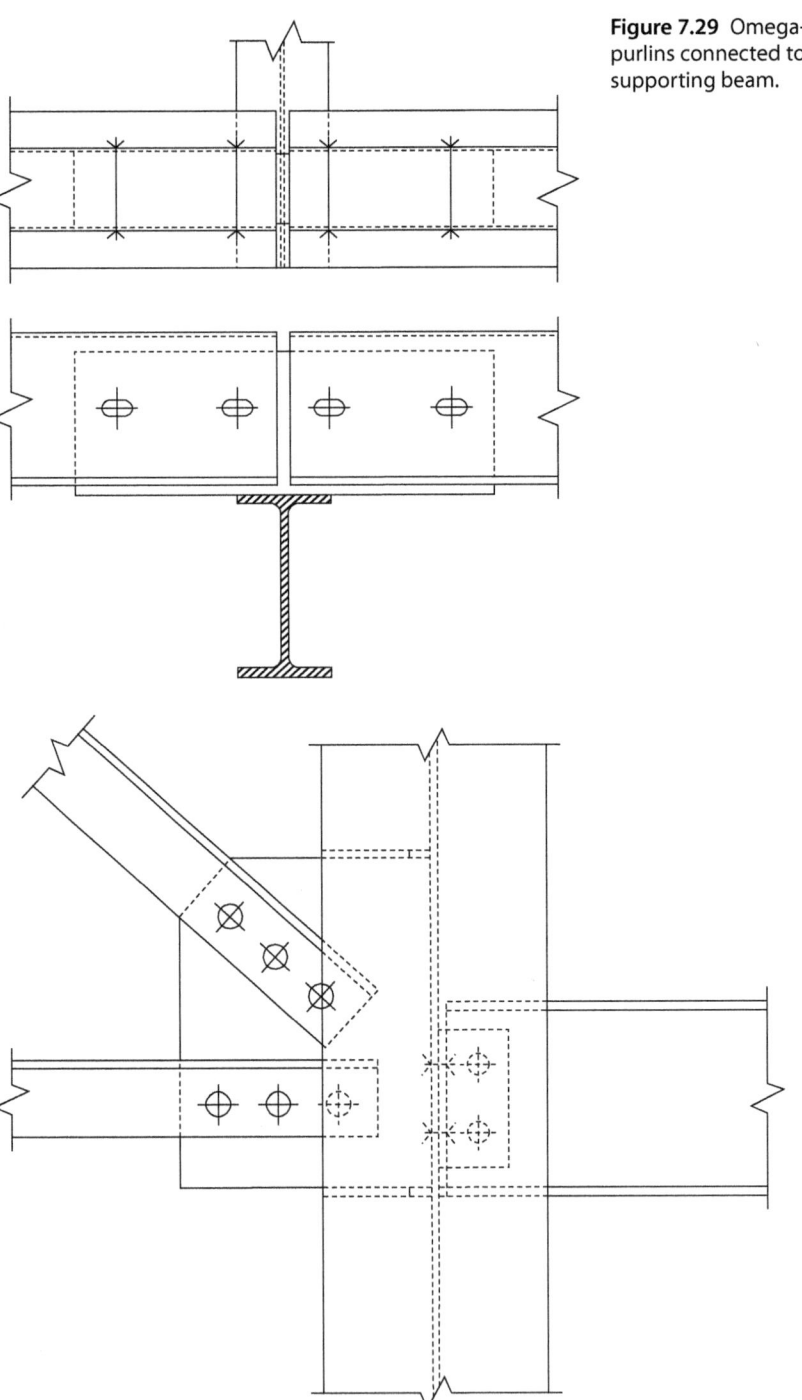

Figure 7.30 Joint on column weak side: on the left angles and on the right-hand side a beam bolted through double angles.

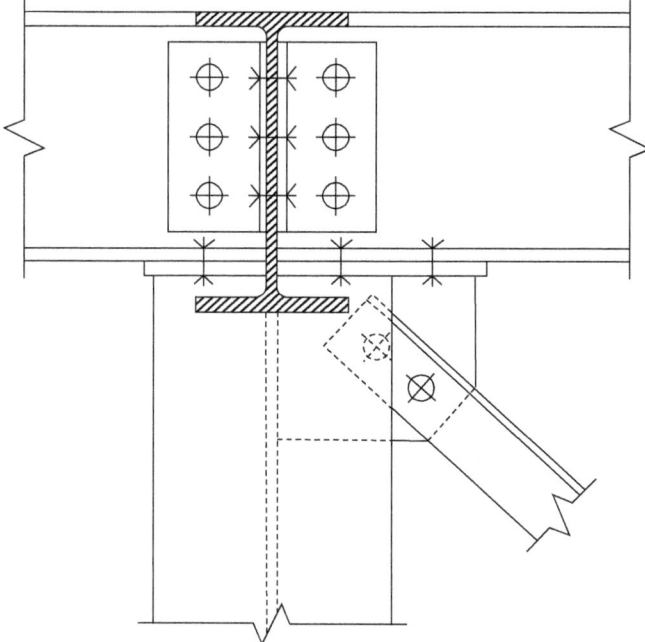

Figure 7.31 Vertical L brace on column; main beam leaning on column (horizontal plates for bolting) and secondary beam bolted to it with double angles.

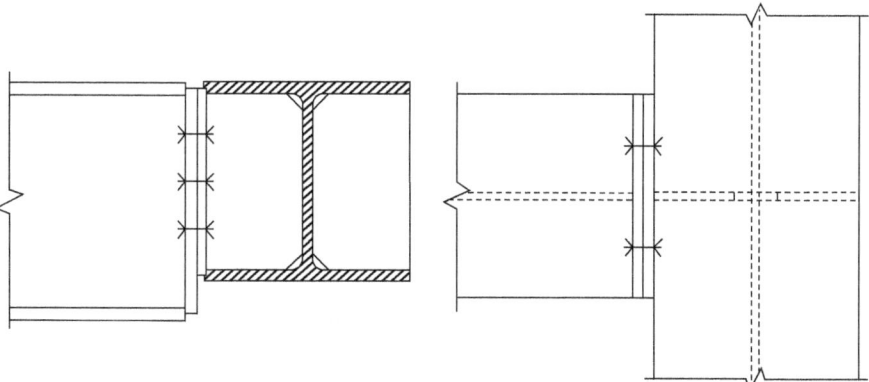

Figure 7.32 End plate between beams of different heights (primary beam actually shallower); stiffeners on both sides of the main beam.

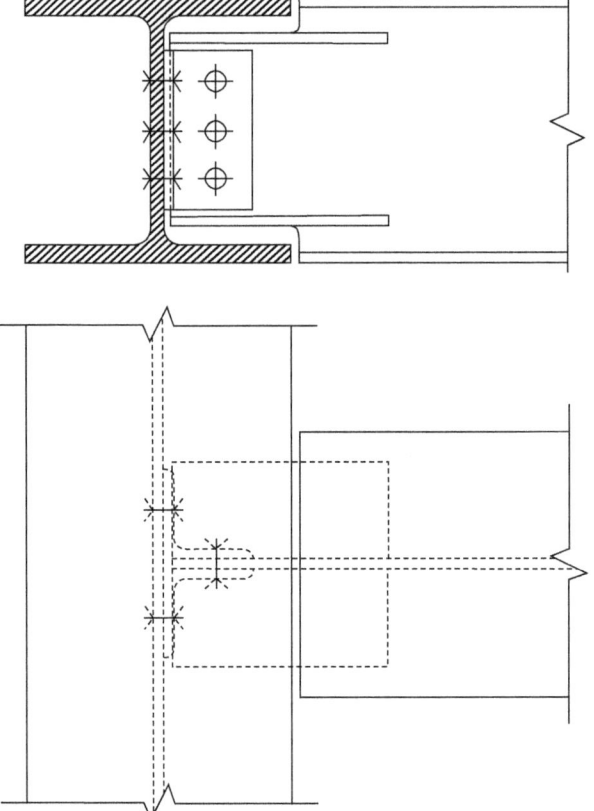

Figure 7.33 Double-angle connection: false flanges added to secondary beam to make up for notches on the flanges.

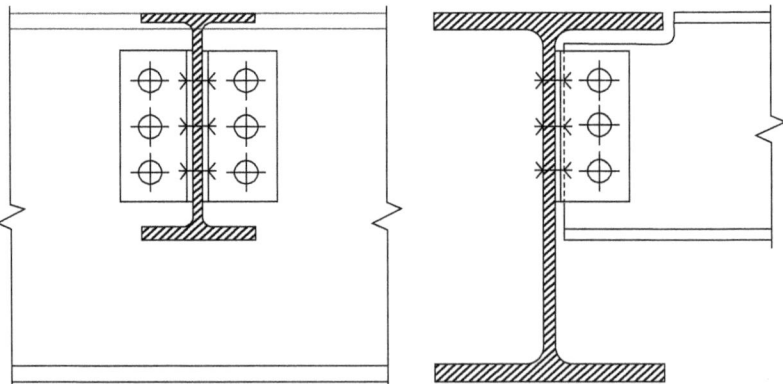

Figure 7.34 Another all-bolted double-angle connection with top notch on secondary beam.

Figure 7.35 Sloped beam stiffened with central plate and bolted to column.

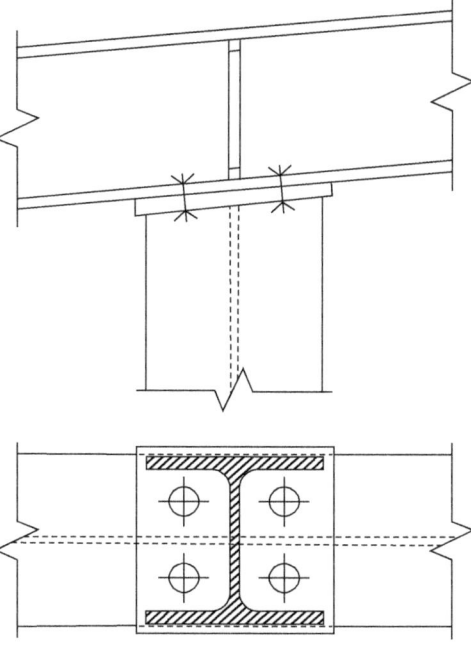

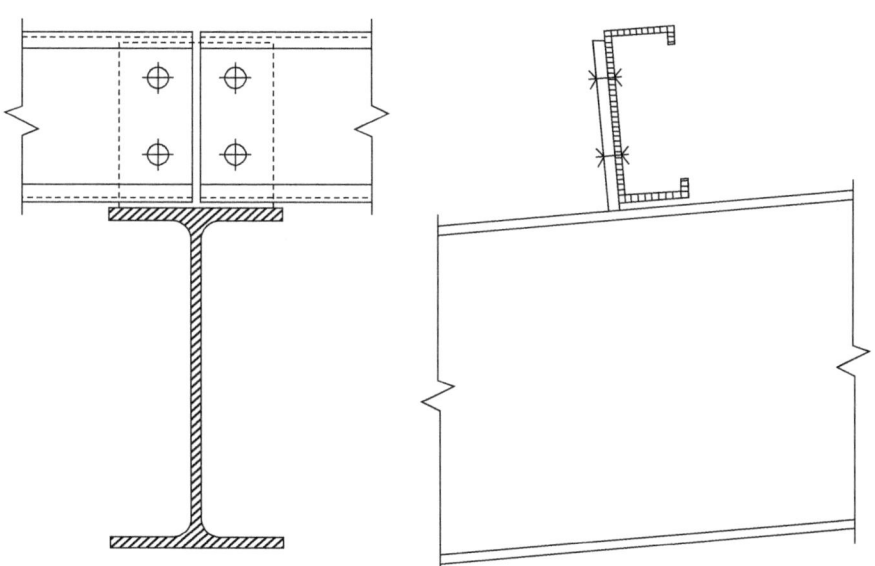

Figure 7.36 Detail of a plate welded to rafter and bolted to C purlins.

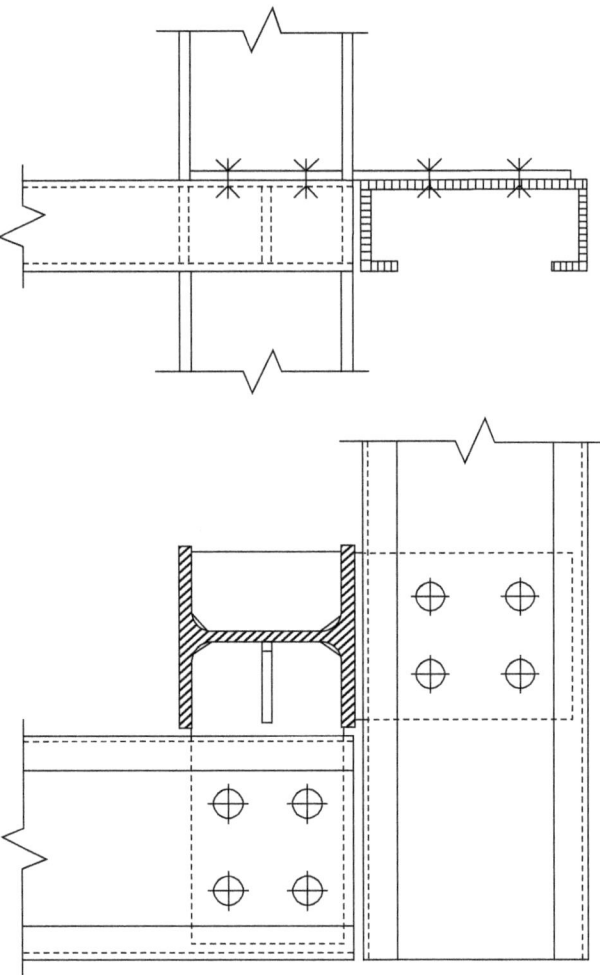

Figure 7.37 Detail of C girts connected to a column in a corner.

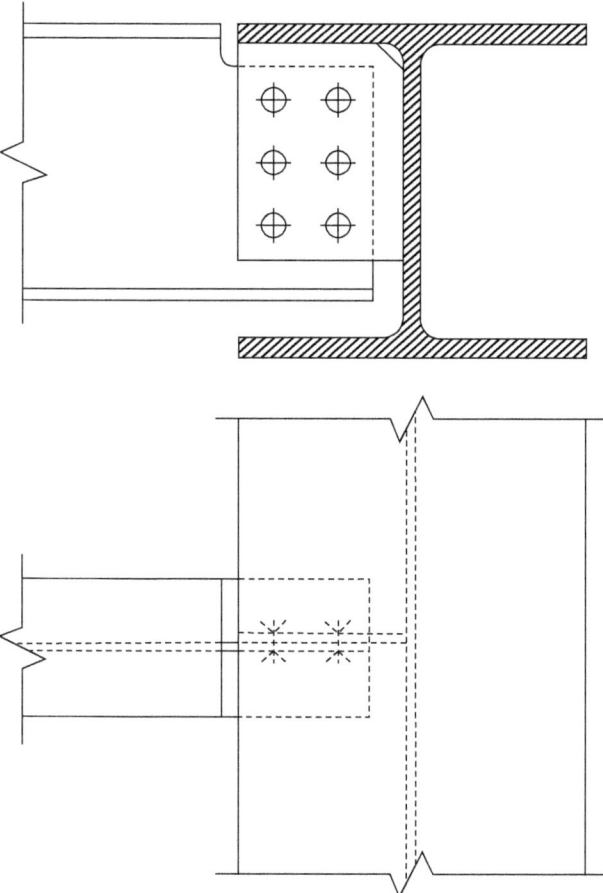

Figure 7.38 Beam-to-beam shear tab (fin plate) with top notch on secondary beam.

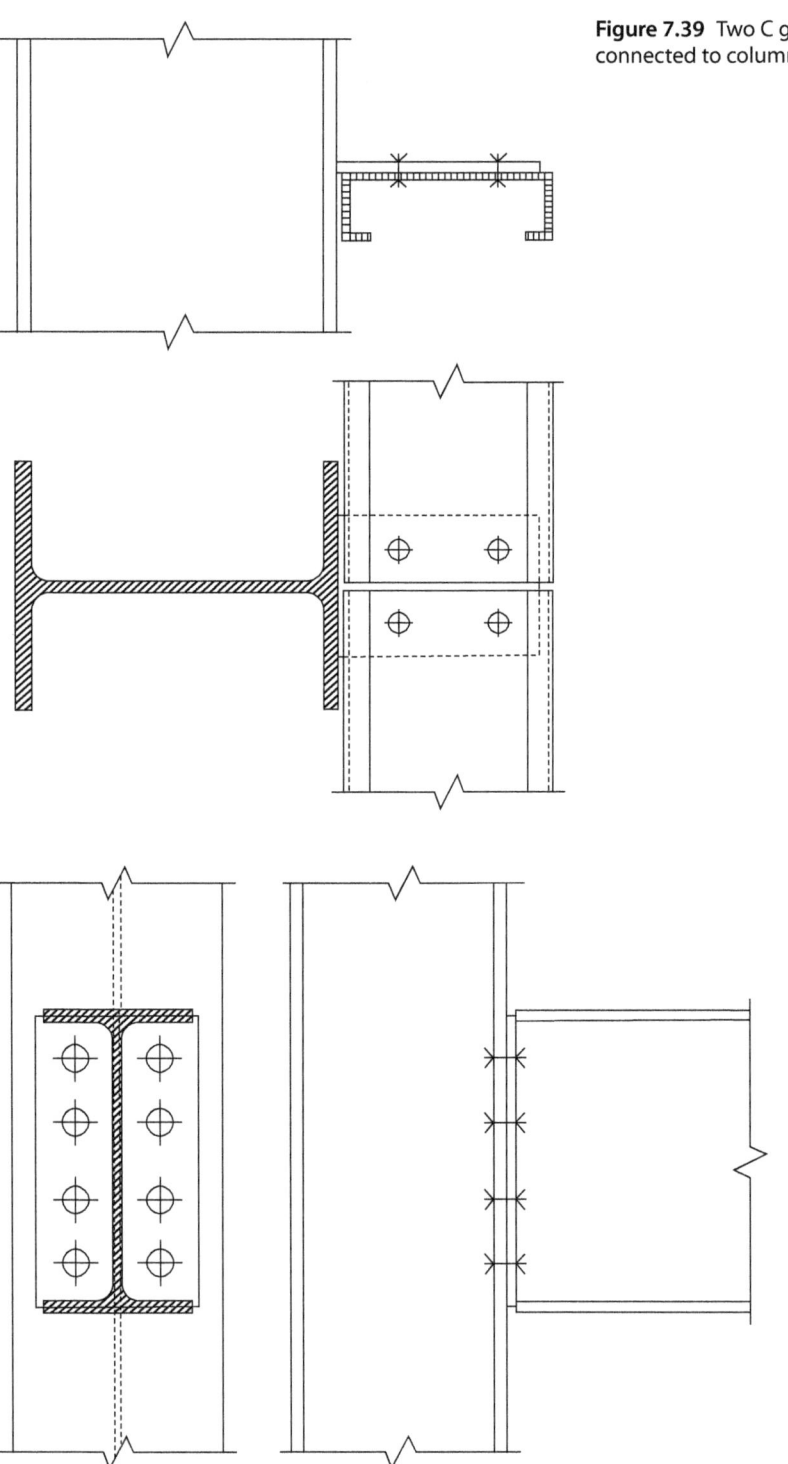

Figure 7.39 Two C girts connected to column.

Figure 7.40 Beam-to-column flexible end plate.

7 Connection Examples | 313

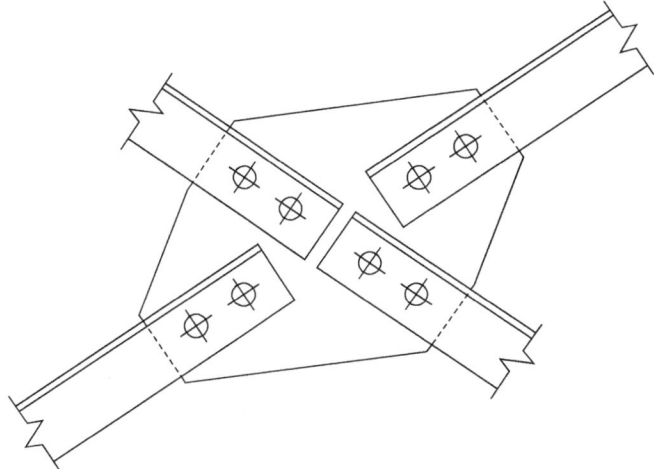

Figure 7.41 Detail at the intersection of X braces in double angles.

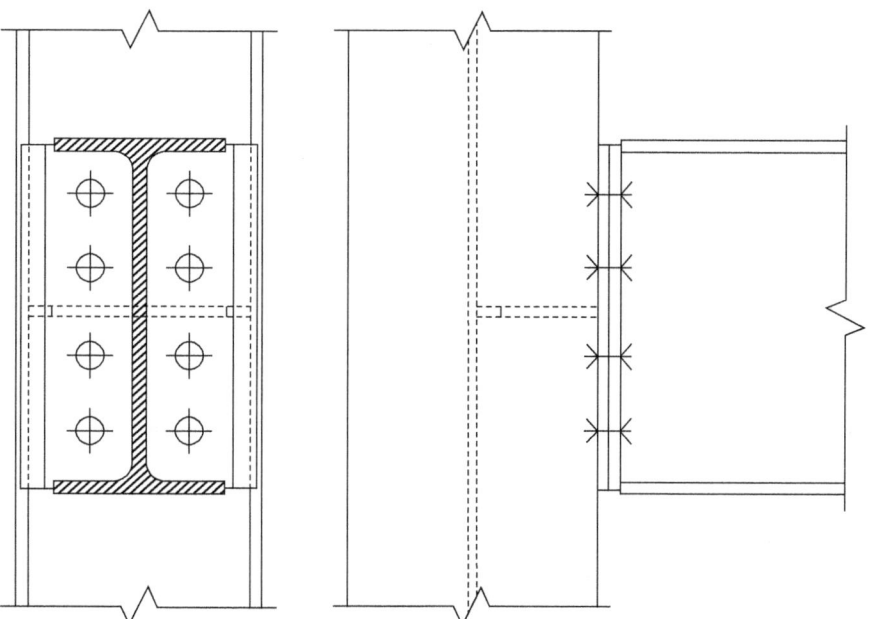

Figure 7.42 Beam connected by a flexible end plate to column weak side (stiffener added).

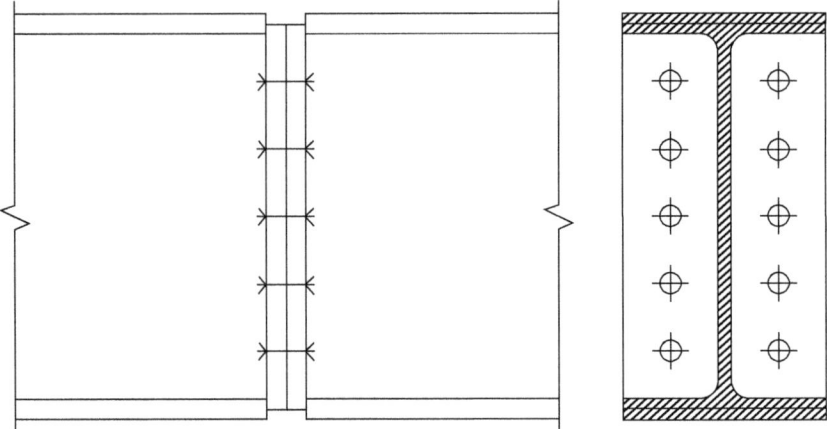

Figure 7.43 End-plate splice between consecutive beams.

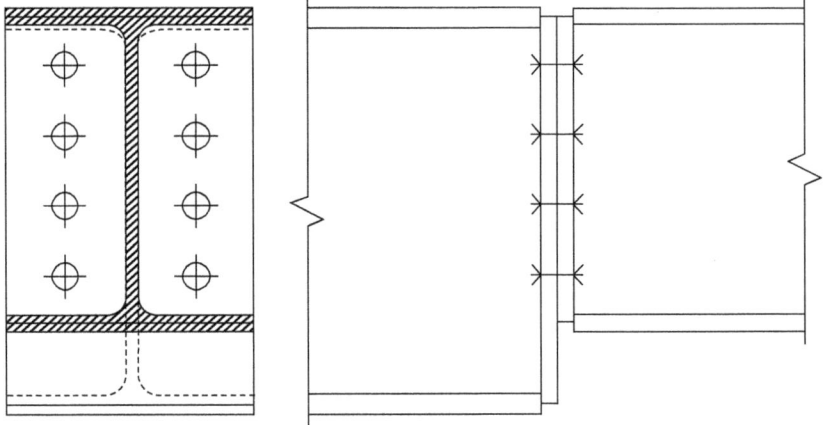

Figure 7.44 End-plate connection between consecutive beams with different depth.

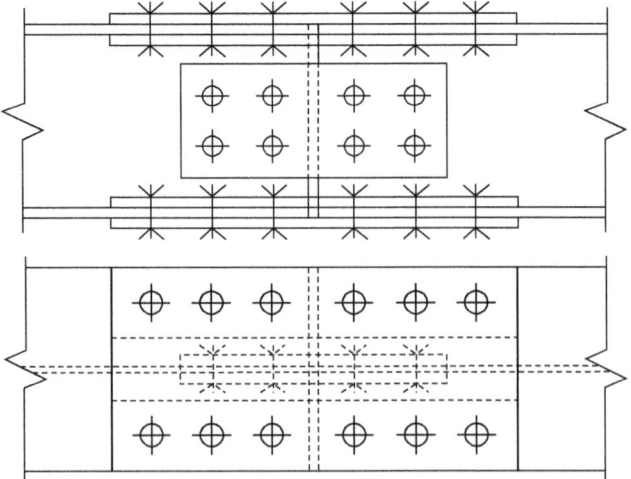

Figure 7.45 All-bolted splice (double plates on web and double plates on flanges).

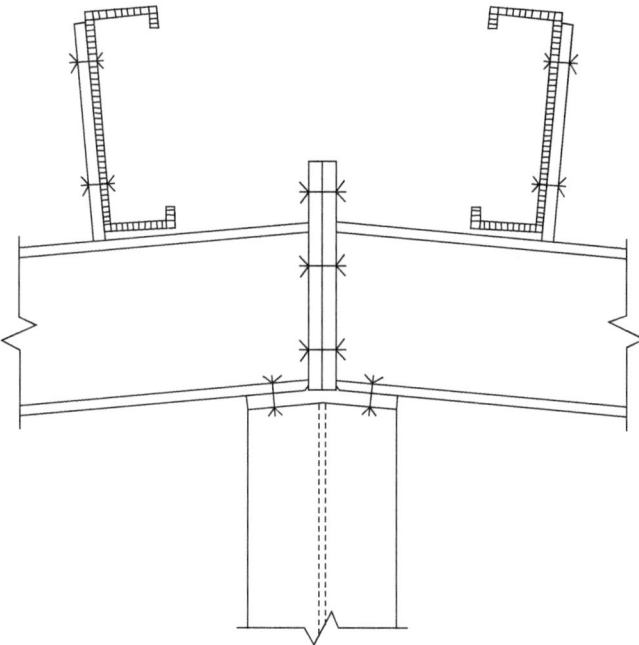

Figure 7.46 Roof detail, C purlins bolted to rafters positioned over central column.

Figure 7.47 All-bolted double angle with reinforcing plate welded to secondary beam to make up for the top notch.

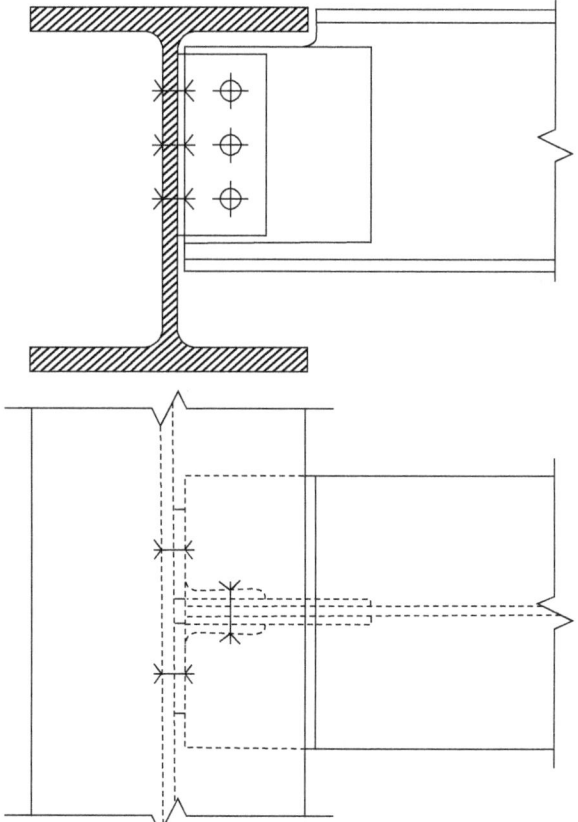

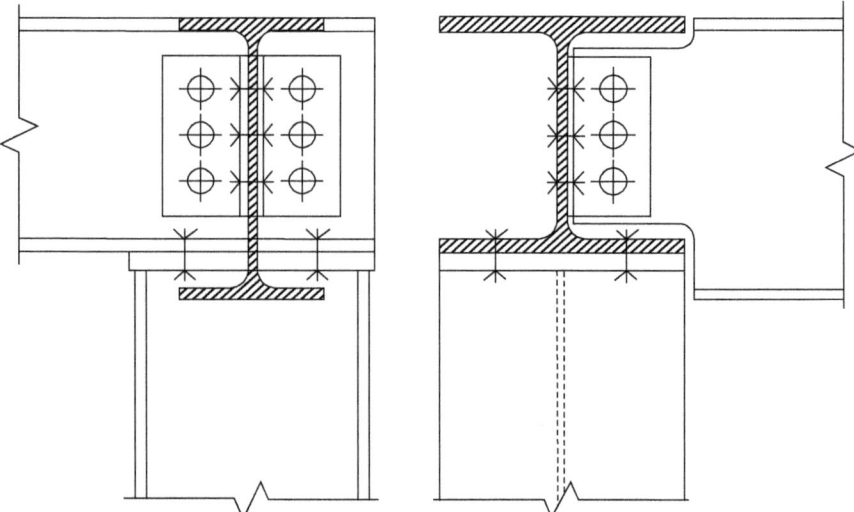

Figure 7.48 Main beam supported by column and notched secondary beam bolted to it with double angles.

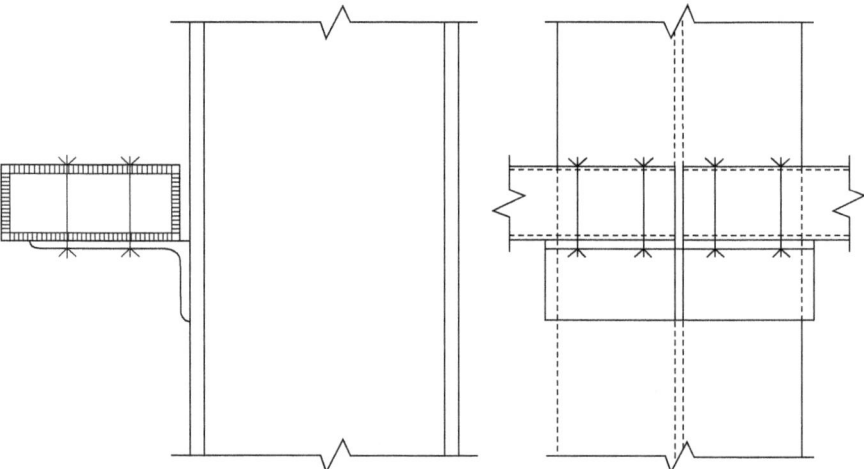

Figure 7.49 Angle welded to column flange to support the RHS (rectangular hollow section) girts.

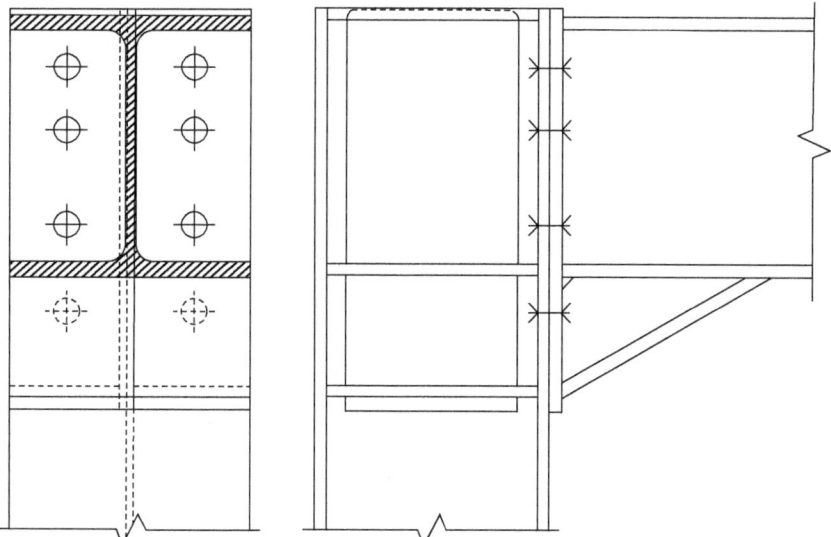

Figure 7.50 Moment connection with bottom haunch, web doubler, and continuity plates.

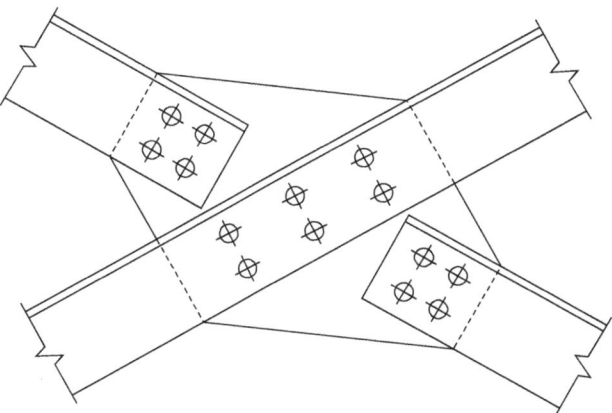

Figure 7.51 Central-plate detail of large double angles in an X brace.

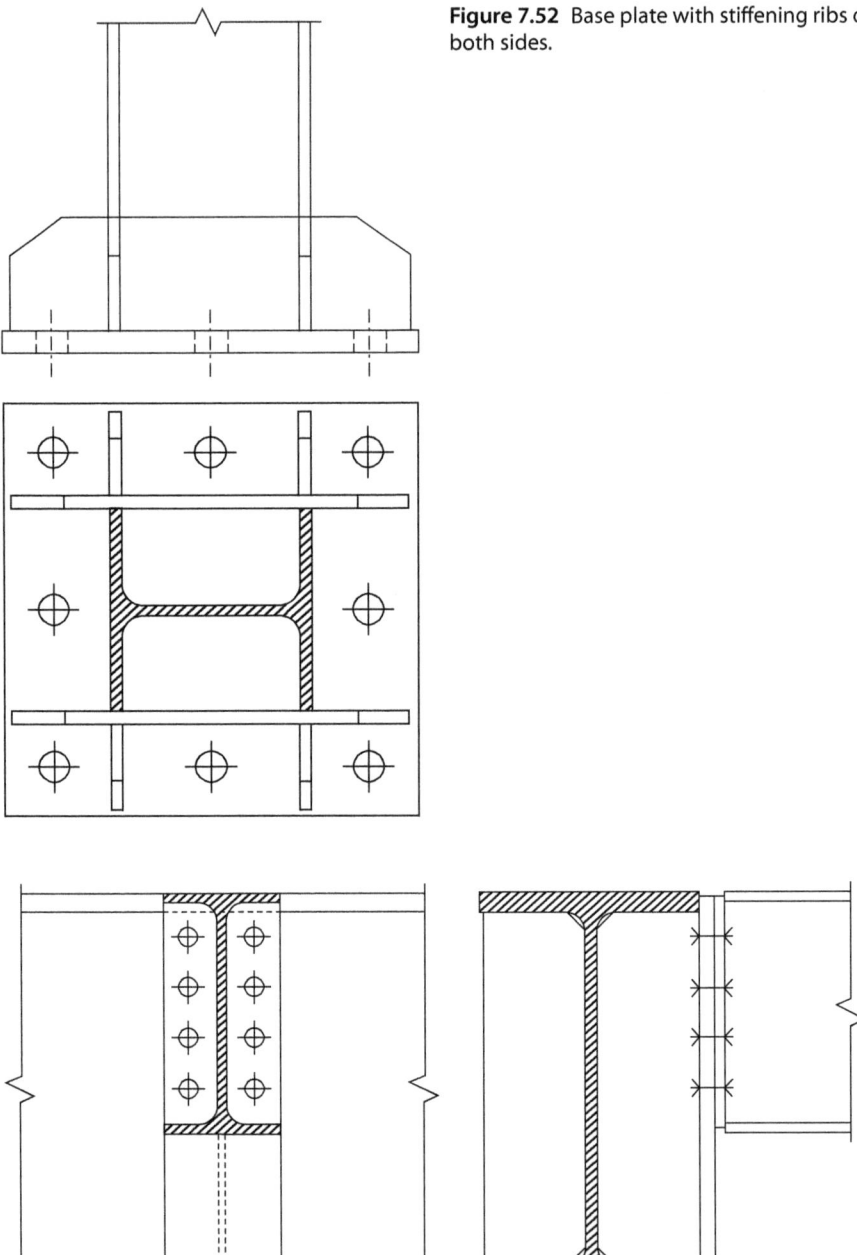

Figure 7.52 Base plate with stiffening ribs on both sides.

Figure 7.53 Flexible end plate between beams of different sizes and stiffener on the opposite side.

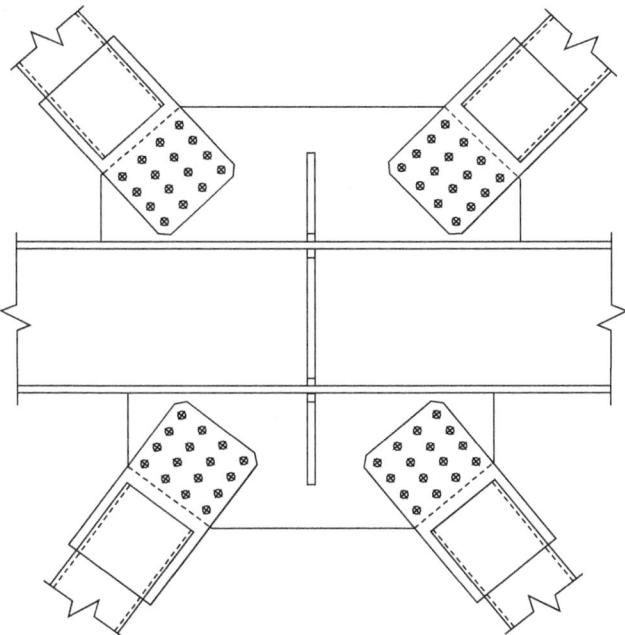

Figure 7.54 Connection between large pipe braces and an intersecting beam; stiffeners on beam web and plates.

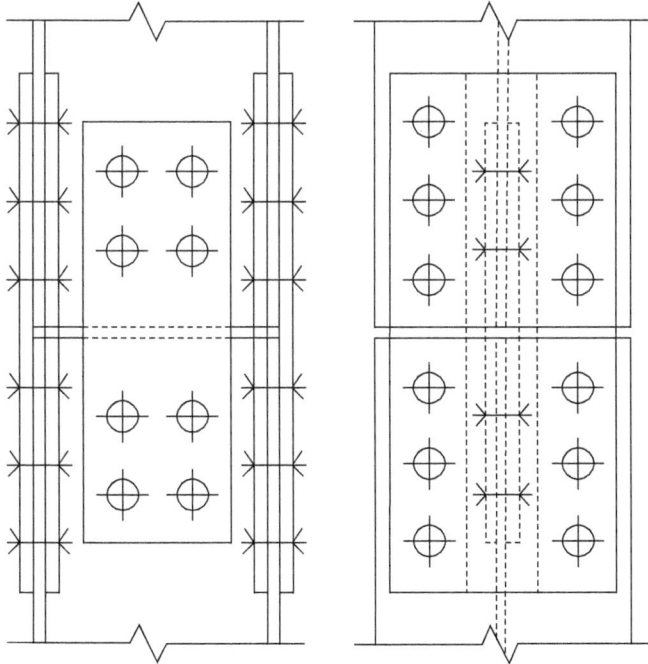

Figure 7.55 Column splice.

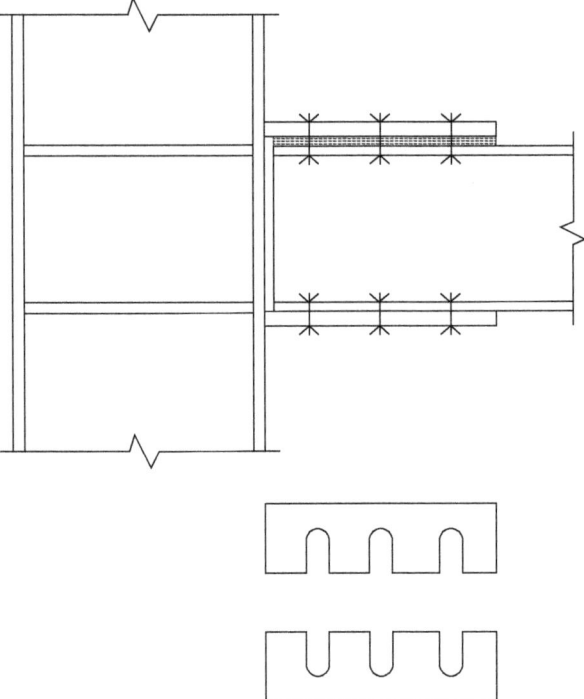

Figure 7.56 Plates welded to column flange and bolted to beam to resist an important bending moment; to help erection there is clearance between top flange and top plate; finger shims will be added during erection.

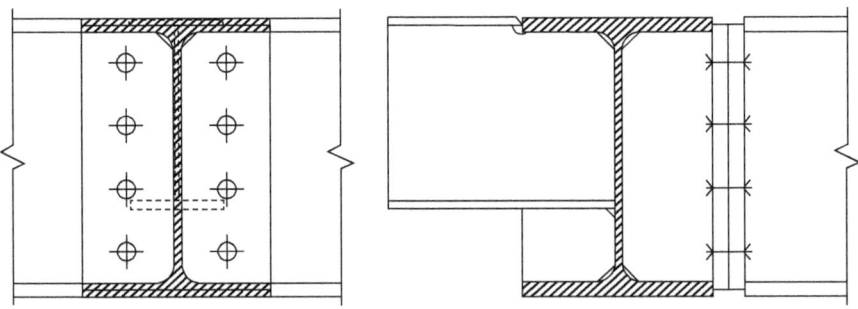

Figure 7.57 End plate between beams; beam stub welded on the opposite side.

7 Connection Examples | 321

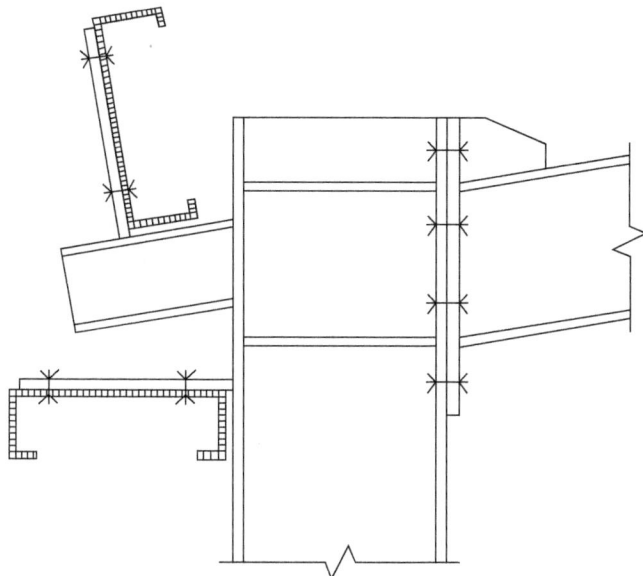

Figure 7.58 Beam-to-column moment connection with continuity plates; welded stub on the left-hand side to support a C purlin; a welded plate supports a C girt.

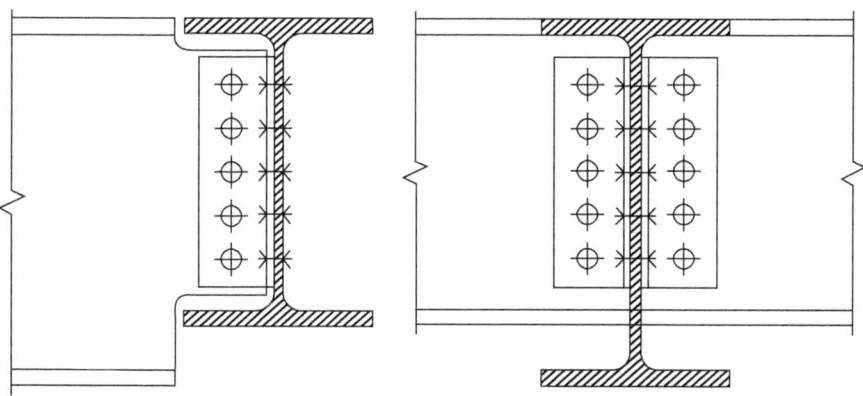

Figure 7.59 Secondary beam (deeper than primary) notched on both flanges and connected by all-bolted double angles.

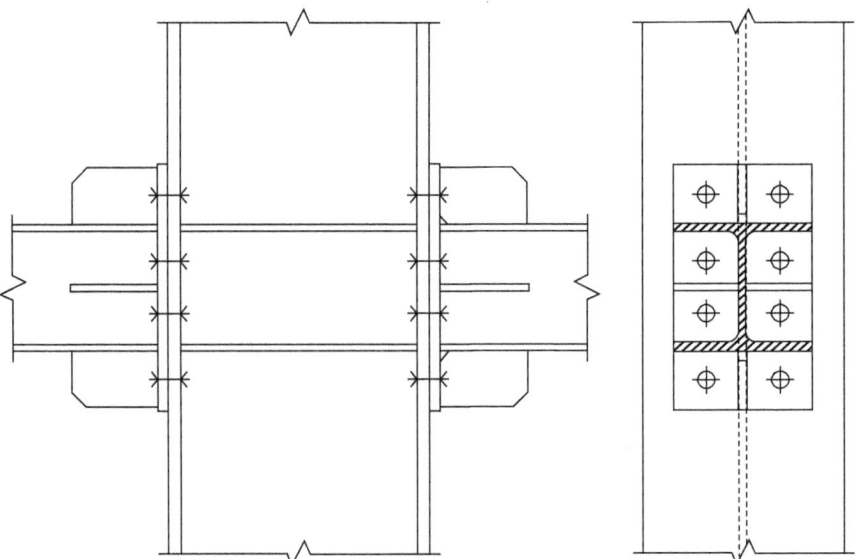

Figure 7.60 Double moment connection; continuity plates in the column and top and bottom stiffeners on beam flanges.

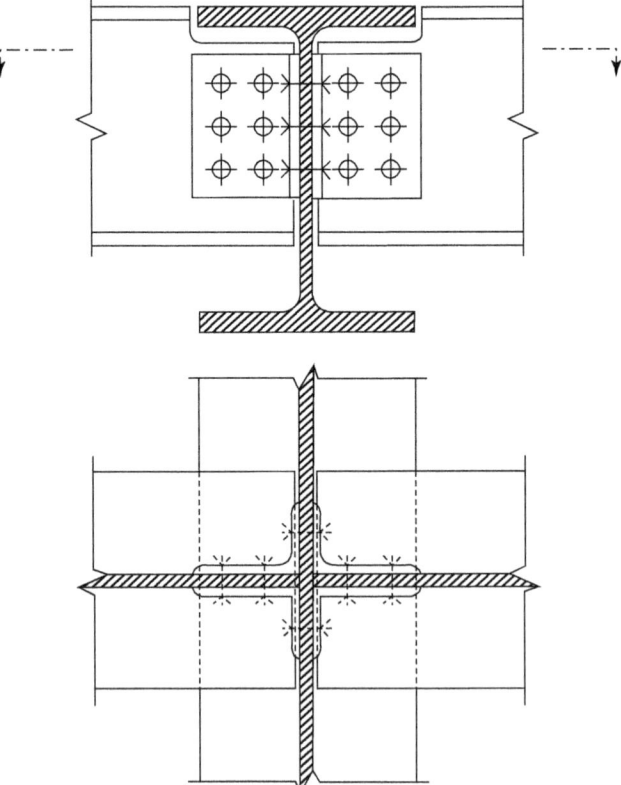

Figure 7.61 Top notched beams connected by double angles.

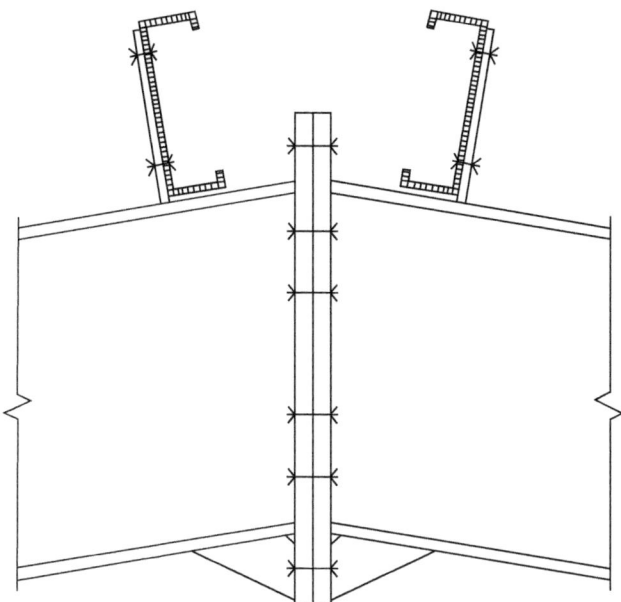

Figure 7.62 Apex connection between rafters and purlins on top.

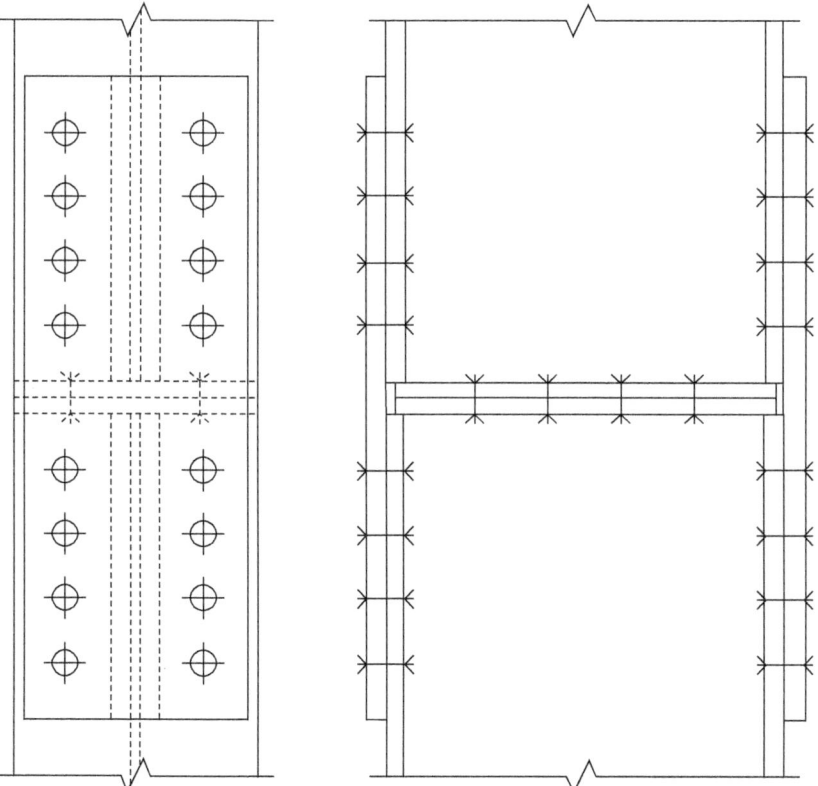

Figure 7.63 Column splice – end plate to support compression and shear; external plates for bending moment.

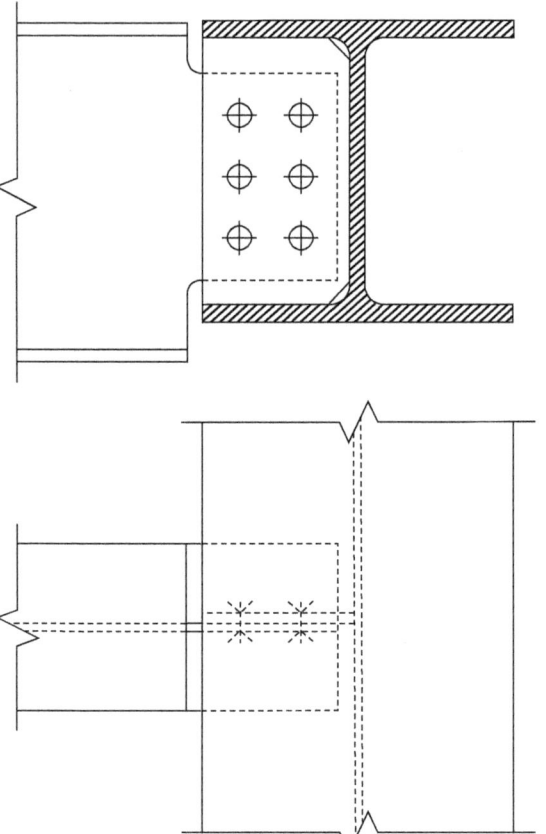

Figure 7.64 Fin plate with notched secondary beam (deeper than supporting beam).

Figure 7.65 Extended shear tab with primary beam positioned over column.

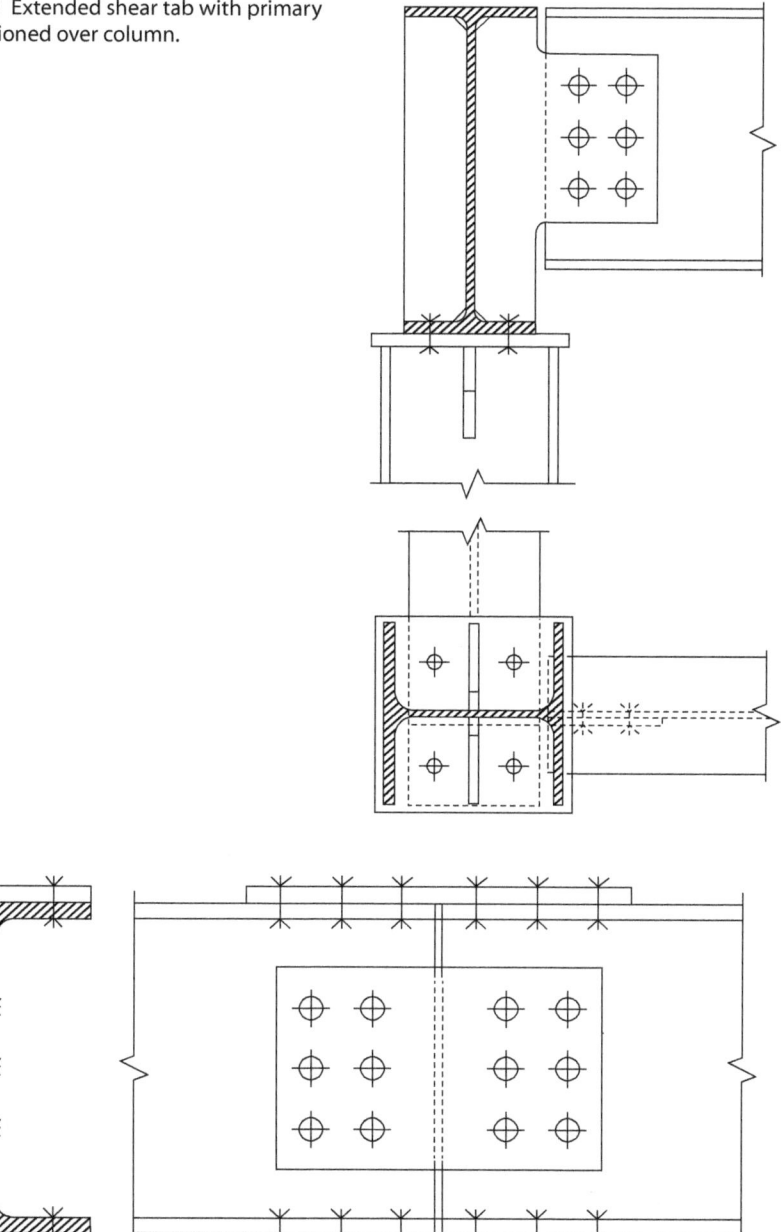

Figure 7.66 Beam splice (single plates on web and flanges).

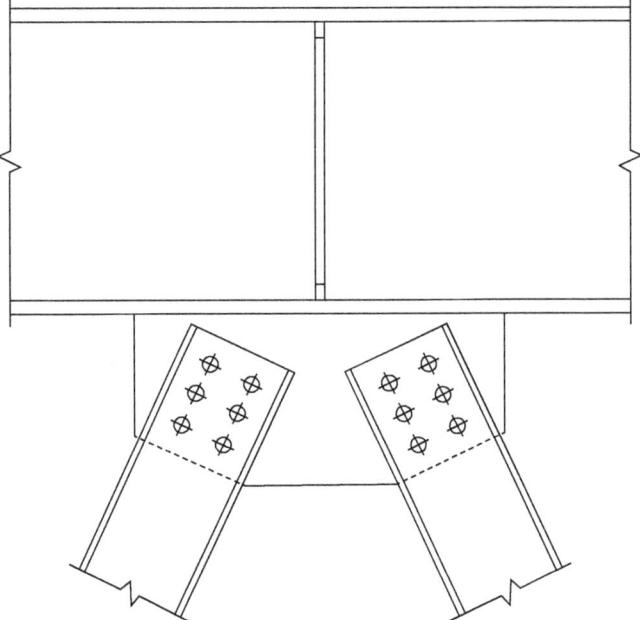

Figure 7.67 Channels framing into a stiffened beam.

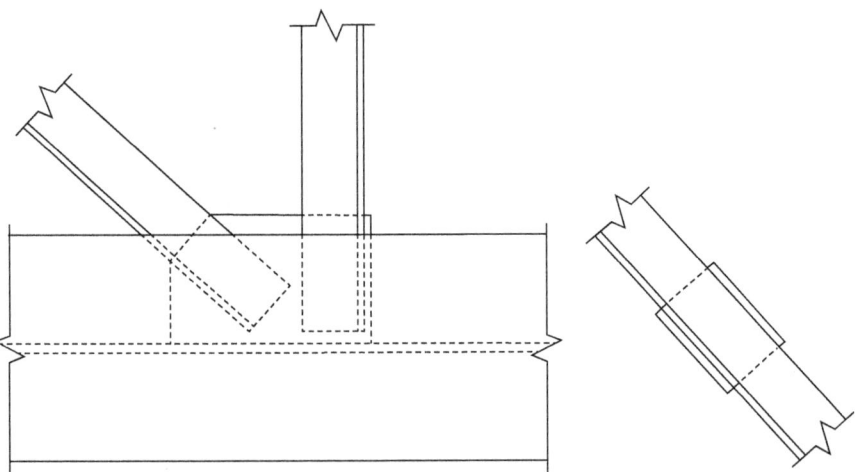

Figure 7.68 Welded truss (bottom chord detail); double angles are interconnected through welded ties.

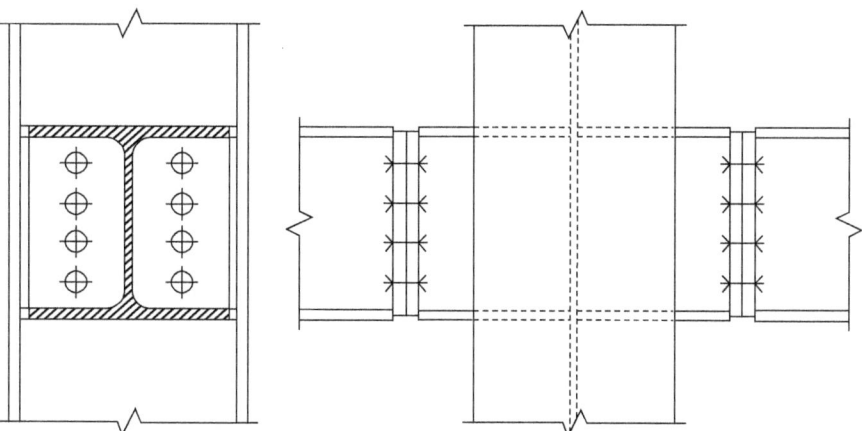

Figure 7.69 Symmetric connection at column – welded beam stubs are bolted by end plates with beams.

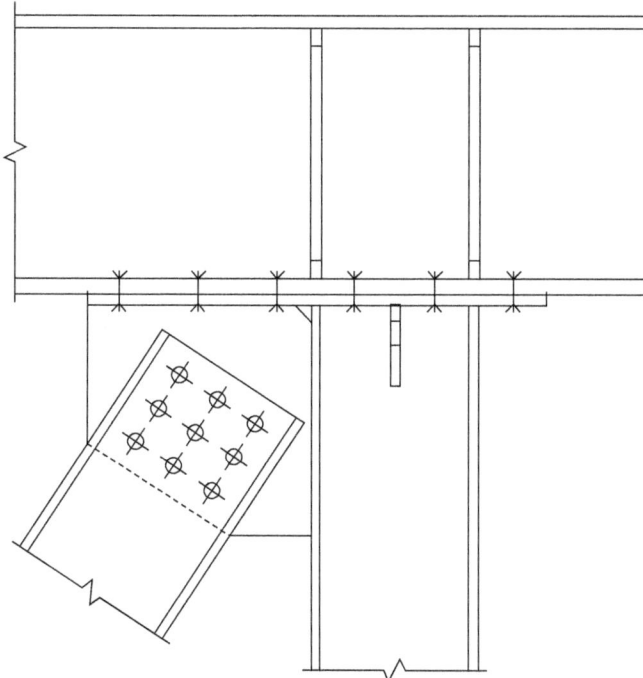

Figure 7.70 Stiffened beam leaning on column; channel brace bolted to gusset welded to column and to top end plate.

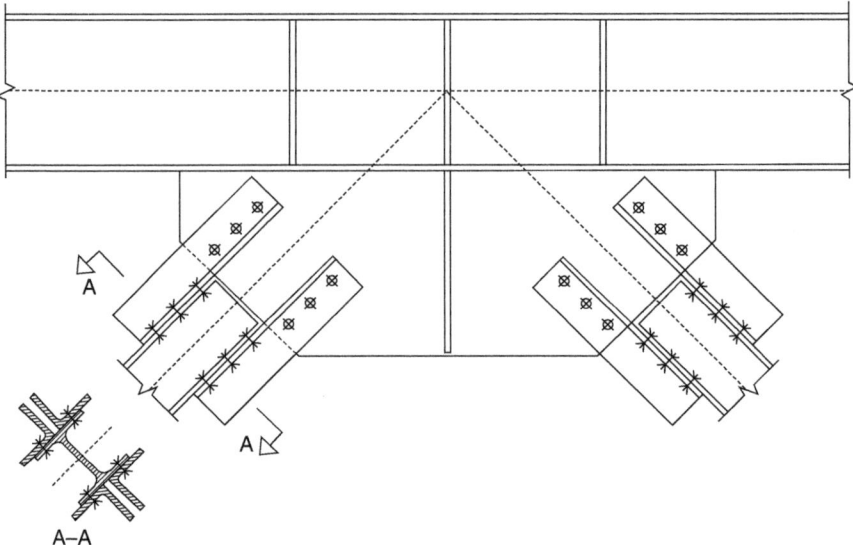

Figure 7.71 H brace flanges connected by bolted angles to the gusset welded to the horizontal beam; stiffeners added to gusset and beam.

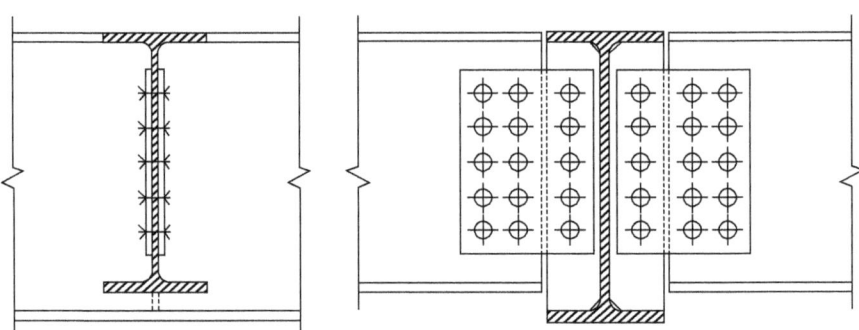

Figure 7.72 Symmetric joints with all-bolted double plates that connect secondary beams.

Figure 7.73 L braces connected to column weak side.

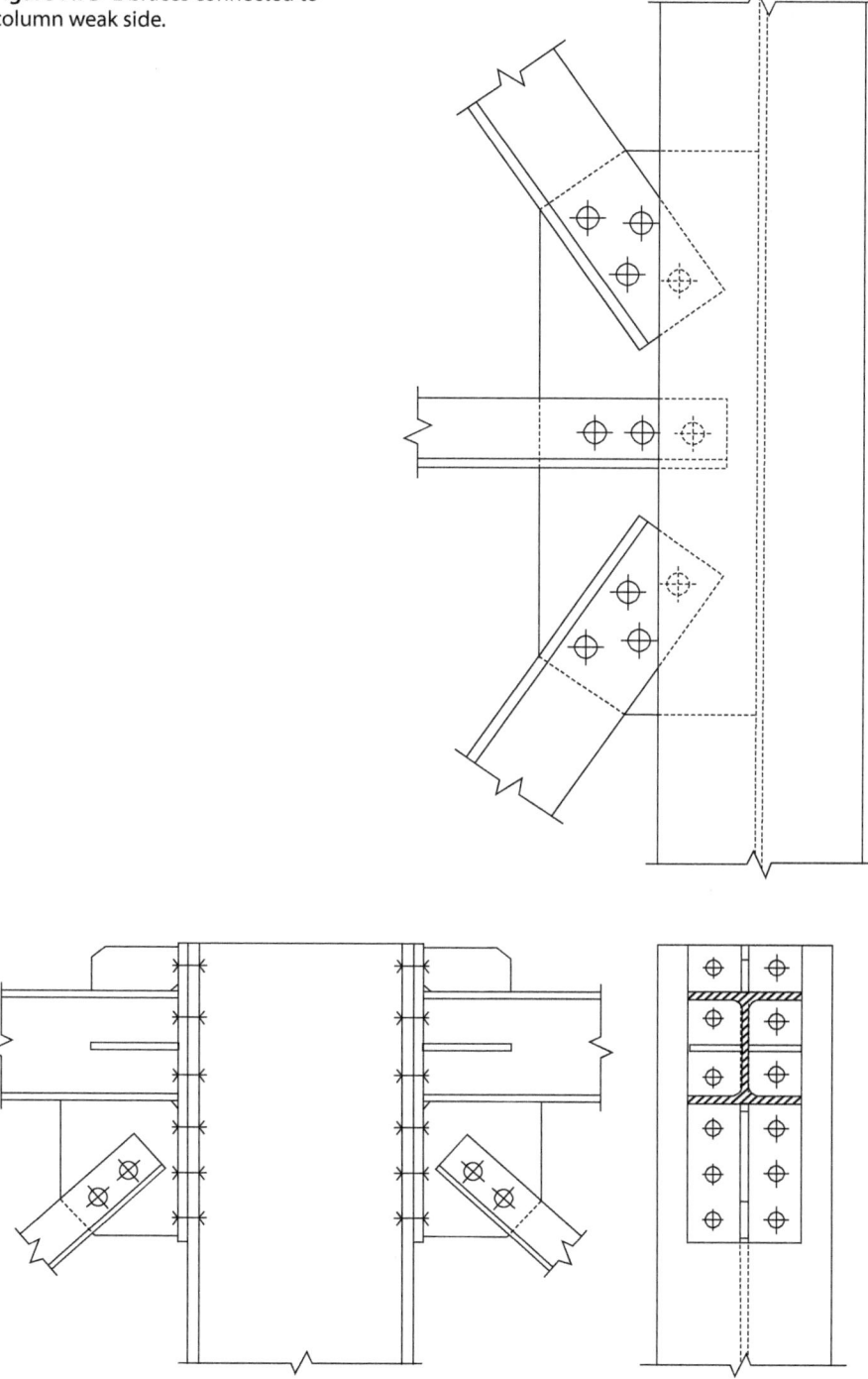

Figure 7.74 Symmetric connection with braces framing into it.

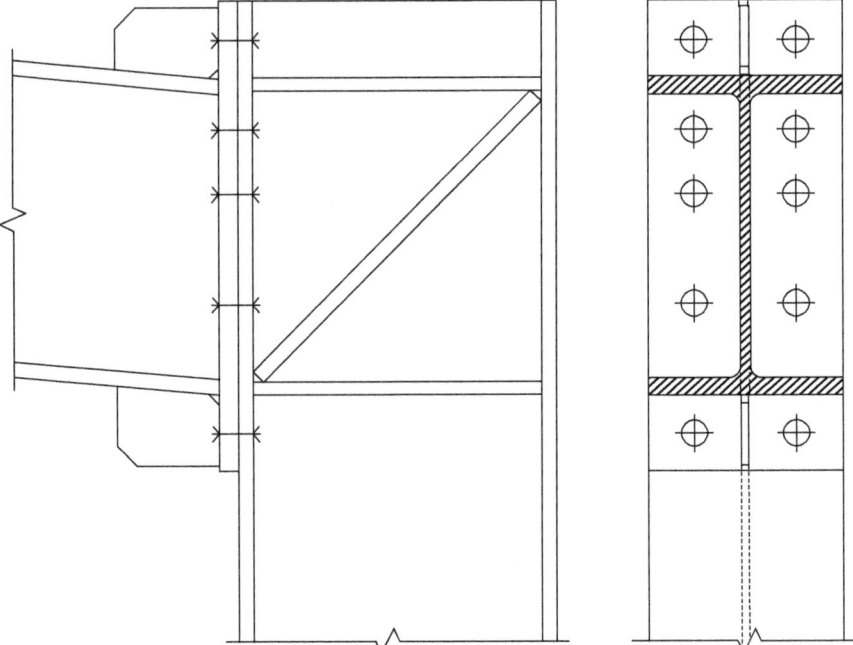

Figure 7.75 Inclined beam bolted through a moment connection to column – various stiffeners are present, including a diagonal stiffener to help web panel shear.

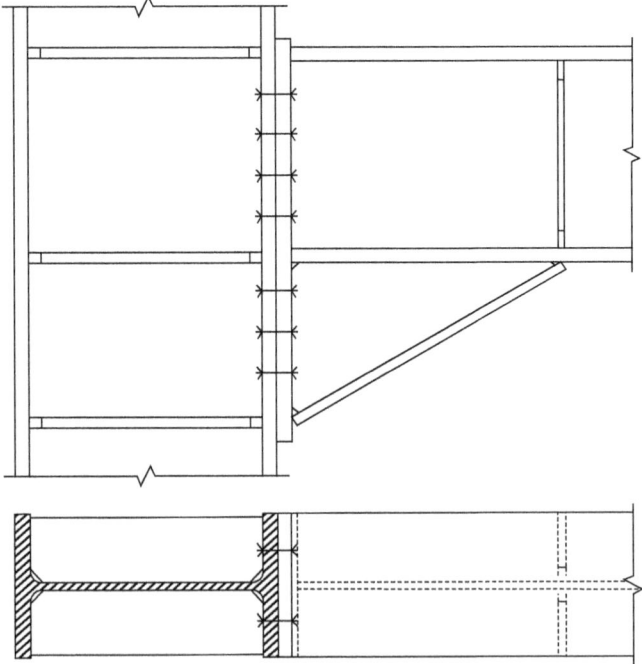

Figure 7.76 Another moment connection with haunch and continuity plates.

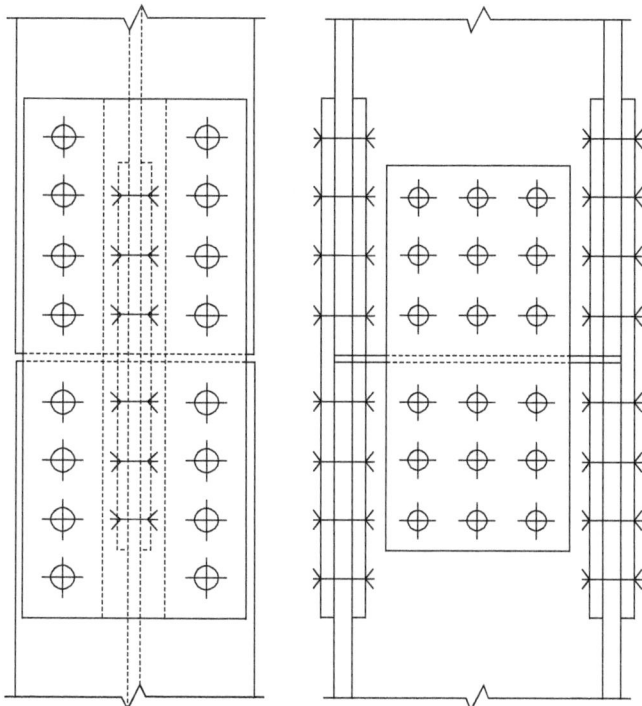

Figure 7.77 Column spliced by double-bolted plates on flanges and web; due to the gap, the compression load is not transferred by contact.

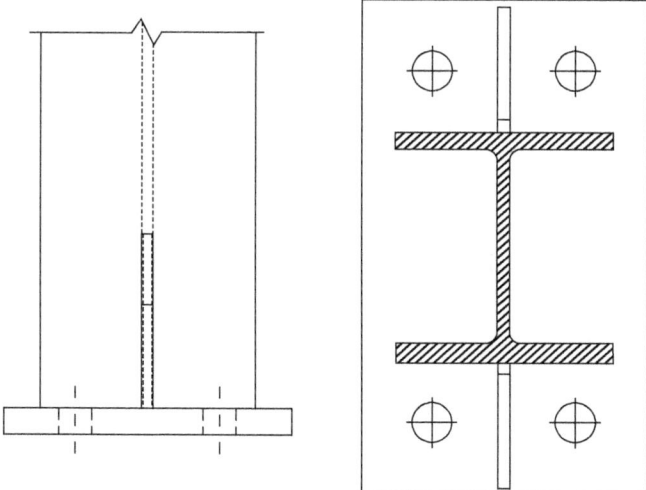

Figure 7.78 Base plate with central stiffener on column strong side.

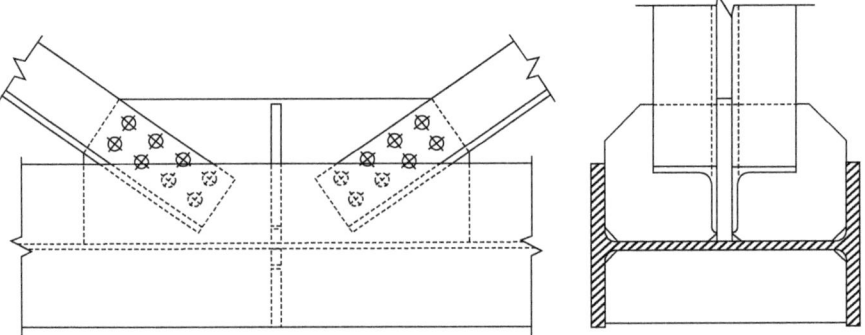

Figure 7.79 Bottom chord of a truss with large bolted double angles; various stiffeners inserted.

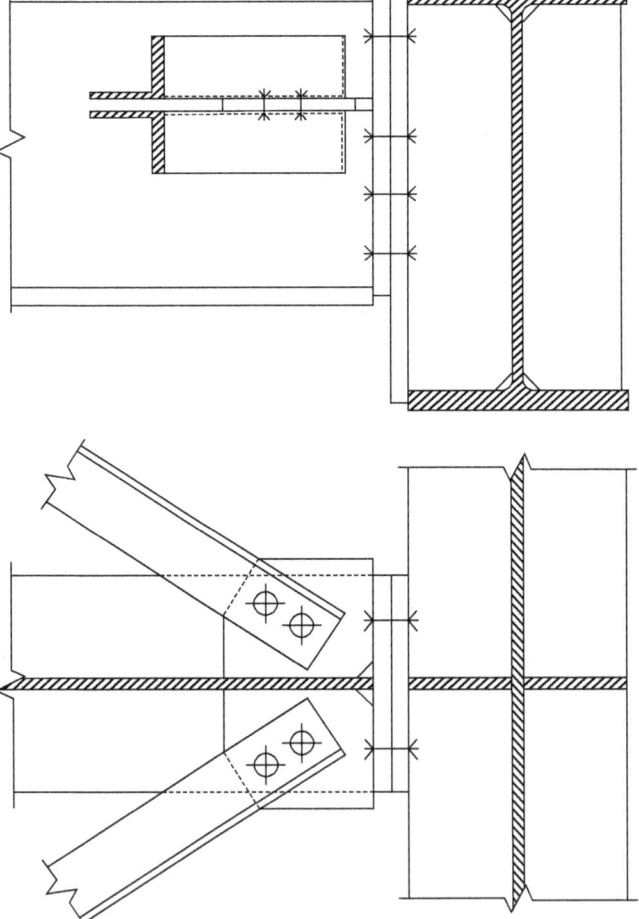

Figure 7.80 Beam-to-beam flexible end plate with horizontal braces in double angle also framing into it.

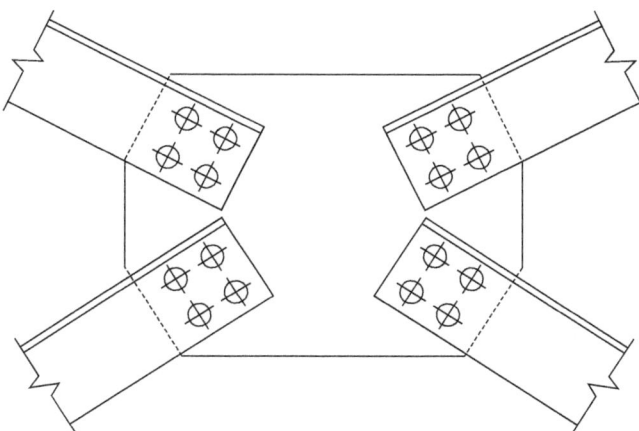

Figure 7.81 Central plate possible detail of X braces designed as L profiles.

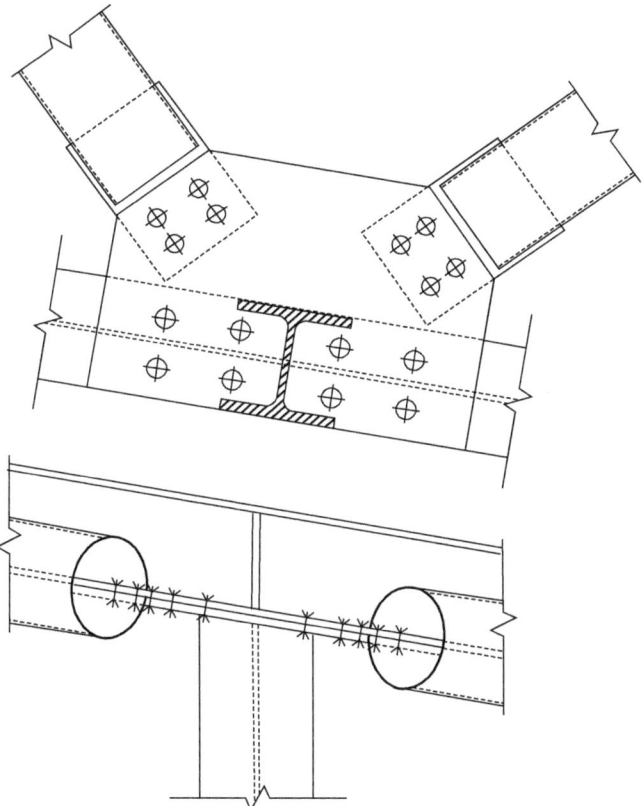

Figure 7.82 Sloping beam bolted on column through a plate also serving as connection for horizontal pipe braces.

334 | *7 Connection Examples*

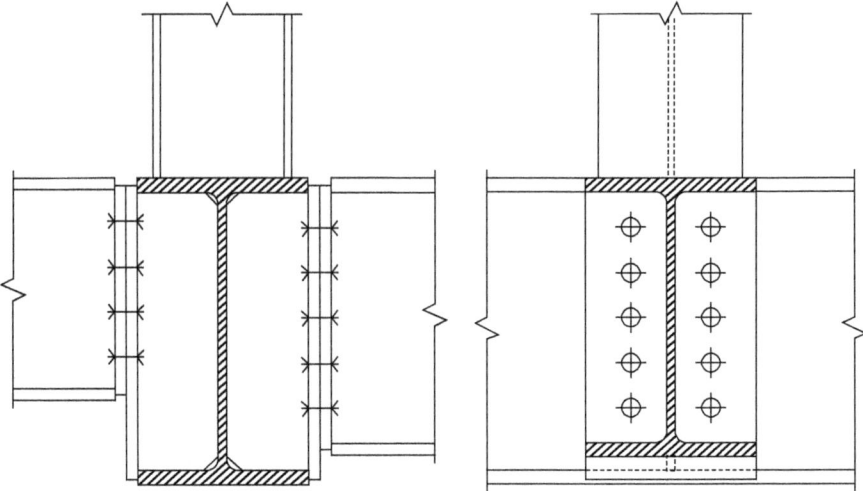

Figure 7.83 Double-sided end-plate connections to a primary beam also supporting a welded stub on top.

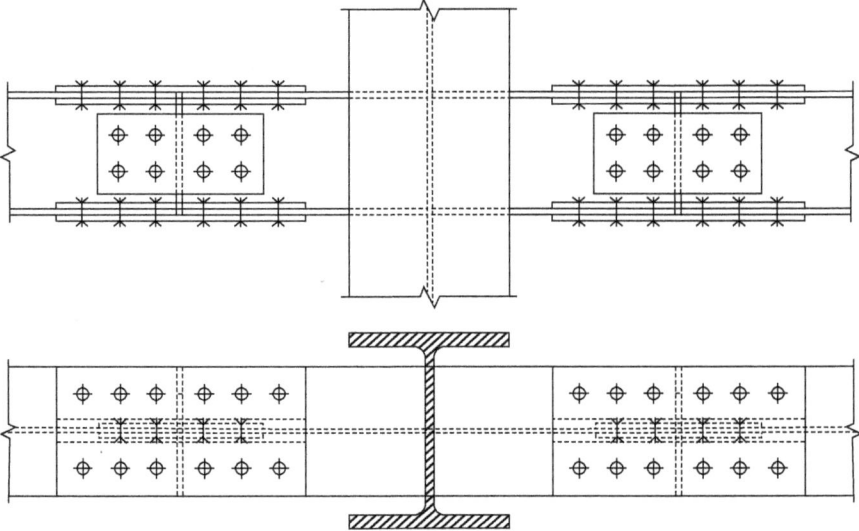

Figure 7.84 Column with beam stubs welded on the weak side that serves to splice beams (likely designed as continuous in the calculation model).

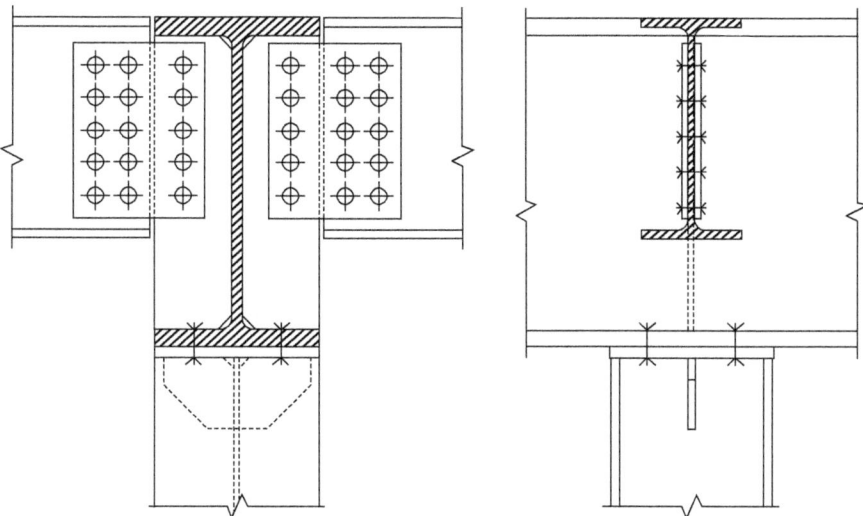

Figure 7.85 Main beam bolted to column (stiffeners added) and connecting secondary beams by all-bolted double plates.

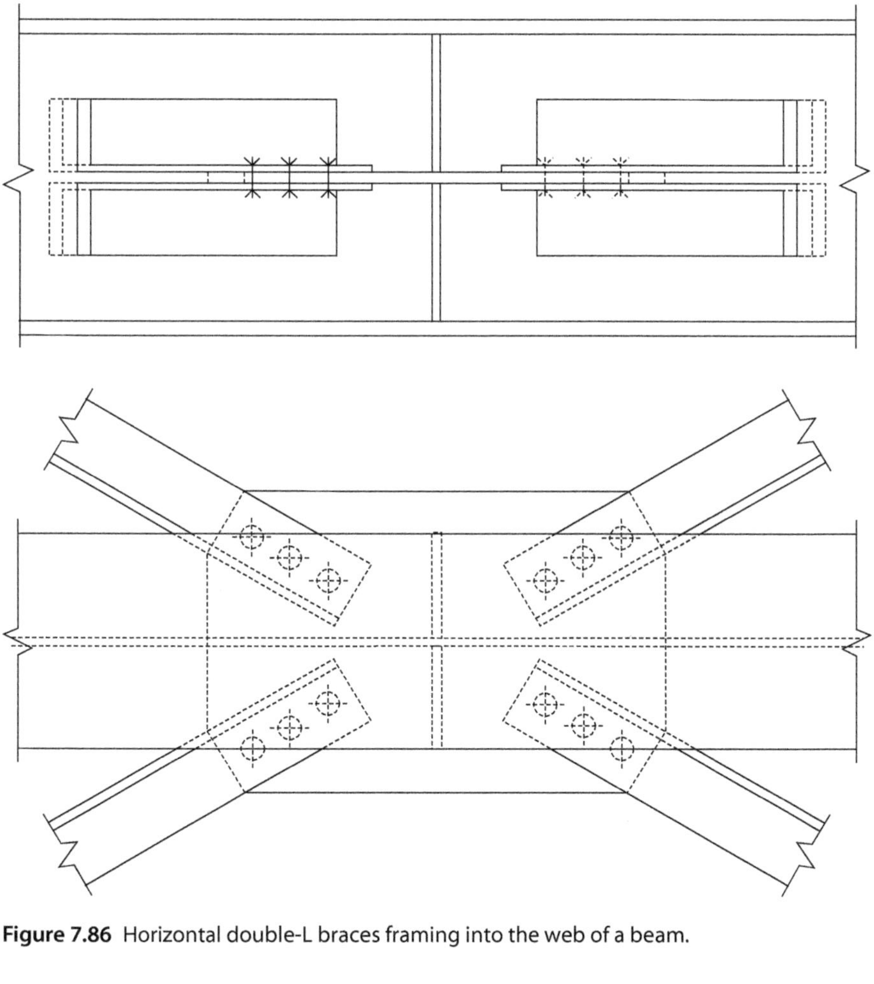

Figure 7.86 Horizontal double-L braces framing into the web of a beam.

Figure 7.87 Lattice tower details.

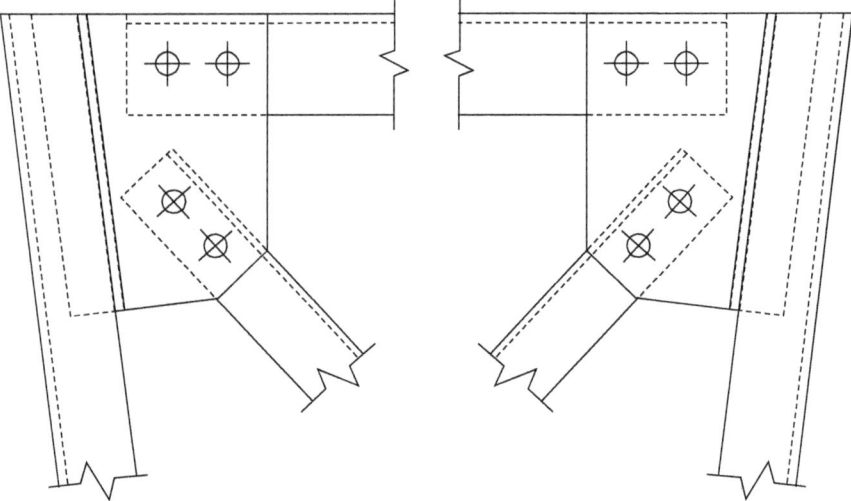

Figure 7.88 Stiffened base plate also connecting a double-U brace.

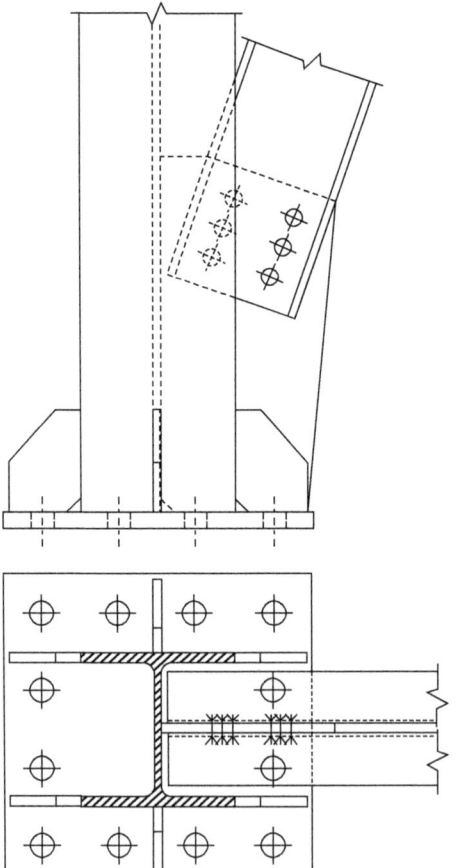

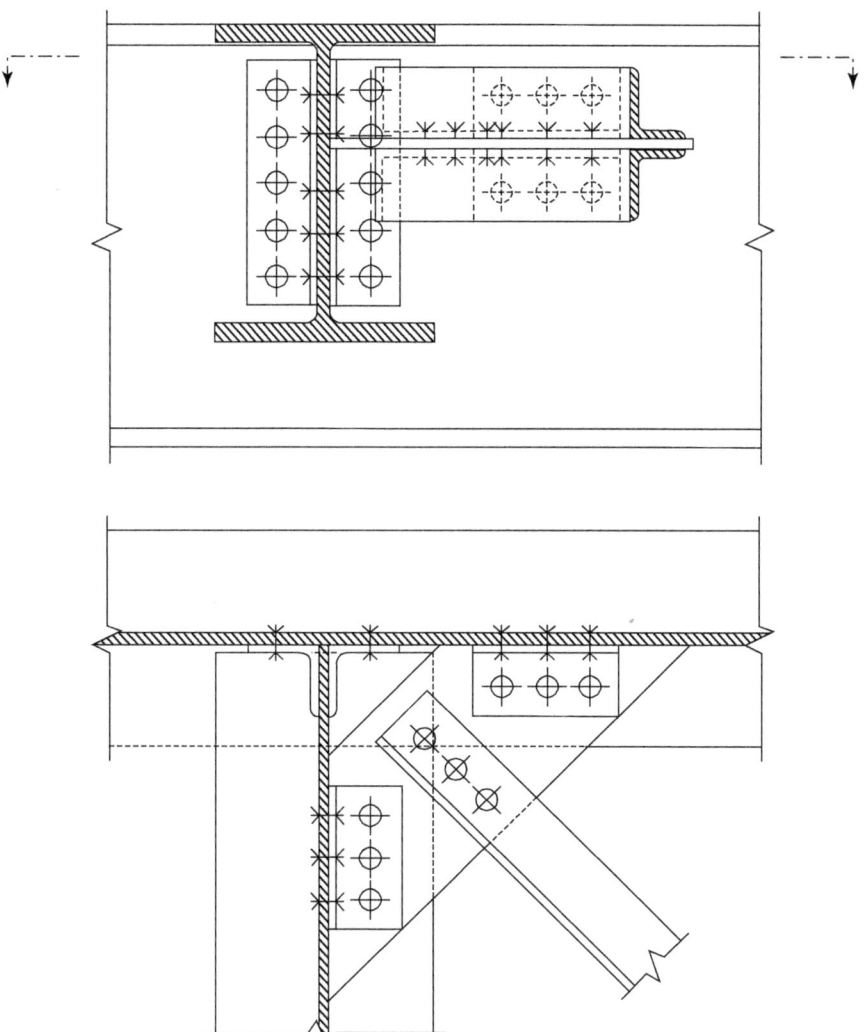

Figure 7.89 Connection without welds – brace, gusset, and beam are all connected through all-bolted double angles.

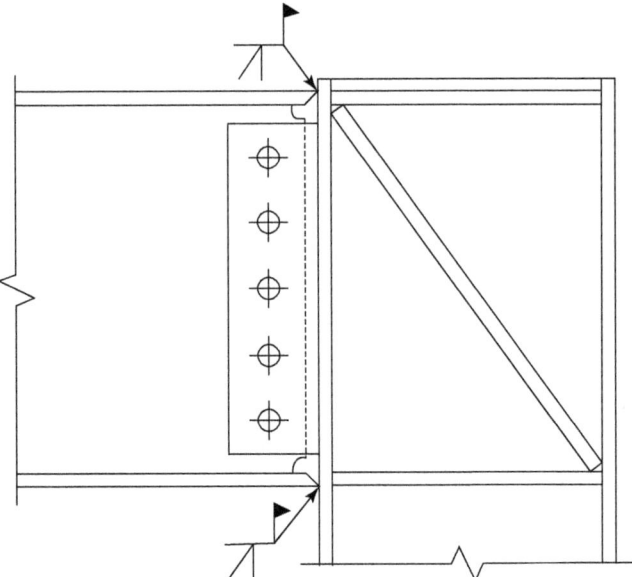

Figure 7.90 On-site welded moment connection – the web is bolted, then full-penetration welds of flanges are executed on-site.

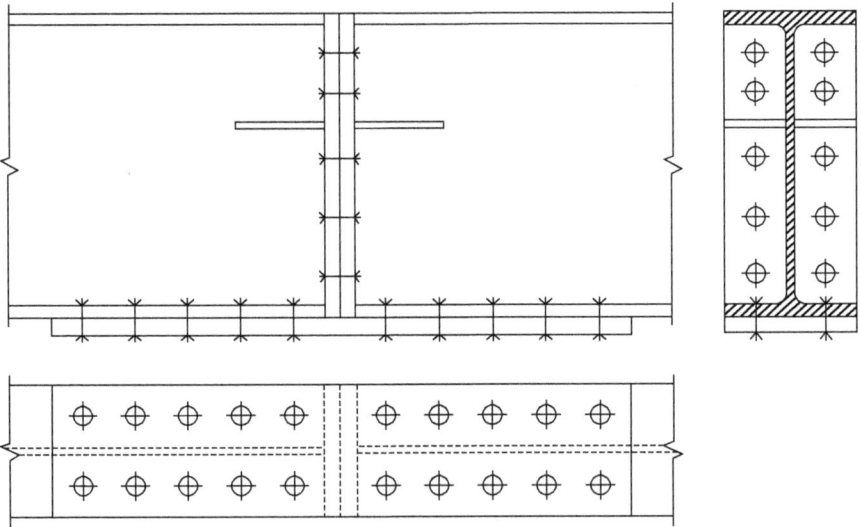

Figure 7.91 Beam splice designed to maintain uniform TOS (top of steel) for a crane girder – external plate only on the bottom flange (the joint is likely in a zone where positive bending is prevalent, i.e. bottom flange has tension) and end plate (with internal stiffeners) for shear and top flange tension.

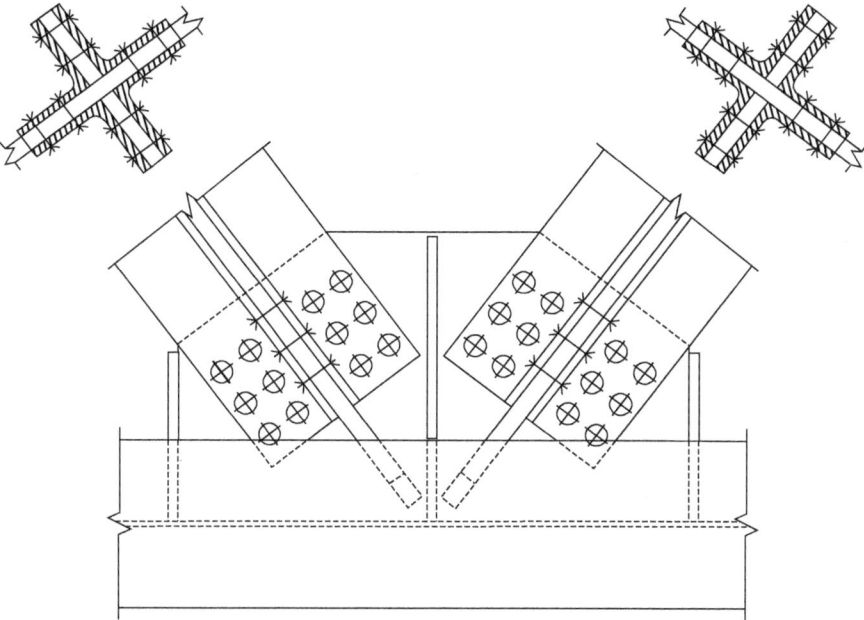

Figure 7.92 Heavy vertical brace made by quadruple angles framing into the stiffened web of a beam.

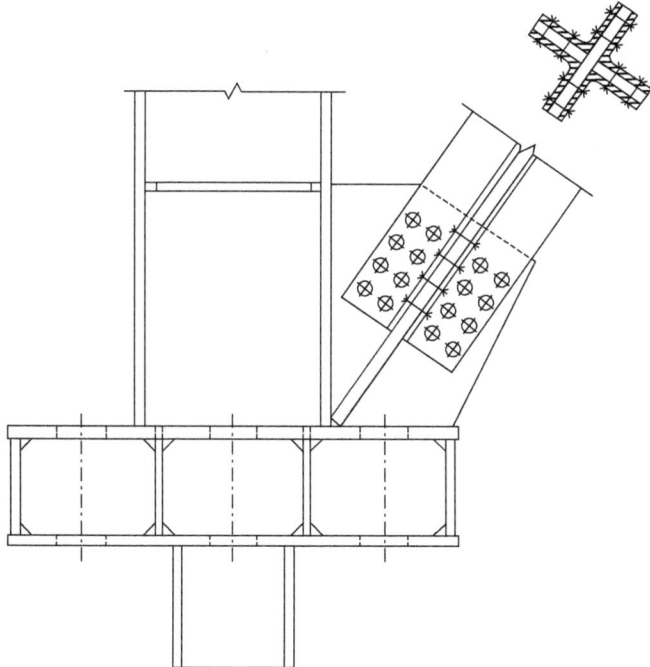

Figure 7.93 Four L braces converging into a heavy base plate realized by a double plate and a shear lug (shear key).

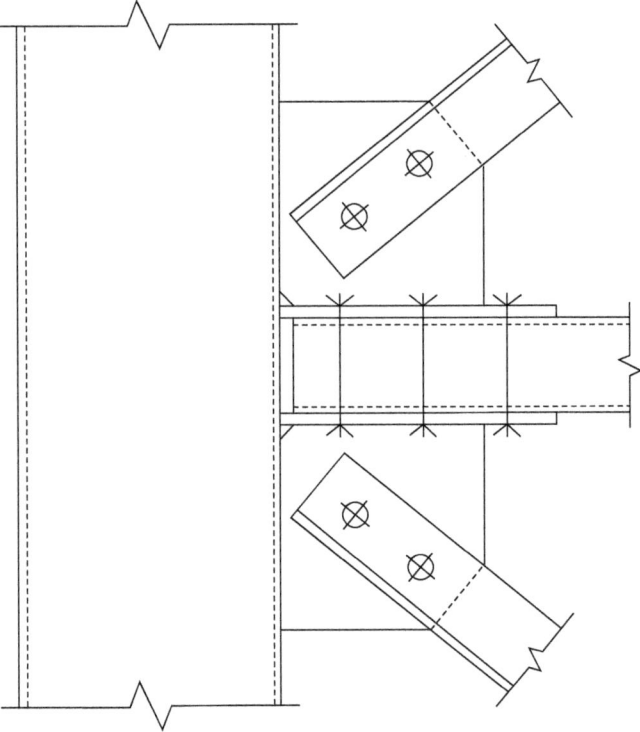

Figure 7.94 Braces bolted to gussets welded to an RHS column; a horizontal RHS is also bolted in the joint.

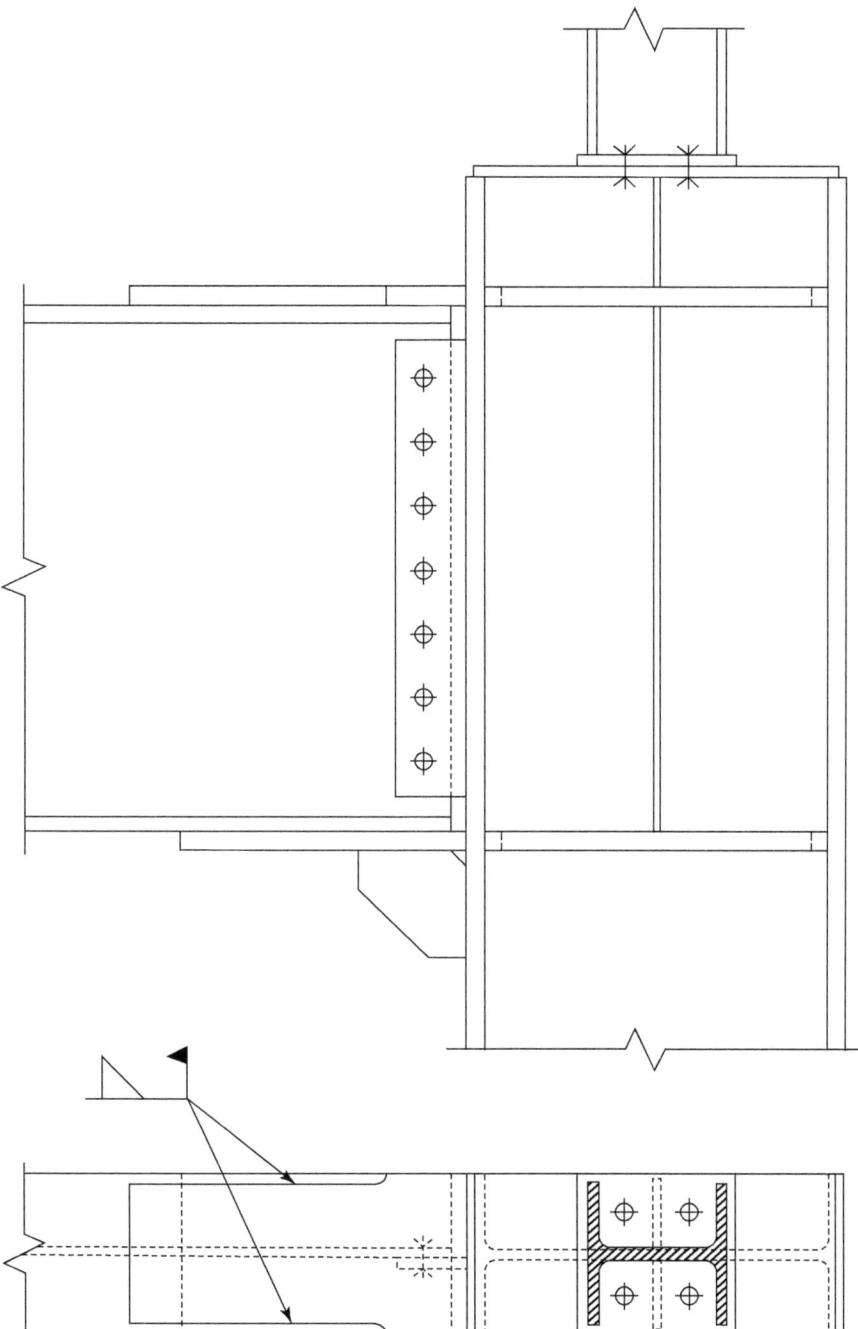

Figure 7.95 Beam-to-column moment connection by bolted web (also good for preassembly during erection) and field welds to beam flanges; continuity plates in the column, also supporting an additional smaller column bolted on top.

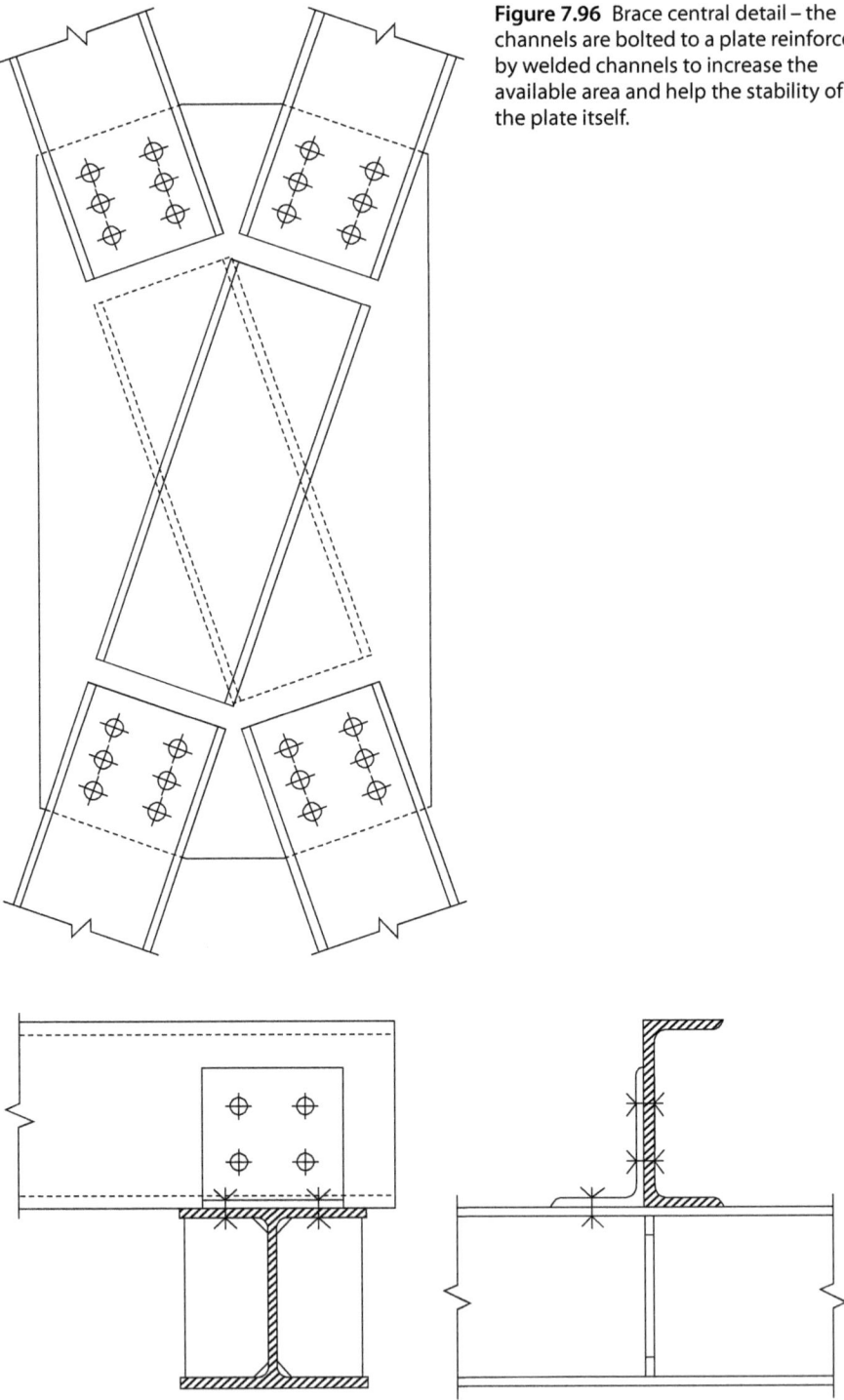

Figure 7.96 Brace central detail – the channels are bolted to a plate reinforced by welded channels to increase the available area and help the stability of the plate itself.

Figure 7.97 Stair U stringer bolted to the supporting beam by a bolted angle.

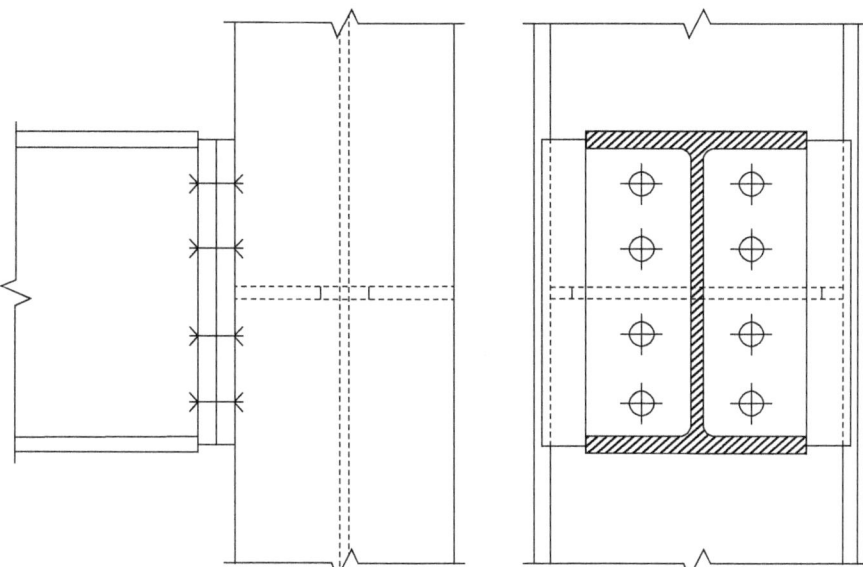

Figure 7.98 End plate for connection to column weak side; stiffeners on both sides of the column.

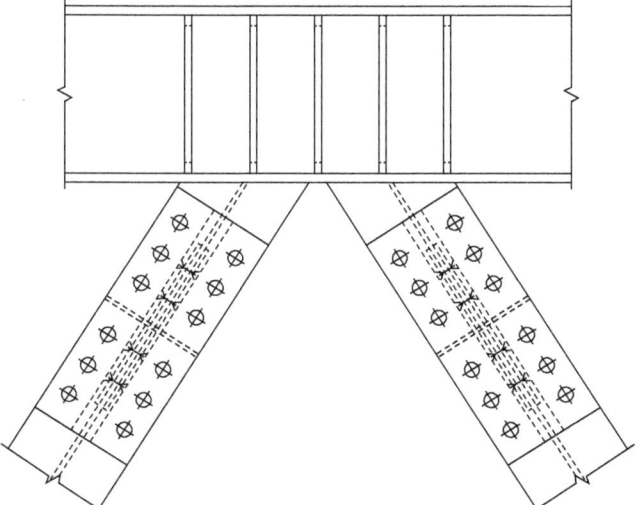

Figure 7.99 Large I-shaped brace splice connected to stiffened beam.

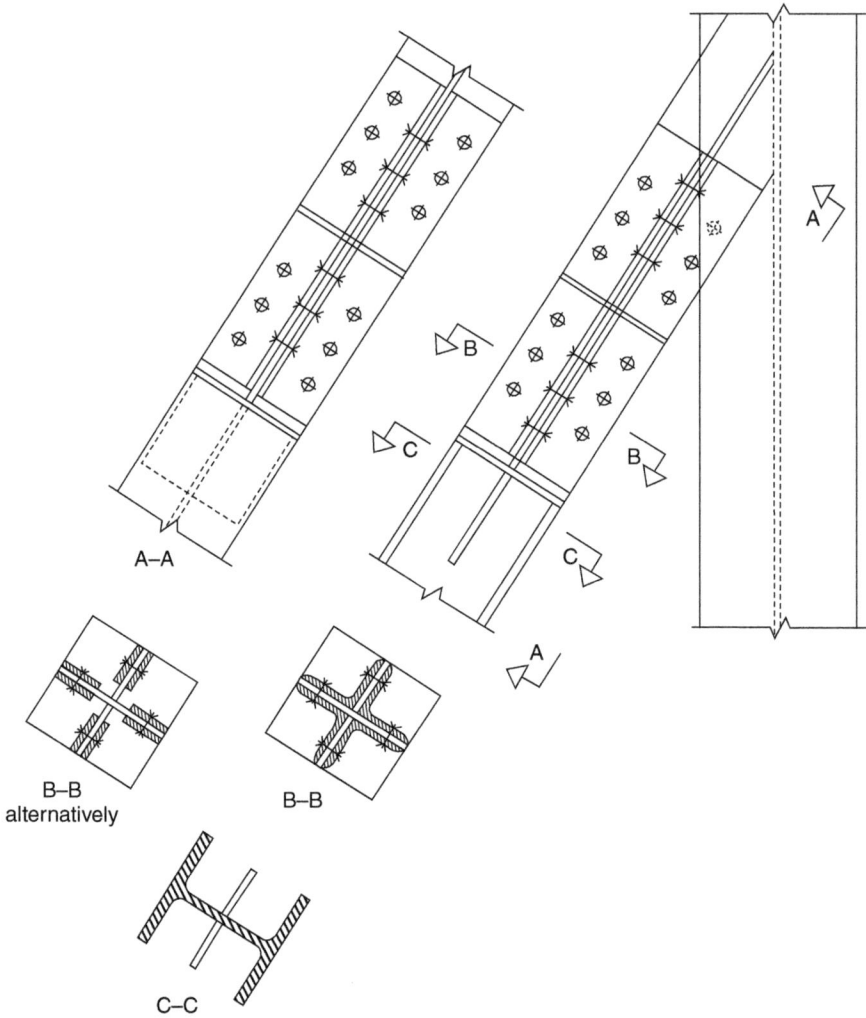

Figure 7.100 Vertical I-shaped brace bolted by four angles (alternatively, four double plates also possible); to allow this kind of connection, a cruciform plate assembly is welded to the column and another to the end of the brace.

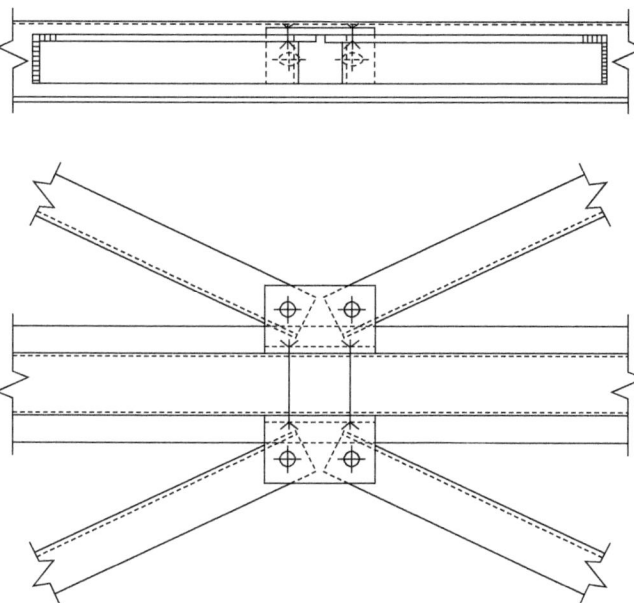

Figure 7.101 Horizontal roof braces connected by bolted angles to omega-shaped purlins.

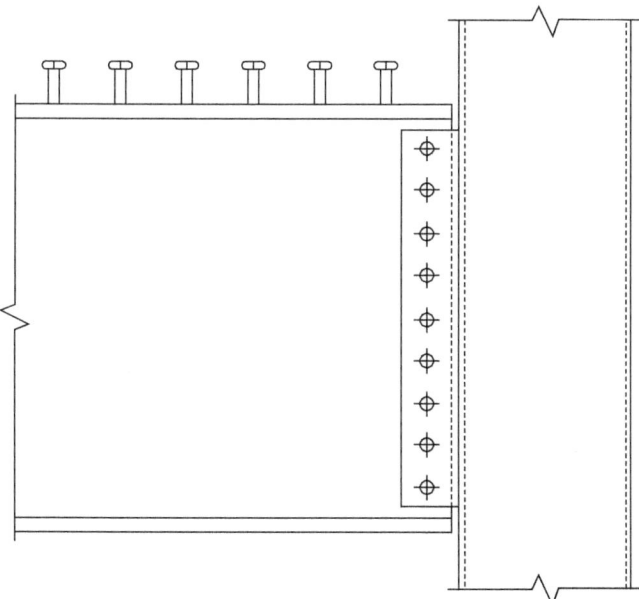

Figure 7.102 Fin plate (shear tab) with Nelson studs welded to the beam.

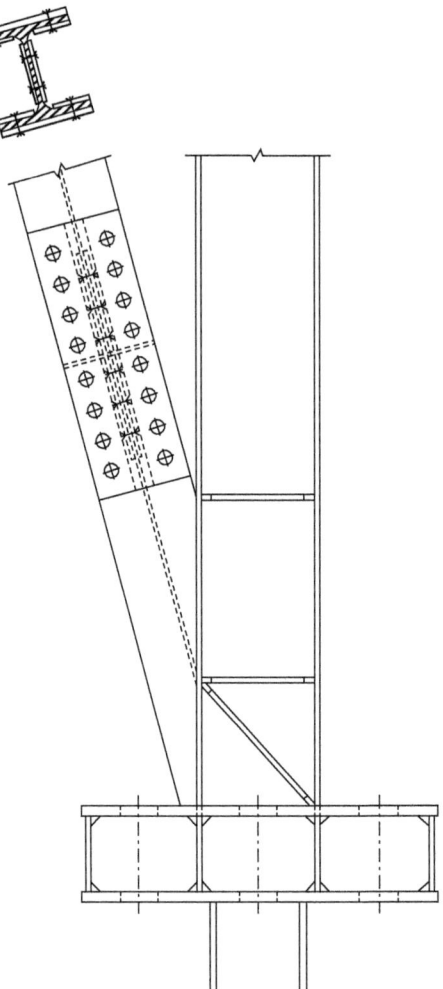

Figure 7.103 I brace spliced to same section stub framing into heavy base plate with shear lug.

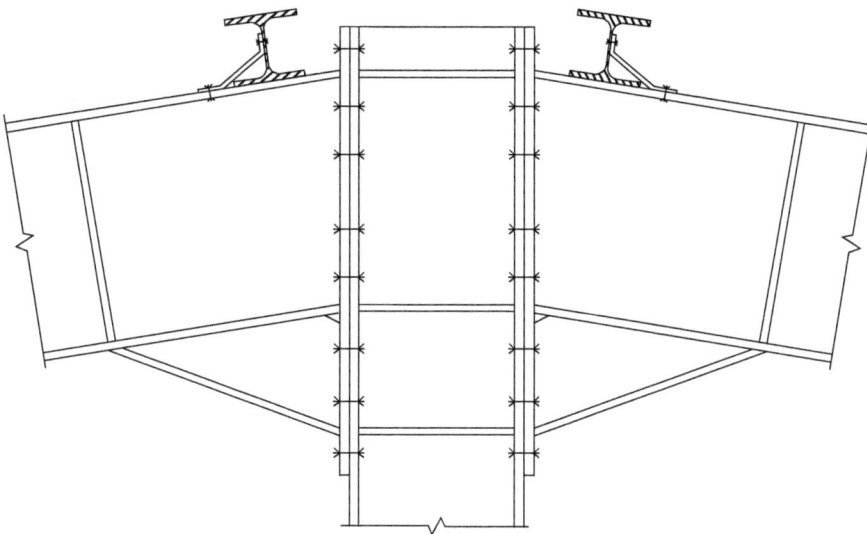

Figure 7.104 Symmetric moment connection with haunches and various stiffeners on column web and beam web; I-shaped purlins bolted on top.

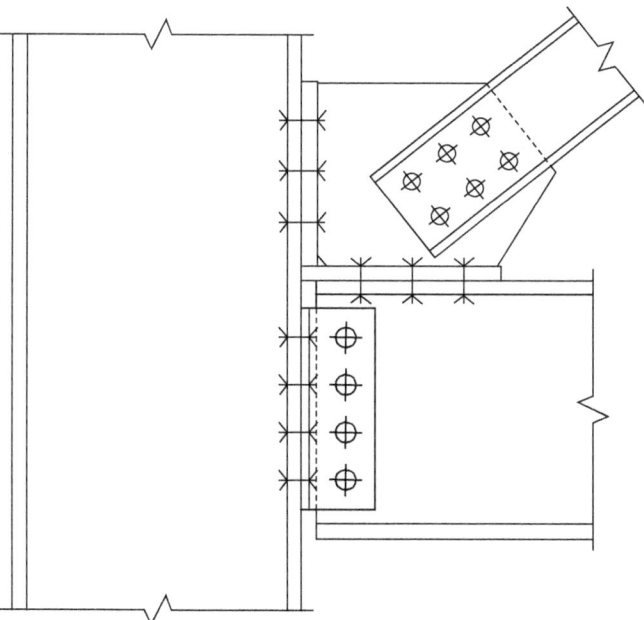

Figure 7.105 Bolted beam-to-column connection where also the gusset plate supporting the brace is field bolted to beam and column.

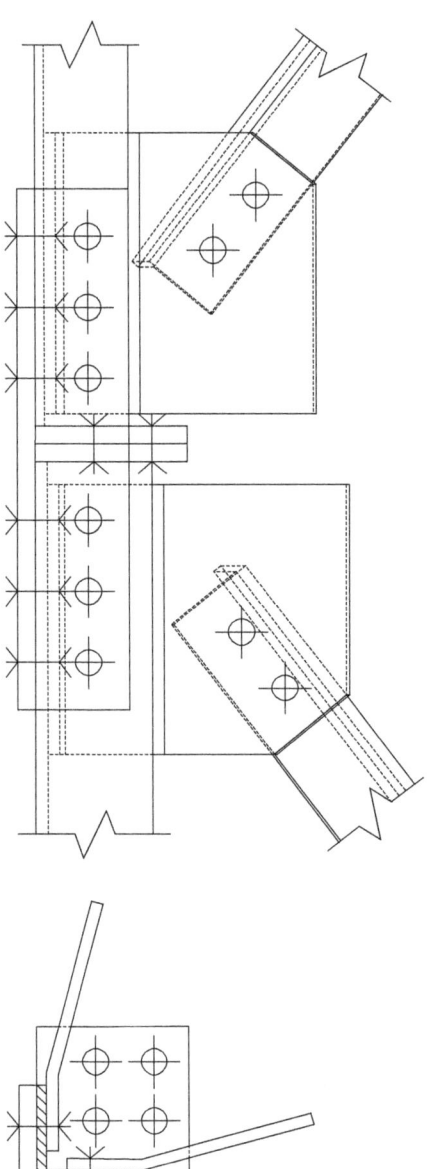

Figure 7.106 Splice in the corner of a lattice tower – end plates connect vertical angles, also connected by external cover plates; bent plates welded to vertical angles connect diagonals.

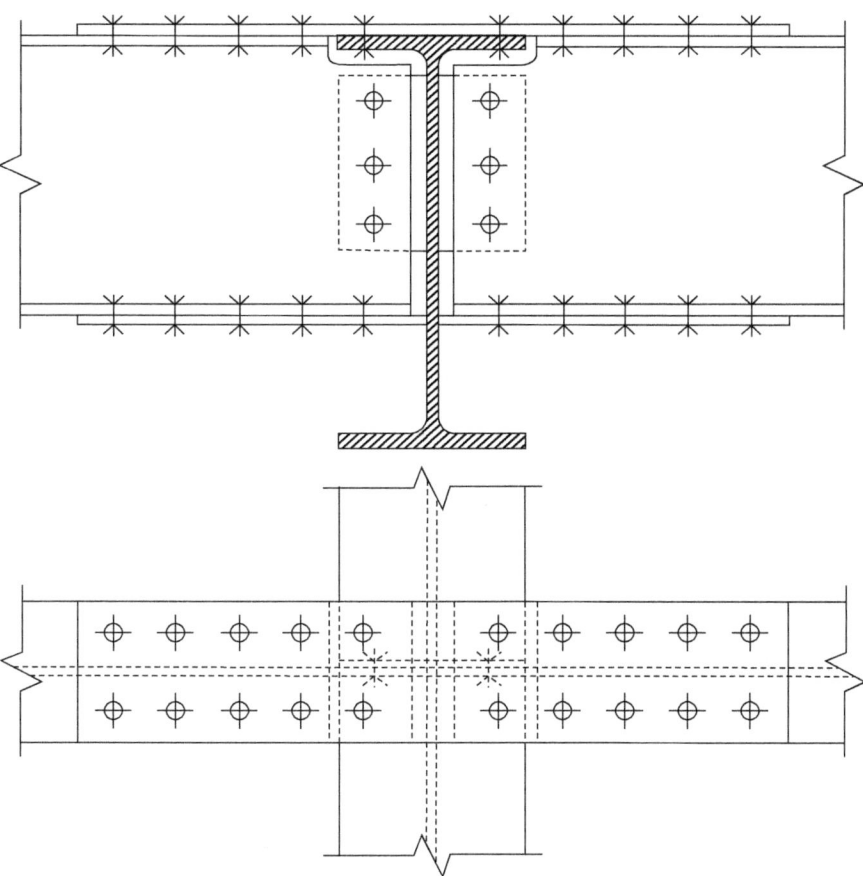

Figure 7.107 Secondary beams connected to main member on web and flanges in order to restore secondary-beam bending resistance.

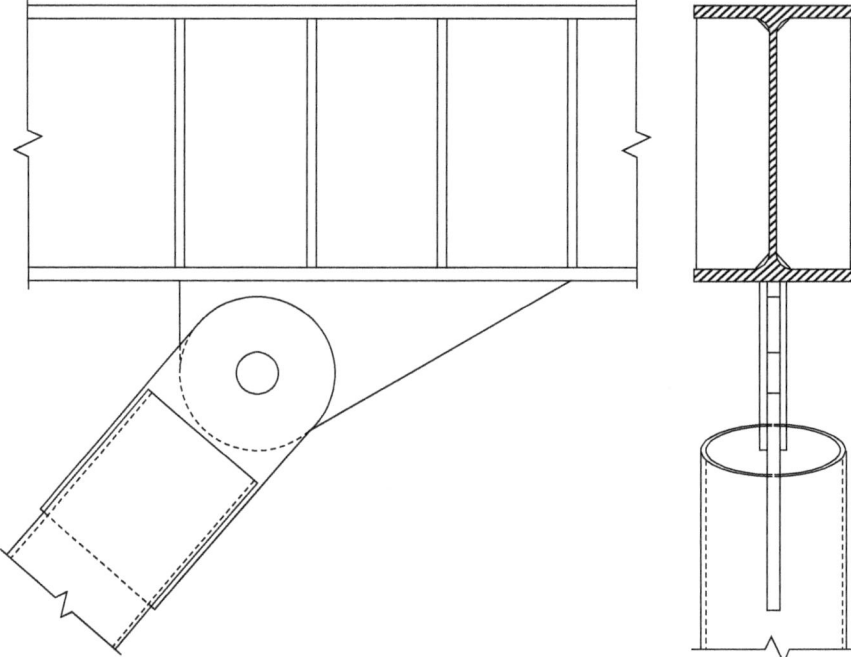

Figure 7.108 Perfect hinge between pipe brace and stiffened beam.

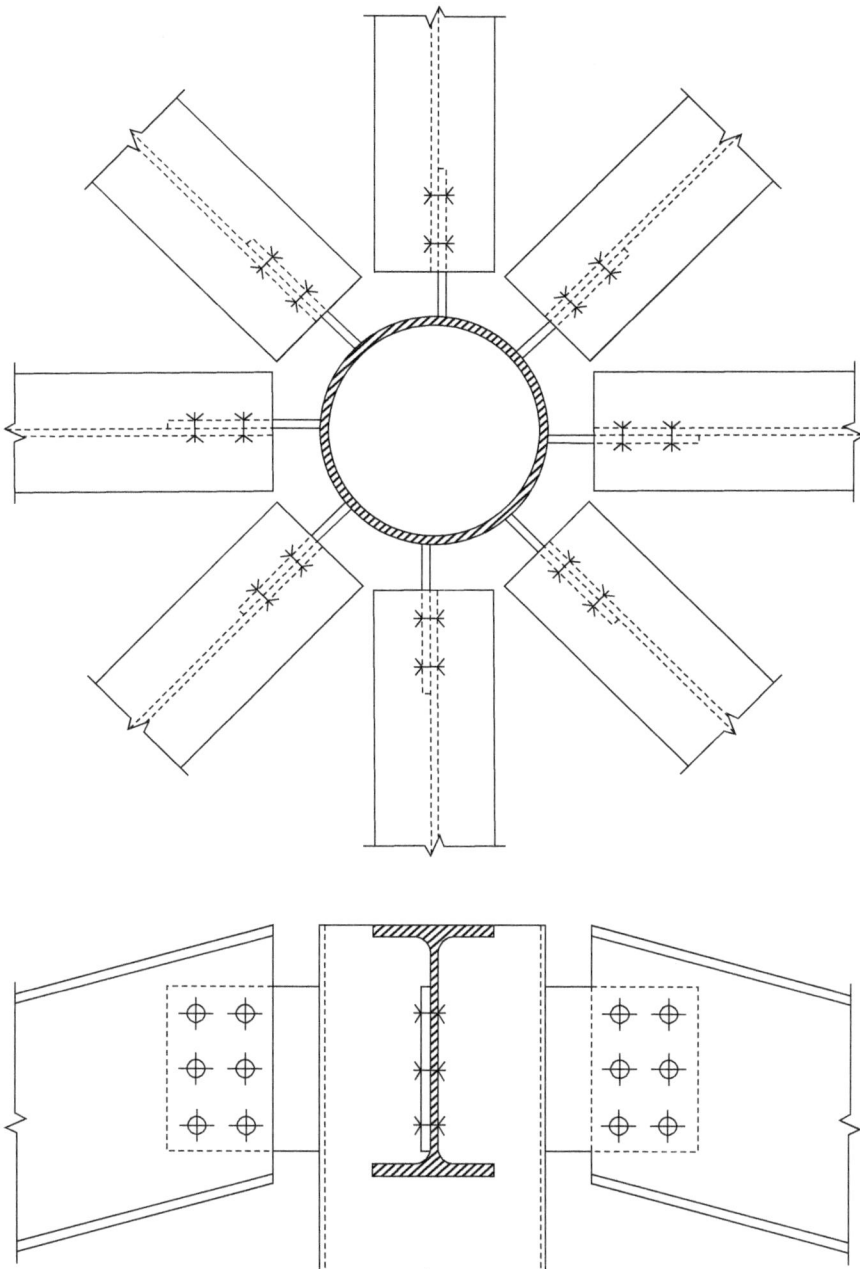

Figure 7.109 Apex connection of multiple beams into a pipe column.

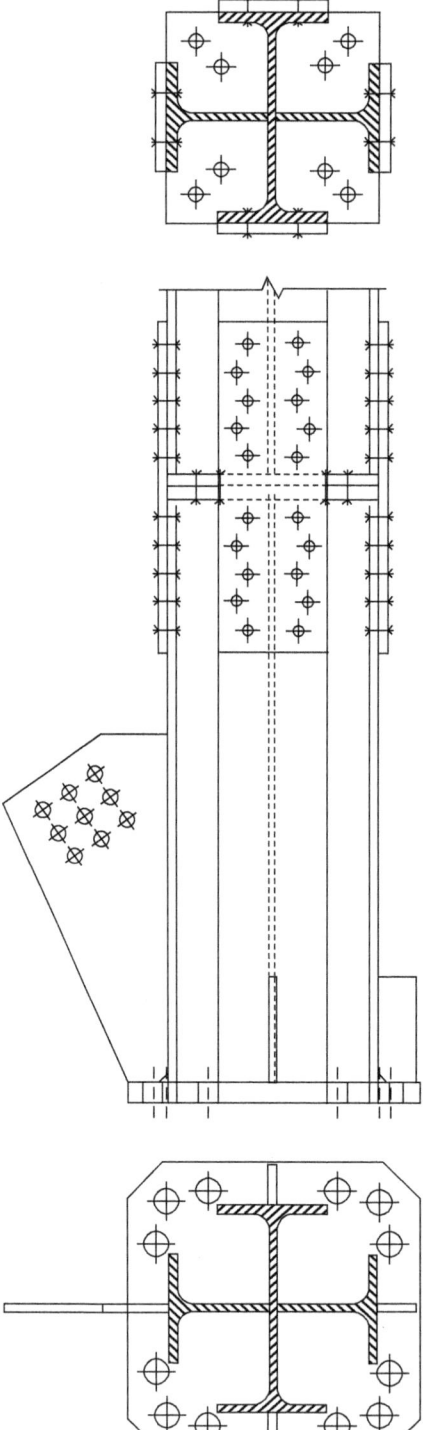

Figure 7.110 Cruciform column splice with end plate and external cover plates; octagonal base plate also taking a gusset plate for braces.

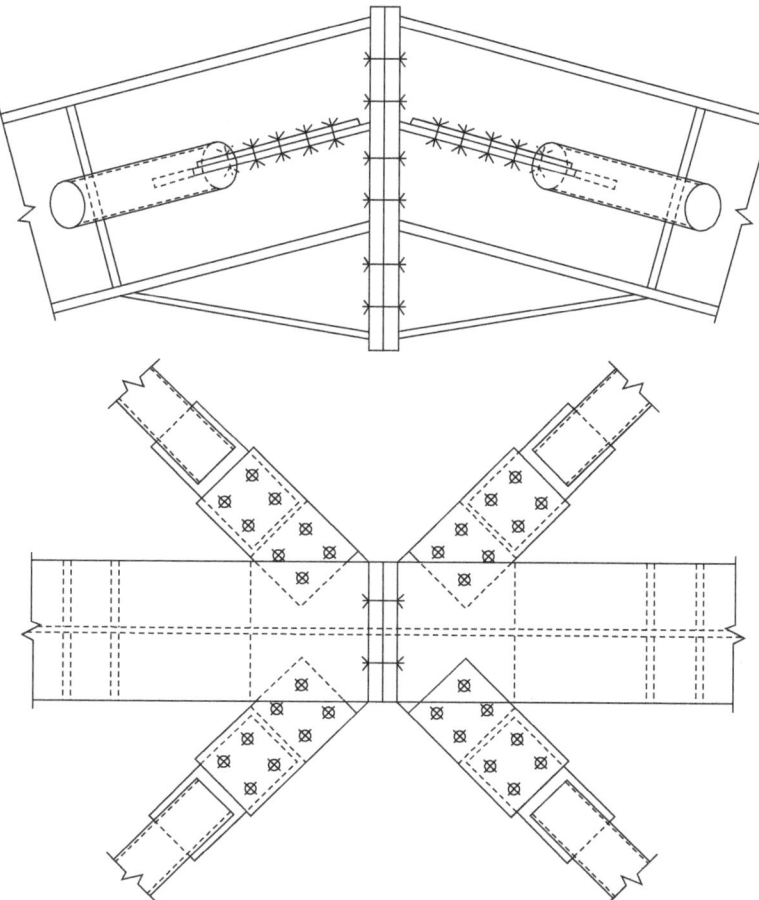

Figure 7.111 Haunched end-plate apex of rafters also taking the loads of plane braces realized with pipes.

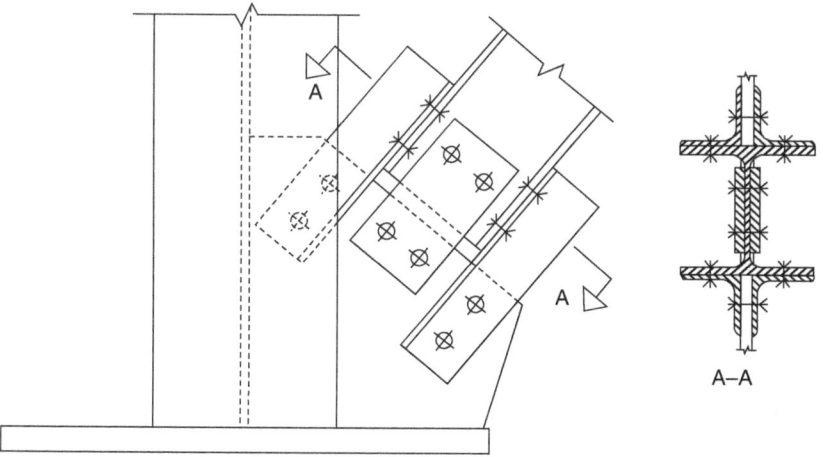

Figure 7.112 H brace connected to base plate gusset by web plates on web and angles on flanges.

Index

b
base plate   9, 125, 244, 282, 284
block shear   49
block tearing *see* block shear   49
bolt hole clearance   33, 254
braces *see* lateral resisting system   2, 217

c
camber   283

d
diffusion angles   23
double angle connection   181, 251
ductility   18, 105, 145, 162, 183, 201, 215, 242

e
eccentricity   22, 115, 120, 155
end plate   71, 81, 120, 175, 189, 217, 242, 251
erection   139, 249, 251, 256, 279, 280, 283
expansion joints   237

f
fatigue   108
fin plate   101, 154, 217, 251

g
galvanization   176, 281
girts *see* purlins   233
grout   137, 143, 145, 148, 256, 284

h
hinge   1, 7, 8, 13, 15, 21, 238, 242, 245

l
lamellar tearing   106
lateral resisting system   2
load path   19, 20

m
moment connection *see* lateral resisting system 2, *see* hinge 8, 40, 82, *see* end plate 189, *see* splice   212

n
notches   257, 264

p
pin *see* hinge   1
plastic hinge *see* hinge   8
portal *see* moment connection   2
pretensioning   24
pretensioning *see* tightening   265
prying action *see* T-stub   61
purlins   233

s
seated connection   230
seismic, 2, *see* ductility   18, 143, 197, 241, 255
shear lag   95, 186
shear tab *see* fin plate   154
shims   201, 280
skewed connections   287

Design and Analysis of Connections in Steel Structures: Fundamentals and Examples,
First Edition. Alfredo Boracchini.
© 2018 Ernst & Sohn Verlag GmbH & Co. KG. Published 2018 by Ernst & Sohn Verlag GmbH & Co. KG.

slenderness   100
spacings and distances   260
splice   212
stiffness   31, 39, 41, 49, 84, 88, 89
structural integrity   103, 162, 171, 178, 183, 201, 215, 251

t
transfer forces   25
T-stub   61, 130, 204

tightening   43
truss   5, 11, 22, 33, 95, 186, 217

w
welds   18, 52, 236, 287

y
yield line   22